COMMUTATIVE NORMED RINGS

COMMUTATIVE NORMED RINGS

By

I. GELFAND D. RAIKOV

AND

G. SHILOV

CHELSEA PUBLISHING COMPANY
BRONX, NEW YORK

The present work is a translation into English of the Russian-language work, published in 1960, KOMMUTATIVNYE NORMIROVANNYE KOL'CA, written by Professors Israel M. Gelfand, Dmitri A. Raikov, and George E. Shilov, to which a supplementary chapter has been added. This added chapter derives from a paper by Professor Shilov (see footnote 1 of Chapter IX for details).

Library of Congress Catalog Card No. 61-15024

Printed in the United States of America

NOTE

The present book gives an account of the theory of commutative normed rings with applications to analysis and topology. The paper by I. N. Gelfand and M. A. Naimark *Normed Rings with an Involution and their Representations,* which is presented here as Chapter VIII, may serve as an introduction to the theory of non-commutative normed rings with an involution.

The book is addressed to mathematicians—students in advanced courses, research students, and scholars—who are interested in functional analysis and its applications.

PREFACE

This book is devoted to an account of one of the branches of functional analysis—the theory of commutative normed rings—and the principal applications of that theory. It is based on our paper written for the *Uspehi Matematičeskih Nauk* in 1940, hard on the heels of the initial period of the development of this theory. The military situation of the time delayed the publication of the paper; it was printed only in 1946, and even then, because of lack of space, in abbreviated form. In the present book certain parts omitted from that paper (concerning harmonic analysis on groups and regular rings) have been restored—more accurately, rewritten—and a number of results have been included that were obtained after the publication of the paper. Furthermore, partly in connection with this, a considerable part of the earlier text has been substantially altered.

The book consists of three parts. Part One, which is concerned with the theory of commutative normed rings, is divided into two chapters, of which the first contains the foundations of the theory and the second deals with more special problems. The most significant novelty here is the extension of the operational calculus to multivalent analytic functions of several variables (§ 13). Part Two, which deals with applications to harmonic analysis, is divided into three chapters. In the first (Chapter III), we discuss the ring of absolutely integrable functions on a line with convolution as multiplication and we find the maximal ideals of this ring, and also of some of its analogues. In the next chapter (Chapter IV) these results are carried over to arbitrary commutative locally compact groups and they are made the foundation of the construction of harmonic analysis and the theory of characters. A new feature here is the construction of an invariant measure on the group of characters and a proof of the inversion formula for Fourier transforms that is not based on theorems on the representation of positive-definite functions or positive functionals; in view of this, the discussion of positive-definite functions is postponed to the very end of the chapter. Fnally, the last chapter of the second part (Chapter V)—the most specialized of all the chapters of the book—is devoted to the investigation of the ring of functions of bounded variation on a line with multiplication defined as convolution; the principal addition to the old text here is the complete description of the maximal ideals of this ring. The third and last part of the book is divided into two chapters devoted to the discussion of two important classes of rings of functions: regular rings (Chapter VI) and rings with uniform convergence (Chapter VII). In the first of these chapters we study essentially the structure of ideals in regular rings; as one of the applications, we prove in a generalized form the well-known Tauberian Theorem of Wiener; the chapter ends with an example, due to Laurant Schwartz, of a ring of functions having closed ideals that cannot be represented as the intersections of maximal ideals. In the second of these chapters (Chapter VII) we discuss the ring $C(S)$ of all bounded continuous complex functions on completely regular spaces S and various of its subrings; the first section here reproduces (although in an entirely new version) results contained in the original paper: the establishment of a natural correspondence between compact extensions of a completely regular space S and symmetric subrings of $C(S)$; the remaining two sections (concerning arbitrary subrings of $C(S)$ and their ideals) contain primarily new results.

Since non-commutative normed rings with an involution are important for group-theoretical applications, the paper by I. M. Gelfand and N. A. Naimark, "Normed Rings with an Involution and their Representations," is reproduced at the end of the book, slightly abridged, in the form of an appendix [Chapter VIII in the present English translation]. The reader who wishes to acquaint himself in more detail with the theory of non-commutative normed rings can find a thorough account of it in the long monograph by Naimark, *Normed Rings.* This monograph also contains an account of the foundations of the theory of commutative normed rings, without, however, touching upon the majority of its analytic applications. The same remark can be made regarding the book by Loomis, *Introduction to Abstract Harmonic Analysis.*

The reader is supposed to have a knowledge of the elements of the theory of normed spaces and of set-theoretical topology. For an understanding of the fourth chapter he also has to know what a topological group is. It stands to reason that the basic concepts of the theory of measure and of the Lebesgue integral are also assumed to be known.

In order not to interrupt the exposition, historico-bibliographical notes are given at the end of the book. The numbers in brackets refer to the list of references that follows the historico-bibliographical notes.

I. M. GELFAND
D. A. RAIKOV
G. E. SHILOV

TABLE OF CONTENTS

PART ONE

PART TWO

PART THREE

NOTE ON THE INTERDEPENDENCE OF THE CHAPTERS

The entire book depends on Chapter I. Chapter II is required only for Chapter VI (which uses § 9) and for § 44.14, of Chapter VII (which is based on § 14); furthermore, in Chapter III there are two references to § 13, but only in that part of the text marked with an asterisk (see footnote 4 of § 1). Chapter III is required for Chapters IV and V (which are based on §§ 16 and 17) and for § 41 of Chapter VI, marked with an asterisk (where there is a reference to § 19). Chapters IV and V are independent; and the chapters that follow them are not based on them. Chapter VI is required for the last two sections of Chapter VII.

Chapter VIII is a modified version of a paper by I. M. Gelfand and M. A. Naimark (see Chapter VIII, footnote 2). Chapter IX is an essentially unaltered version of a paper by G. E. Shilov (see Chapter IX, footnote 1). Chapter IX does not constitute part of the original Russian text and was added in translation, in pursuance of a suggestion for which the publishers wish to thank Professor John Lindberg, of Syracuse University and Mr. Roger S. Thompson, Physics Librarian at Argonne National Laboratory.

PART ONE

CHAPTER I

THE GENERAL THEORY OF COMMUTATIVE NORMED RINGS

§ 1. The Concept of a Normed Ring

1. DEFINITION 1: A *normed ring*[1] is a complex Banach space in which an associative multiplication is defined that is permutable with the multiplication by complex numbers, distributive with respect to addition, and continuous in each factor.

Henceforth we shall assume that the *multiplication is commutative.*

Every normed ring that does not contain a unit element e of multiplication can be completed to a normed ring with unit element by formally adjoining such a unit, i.e., by constructing the ring of formal sums $\lambda e + x$, where λ ranges over all complex numbers, x ranges over all the elements of the ring in question, and e is the unit to be adjoined; the operations in the extended ring are defined in the natural way:

$$(\lambda e + x) + (\mu e + y) = (\lambda + \mu) e + (x + y),$$
$$\mu (\lambda e + x) = \mu\lambda e + \mu x,$$
$$(\lambda e + x)(\mu e + y) = \lambda\mu e + (\mu x + \lambda y + xy),$$

and the norm is given by the formula

$$\| \lambda e + x \| = | \lambda | + \| x \|.$$

Therefore, in the construction of the general theory of normed rings we can confine ourselves to the study of *normed rings with unit element,* and this we shall do from now on.

2. Let us give some examples of normed rings.

1. Let C be the space of all complex functions that are defined and continuous on the interval $[0, 1]$, with the norm given by $\| x \| = \max_{0 \leqq t \leqq 1} | x(t) |$.

[1] In another terminology: a *Banach algebra.* Henceforth, the term 'ring' shall be understood to mean an algebra over the field of complex numbers.

C is a normed ring (with the unit element $x(t) \equiv 1$) under ordinary multiplication (it is obvious that it satisfies all the conditions of Definition 1).

2. Let D_n be the space of all complex functions that are defined on the interval $[0, 1]$, have a continuous n-th derivative on the interval, and have the norm given by

$$\|x\| = \sum_{k=0}^{n} \max_{0 \leq t \leq 1} |x^{(k)}(t)|. \tag{1}$$

D_n is a normed ring (with unit element $x(t) \equiv 1$) under ordinary multiplication (which is easily verified to be continuous in the norm (1) with respect to both factors and obviously also satisfies the remaining conditions of Definition 1).

3. Let W be the space of all complex functions of a real variable that can be developed in an absolutely convergent trigonometric series, with the norm given by

$$\| z \| = \left\| \sum_{n=-\infty}^{\infty} c_n \exp(int) \right\| = \sum_{n=-\infty}^{\infty} | c_n |. \tag{2}$$

(The sum of this series $z = z(t)$ is then uniquely determined.) If

$$x(t) = \sum_{k=-\infty}^{\infty} a_k \exp(ikt) \in W \quad \text{and} \quad y(t) = \sum_{k=-\infty}^{\infty} b_l \exp(ilt) \in W,$$

then we also have $z(t) = x(t)y(t) \in W$. For the product of the absolutely convergent series $\sum_{k=-\infty}^{\infty} a_k \exp(ikt)$ and $\sum_{l=-\infty}^{\infty} b_l \exp(ilt)$ is the absolutely convergent series $\sum_{n=-\infty}^{\infty} c_n \exp(int)$, where

$$c_n = \sum_{k+l=n} a_k b_l = \sum_{l=-\infty}^{\infty} a_{n-l} b_l.$$

Moreover, since

$$\begin{aligned} \|z\| = \sum_{n=-\infty}^{\infty} |c_n| &\leq \sum_{n=-\infty}^{\infty} \sum_{l=-\infty}^{\infty} |a_{n-l}| |b_l| = \\ &= \sum_{l=-\infty}^{\infty} \left(\sum_{n=-\infty}^{\infty} |a_{n-l}| \right) |b_l| = \\ &= \sum_{k=-\infty}^{\infty} |a_k| \sum_{l=-\infty}^{\infty} |b_l| = \|x\| \|y\|, \end{aligned}$$

multiplication is continuous in the norm (2) with respect to both factors. Hence W is a normed ring (with unit element $x(t) \equiv 1$) under the usual multiplication.

4. Let $I^{(n)}$ be the ring with unit element generated by the differentiation operator D in the space of polynomials of degree less than or equal to n in a single variable, with complex coefficients. Since $D^{n+1}=0$, the elements of this ring are the totality of all possible polynomials $\sum_{k=0}^{n} a_k D^k$, where a_k is an arbitrary complex number and D^0 is the unit operator. We put

$$\left\| \sum_{k=0}^{n} a_k D^k \right\| = \sum_{k=0}^{n} |a_k|.$$

Then $I^{(n)}$ is a normed ring under the multiplication of operators, with the unit element $e=D^0$.

5. Let $L^1(0, 1)$ be the space of all absolutely integrable measurable complex functions on the interval $[0, 1]$, with norm given by $\|x\| = \int_0^1 |x(t)|\,dt$. By means of Fubini's theorem on the relation between the Lebesgue double integral and repeated integrals it can be shown that for any two functions $x(t)$ and $y(t)$ belonging to $L^1(0, 1)$ their 'convolution'

$$(x * y)(t) = \int_0^t x(t-\tau)\, y(\tau)\, d\tau \qquad (0 \leqq t \leqq 1) \tag{3}$$

exists for almost all t and belongs to $L^1(0, 1)$, and the operation of 'convolution' (3) is associative.[2] This operation is obviously bilinear, and the substitution $\tau \to t-\tau$ shows that it is also commutative. Moreover (again on the basis of Fubini's Theorem),

$$\begin{aligned}
\|x * y\| &= \int_0^1 \left| \int_0^t x(t-\tau)\, y(\tau)\, d\tau \right| dt \leqq \\
&\leqq \int_0^1 \left(\int_0^t |x(t-\tau)|\,|y(\tau)|\, d\tau \right) dt = \\
&= \int_0^1 \left(\int_\tau^1 |x(t-\tau)|\, dt \right) |y(\tau)|\, d\tau = \\
&= \int_0^1 \left(\int_0^{1-\tau} |x(t)|\, dt \right) |y(\tau)|\, d\tau \leqq \int_0^1 |x(t)|\, dt \int_0^1 |y(\tau)|\, d\tau = \\
&= \|x\|\,\|y\|,
\end{aligned}$$

[2] For further details, see § 16.

and hence it follows that convolution is continuous with respect to both 'factors.' Thus, $L^1(0, 1)$ is a normed ring under convolution. It is easy to see that it does not contain a unit element.[3] *We shall denote by I the normed ring that is obtained as the result of formally adjoining a unit element to* $L^1(0, 1)$.

6. Let A be the space of all functions of a complex variable ζ that are defined and continuous in the circle $|\zeta| \leqq 1$ and regular throughout the interior of this circle, with the norm given by $\| x \| = \max_{|\zeta| \leqq 1} | x(\zeta) |$. A is a normed ring (with unit element $x(\zeta) \equiv 1$) under ordinary multiplication (it obviously satisfies all the conditions of Definition 1).

As we have shown above, in the rings of Examples 3 and 5, the norm has the following property:

$$\| xy \| \leqq \| x \| \, \| y \|. \tag{4}$$

Clearly, the same inequality is satisfied by the norm in the rings of Examples 1, 4, and 6. However, in Example 2 the inequality (4) is not, in general, satisfied for $n \geqq 2$; for example, we have for $x(t) \equiv t$:

$$\| x(t) \| = 2, \quad \| x^2(t) \| = 5 > \| x(t) \|^2.$$

But if we choose the norm

$$\| x(t) \| = \sum_{k=0}^{n} \frac{\max_{0 \leqq t \leqq 1} | x^{(k)}(t) |}{k!}, \tag{5}$$

in D_n instead of (1) then, as is easily verified, the inequality (4) holds. Furthermore, the norms (1) and (5) are *topologically equivalent.* It is a general property of normed rings that such a re-norming is always possible.

3. THEOREM 1: *For every normed ring R we can find a ring R' that is topologically and algebraically isomorphic to it and has the properties*

$$\| xy \| \leqq \| x \| \, \| y \| \quad \textit{and} \quad \| e \| = 1. \tag{*}$$

***4.** *Proof:*[4] Every element x of R generates the corresponding operator A_x of multiplication by x: $A_x y = xy$. By Definition 1, this operator is

[3] See § 16. The absence of a unit element in $L^1(0, 1)$ also follows from the fact that this ring consists of generalized nilpotent elements (see § 4.9).

[4] Subsections marked with an asterisk may be omitted on a first reading. In the original Russian text, the material of such subsections was printed in a smaller type than the main body of the text.

linear. In the ring Q of all linear operators mapping the Banach space R into itself, the operators A_x form a subring R' with unit element (the unit element being the unit operator E generated by the unit element e of the ring R).

Let us show that R' is a normed ring under the norm $\| A_x \| = \sup_{\|y\| \leqq 1} \| xy \|$. In the proof we only require that R' is complete, i.e., that R' is closed in Q.

By the associativity of multiplication we have

$$A_x(yz) = x(yz) = (xy)z = A_x y \cdot z.$$

It is not difficult to see that this property is characteristic for the operators of the ring R'. For, if for an operator A and for arbitrary y and z the equation $A(yz) = Ay \cdot z$ holds, then, putting $Ae = x$, we have:

$$Ay = A(ey) = Ae \cdot y = xy,$$

i.e., A is the operator of multiplication by x.

Suppose now that the operators $A_n \in R'$ strongly converge to some operator A, i.e., that $A_n x$ converges to Ax in the norm of the space R for every $x \in R$. By the continuity of multiplication with respect to the first factor, we then have: $A(xy) = \lim A_n(xy) = \lim A_n x \cdot y = Ax \cdot y$ and hence, by what we have just proved, A is also in R'. Thus, R' is closed in Q not only in the sense of uniform convergence of operators, but also in that of strong convergence.

Obviously, the rings R and R' are algebraically isomorphic. Let us show that they are also topologically isomorphic. We have:[5]

$$\| A_x \| = \sup_{\|y\| \leqq 1} \| xy \| \geqq \left\| x \frac{e}{\| e \|} \right\| = \frac{\| x \|}{\| e \|}$$

or

$$\| x \| \leqq \| e \| \ \| A_x \|. \tag{6}$$

Thus, the mapping $A_x \rightarrow x$ of the space R' onto the space R is continuous; but since both these spaces are complete, we have, by the well-known theorem of Banach,[6] that the inverse mapping $x \rightarrow A_x$ is also continuous. We have thus proved that the rings R and R' are topologically isomorphic and, with this, we have also proved the theorem, because the norm in R' has the

---

[5] It is easy to verify that $e \neq 0$, and therefore $\| e \| > 0$.

[6] See A. N. Kolmogorov and S. V. Fomin, *Elements of the Theory of Functions and Functional Analysis,* Vol. I, Rochester, 1957, pp. 99-101, or L. A. Liusternik and W. I. Sobolev, *Elements of Functional Analysis,* F. Ungar, New York, 1961, pp. 88-90.

property (*). At the same time, we have obtained the result that *every normed ring is topologically and algebraically isomorphic to a normed ring of operators in a Banach space.* //[7]

Note: If the condition (*) is satisfied in the ring R, then R and R' are *isometric*. For in this case, (6) gives: $\| x \| \leqq \| A_x \|$. On the other hand, by (4) we have

$$\| A_x \| = \sup_{\| y \| \leqq 1} \| xy \| \leqq \| x \| \sup_{\| y \| \leqq 1} \| y \| = \| x \|.$$

Combining the two inequalities, we obtain: $\| A_x \| = \| x \|$.

5. COROLLARY 1: *The product xy is continuous with respect to both factors.*

DEFINITION 2: A series $x_1 + x_2 + \ldots + x_n + \ldots$ is called *absolutely convergent* if the series

$$\| x_1 \| + \| x_2 \| + \ldots + \| x_n \| + \ldots$$

converges.

Obviously, every absolutely convergent series in R converges.

COROLLARY 2: *Absolutely convergent series formed from elements of a normed ring can be added and multiplied like absolutely convergent numerical series.*

Henceforth *the norm will always be assumed to satisfy the condition* (*).

§ 2. Maximal Ideals

1. LEMMA: *The set O of all elements x of a normed ring R for which there exists an inverse element x^{-1} is open, and x^{-1} is a continuous function of x on O.*

Proof: First of all, *every element x for which $\| e - x \| < 1$ has an inverse element x^{-1}.* For let us consider the series

$$e + (e - x) + (e - x)^2 + \ldots . \tag{1}$$

Since $\| (e - x)^n \| \leqq \| e - x \|^n$, this series converges absolutely and it therefore represents an element of R. Upon multiplying it by $x = e - (e - x)$ and making use of Corollary 2 to Theorem 1 of § 1.5, we obtain

$$e + (e - x) + (e - x)^2 + \ldots - (e - x) - (e - x)^2 - \ldots = e.$$

Therefore the sum of the series (1) is the inverse element x^{-1}, whose existence we wished to show.

---

[7] This mark signifies the conclusion of a proof.

Now let x be an arbitrary element of O. We denote by $U_0(e)$ the neighborhood $\| e-y \| < 1$ of the unit element e, which by what we have proved belongs to O. Since $xx^{-1}=e$, by the continuity of multiplication there exists a neighborhood $U(x)$ of x such that $U(x)x^{-1} \subset U_0(e)$. Hence, for arbitrary $z \in U(x)$, zx^{-1} has the inverse element $(zx^{-1})^{-1}$:

$$zx^{-1}(zx^{-1})^{-1} = e.$$

But then $x^{-1}(zx^{-1})^{-1}$ is the inverse element of z, i.e., when x is contained in O, its neighborhood $U(x)$ is also contained in O.

Suppose, finally, that $x_n \rightarrow x \in O$. Then $z_n = x_n x^{-1} \rightarrow xx^{-1} = e$; and therefore, as follows from the expression (1) for the inverse element, we also have

$$x x_n{}^{-1} = z_n{}^{-1} = e + (e - z_n) + (e - z_n)^2 + \ldots \rightarrow e.$$

Multiplying both sides of this limit relation by x^{-1}, we obtain $x_n{}^{-1} \rightarrow x^{-1}$. //

DEFINITION 1: A set I of elements of a ring R is called an *ideal* if it has the following properties:

a) if $x \in I$ and $y \in I$, then $x + y \in I$;

b) if $x \in I$, then $zx \in I$ for all $z \in R$.

An ideal I of a ring R is called a *proper ideal* if, in addition,

c) $I \neq R$.

As an example of a proper ideal of the ring C of Example 1 of § 1, we can take the set of all functions of C that are equal to zero on the interval $[0, 1/2]$.

An element of a ring R that has an inverse element cannot be contained in any proper ideal. For if $x \in I$, then the existence of x^{-1} would entail for every element z of R that $z = (zx^{-1})x \in I$, i.e., I would coincide with R.

On the other hand, every element x that does not have an inverse is contained in some proper ideal, viz., the totality of all elements zx, where z ranges over the whole of R.

Thus, *for an element of a ring R to have an inverse it is necessary and sufficient that it does not belong to any proper ideal.* In particular, *if the ring R contains no proper ideals other than the zero ideal* (consisting only of the element 0), *then R is a field.*

It is easy to verify that the closure $\bar{I}$ of an ideal I satisfies the conditions a) and b) of Definition 1. Moreover, since every proper ideal I is contained in the set $R \setminus O$, which by the lemma is closed, $\bar{I}$ is also contained in $R \setminus O$ and hence does not coincide with R. Thus, *the closure of a proper ideal is itself a proper ideal.*

DEFINITION 2: A *maximal ideal* is a proper ideal that is not contained in any other proper ideal.

***2.** Let us find all the maximal ideals of the ring C of Example 1 of § 1.

The set of all functions of C that vanish at an arbitrary fixed point of an interval in $[0,1]$ is a maximal ideal of C.

The set M_τ of all functions $x(t) \in C$ for which $x(\tau)=0$ is a proper ideal of C. Let $y(t)$ be any function of C not belonging to M_τ. What we have to show is that there exists no proper ideal containing M_τ and $y(t)$. But this follows from the fact that every function $z(t) \in C$ can be represented in the form

$$z(t)=\frac{z(\tau)}{y(\tau)}\, y(t)+\left(z(t)-\frac{y(t)}{y(\tau)}\, z(\tau)\right),$$

where

$$z(t)-\frac{y(t)}{y(\tau)}\, z(\tau) \in M_\tau$$

and the first summand is a multiple of $y(t)$.[8]

Now let M be any maximal ideal of C. We shall show that all the functions that occur in this maximal ideal vanish at some fixed point of the interval $[0,1]$. Indeed, if this were not so, then for every point $\tau \in [0,1]$ we could find a function $x_\tau(t) \in M$ such that $x_\tau(\tau) \neq 0$ and hence

$$|x_\tau(t)|>\delta_\tau>0$$

in some interval containing τ. By the Heine-Borel Theorem, there exists a finite number of such intervals covering the whole interval $[0,1]$. Let $\tau_1, \ldots, \tau_n$ be the points corresponding to each of these intervals. The function

$$x(t)=x_{\tau_1}(t)\,\overline{x_{\tau_1}(t)}+\ldots+x_{\tau_n}(t)\,\overline{x_{\tau_n}(t)}=|x_{\tau_1}(t)|^2+\ldots+|x_{\tau_n}(t)|^2$$

is contained in M. But on the other hand, $x(t)>\min\limits_{1 \leqq k \leqq n} \delta^2_{\tau_k}>0$ and hence the function $1/x(t)$ exists in C, so that in this case $x(t)$, as we have seen, cannot belong to any proper ideal; in particular, it cannot belong to the maximal ideal M. This contradiction shows that there exists a point τ such that $x(\tau)=0$ for all $x(t) \in M$. But then M, being maximal, is the ideal M_τ consisting of *all* the functions of C that vanish at the point τ. //

In exactly the same way, we can satisfy ourselves that the set of all absolutely convergent series $\sum\limits_{n=-\infty}^{\infty} c_n \exp(int)$ whose sums vanish at some point

[8] In the same way, one can see that in *every* ring of functions (with the usual algebraic operations) the collection of all functions that vanish at any given point forms a maximal ideal.

τ forms a maximal ideal in the ring W of Example 3 of § 1. But when we wish to repeat for W the same proof as above of the converse statement, we reach a point in the proof where we must draw the conclusion, from the fact that the sum of the absolutely convergent series $S(t) = \sum_{n=-\infty}^{\infty} c_n \exp(int)$ is different from zero for all t, that $1/S(t)$ also belongs to W, i.e., can be expanded in an absolutely convergent trigonometric series. This conclusion is a correct one and constitutes a theorem of Wiener ([32], p. 14; [33], p. 91),[9] but when we give a proof later on it will be on the basis of the very fact that all the maximal ideals of W are of the given form.

Note: Let R be an arbitrary normed ring formed from (not necessarily all) continuous functions $x(t)$, given on a compactum (i.e., a compact Hausdorff space) S with the usual addition and multiplication. C and D_n (Example 2 of § 1) are of this type; so is W, if the functions occurring in it are assumed to be given not on a line, but on a circle of radius 1. When we run over the arguments given above, we see that the set of all functions of R that vanish simultaneously at some point of the set S is always a maximal ideal of R;[1] for the converse statement to be true it is sufficient that R should have the following properties:

a) if $x(t)$ is in R, its complex conjugate $\overline{x(t)}$ is also in R;

b) if $x(t)$ is in R and if it does not vanish anywhere on S, $1/x(t)$ is also in R.

In order to establish a one-to-one correspondence between maximal ideals and the points of S, it is also necessary that R should satisfy the following 'separability condition':

c) for any two distinct points t_1, t_2 of S there exists a function $x(t)$ in R such that $x(t_1) \neq x(t_2)$.

The condition b) is not only sufficient (in conjunction with condition a)) but is also necessary. The condition a) is not necessary, as we shall see in § 4, when we determine the maximal ideals of the ring A of Example 6 of § 1.

Since D_n has the properties a), b), and c), we conclude that *the maximal ideals of the ring D_n are the sets of all functions of D_n that vanish at an arbitrary fixed point of the interval* $[0, 1]$.

Furthermore, the ring $C(S)$ of all continuous complex functions on S, which obviously satisfies the conditions a) and b), also satisfies the condition of separability c) by the normality of the space S and the well-known theorem

---

[9] See also A. Zygmund, *Trigonometric Series*, Vol. I, Cambridge, 1959, pp. 245-247.

[1] And this is independent of the compactness of S and the continuity of $x(t)$ (see Footnote 8).

of Uryson.[2] Thus, *the maximal ideals of the ring $C(S)$ are the sets of all functions of $C(S)$ that vanish at an arbitrary fixed point of the compactum S.*

3. Obviously, *every maximal ideal is closed*: otherwise it would be contained properly in its closure and hence would not be maximal.

THEOREM 1: *Every proper ideal I is contained in a maximal ideal.*

Proof: We prove the theorem by transfinite induction. Let

$$x_1, x_2, \ldots, x_\omega, \ldots, x_\alpha, \ldots$$

be a well-ordered transfinite sequence of all the elements of R. With every proper and non-maximal ideal I we associate a proper ideal $I^+ \supset I$ in the following way: Since I is not maximal, there exist elements $x \in R \setminus I$ having the property that the set of all elements of the form $j + rx$, where $j \in I$, $r \in R$, forms a proper ideal; let x_I be the first of such elements in the sequence $\{x_\alpha\}$; then we put $I^+ = \{j + rx_I\}$. We now construct a transfinite sequence of proper ideals I_α in the following way: We put $I_0 = I$ and suppose that the I_α have already been constructed for all $\alpha < \beta$; if β is of the first class, i.e., if $\beta - 1$ exists, then we put $I_\beta = I^+_{\beta-1}$; if β is of the second class, then we put $I_\beta = \bigcup_{\alpha<\beta} I_\alpha$. This sequence has a cardinal number not exceeding the cardinal number of R and it therefore has a last term that is then a maximal ideal containing I. //

In conjunction with the necessary and sufficient condition mentioned above for the existence of an inverse element, this theorem now yields the following.

THEOREM 2: *For an element of a commutative normed ring R to have an inverse it is necessary and sufficient that it is not contained in any maximal ideal. In particular, if the ring does not contain any non-zero maximal ideal, then it is a field.*

4. Elements $x, y \in R$ are called *congruent modulo the ideal I* if $x - y \in I$. Since the relation of congruence is reflexive, symmetric, and transitive, R splits into classes of congruent elements. Defining the sum (product) of classes X, Y as the class containing the sums (products) of elements x, y from X, Y and denoting by λX (where λ is a complex number) the class formed by the elements λx ($x \in X$), we obtain the ring R/I of residue classes of R with respect to I. The zero element of this *residue-class ring* is the

---

[2] See, for example P. S. Aleksandrov, *Introduction to the General Theory of Sets and Functions*, Moscow-Leningrad, 1948, p. 305 [in Russian; English translation in prep.]. (Or see J. L. Kelley, *General Topology*, New York, 1955, p. 115; H. L. Royden, *Real Analysis*, New York, 1963, p. 130.)

class formed by all the elements $x \in I$ and the unit element E is the class containing the unit element e of R.

In R/I we introduce the norm

$$\| X \| = \inf_{x \in X} \| x \|. \tag{2}$$

THEOREM 3: *If I is a closed proper ideal, then R/I is a normed ring.*

Proof:

1) $\| \lambda X \| = |\lambda| \, \| X \|$.

Obvious.

2) $\| X + Y \| \leqq \| X \| + \| Y \|$.

We have:

$$\|X+Y\| = \inf_{z \in X+Y} \|z\| = \inf_{x \in X,\, y \in Y} \|x+y\| \leqq$$
$$\leqq \inf_{x \in X,\, y \in Y} \{\|x\| + \|y\|\} = \inf_{x \in X} \|x\| + \inf_{y \in Y} \|y\| = \|X\| + \|Y\|.$$

3) $\| XY \| \leqq \| X \| \, \| Y \|$.

We have:

$$\|XY\| = \inf_{z \in XY} \|z\| \leqq \inf_{x \in X, y \in Y} \|xy\| \leqq \inf_{x \in X,\, y \in Y} \|x\| \, \|y\| =$$
$$= \inf_{x \in X} \|x\| \inf_{y \in Y} \|y\| = \|X\| \, \|Y\|.$$

4) $\| E \| = 1$.

Since $e \in E$, we have $\| E \| \leqq 1$. Let y be an arbitrary element of E. We have: $y = e + x$, where $x \in I$. Now if $\| y \|$ were less than 1, then, by what has been proved at the beginning of this section, x would have an inverse and hence could not belong to the proper ideal I. Thus, $\| E \| \geqq 1$ and hence $\| E \| = 1$.

5) If $\| X \| = 0$, then X is the zero class.

By (2), there exists a sequence $x_n \in X$ such that $x_n \to 0$ for $n \to \infty$. Let x be an arbitrary element of X. We have $x - x_n \in I$, and since $x_n \to 0$, we have $x = \lim_{n \to \infty} (x - x_n) \in \bar{I}$. But, by assumption, the ideal I is closed: $\bar{I} = I$. Thus X coincides with I, i.e., it is the zero class.

6) R/I is complete in the norm (2).

Let $\{X_n\}$ be a fundamental sequence of classes, that is, $\| X_n - X_m \| \to 0$ for $m, n \to \infty$. Then we can choose a subsequence $\{X_{n_k}\}$ from it such that the series $\sum_k \| X_{n_{k+1}} - X_{n_k} \|$ converges. By (2), for an arbitrary element $x_1 \in X_{n_1}$ we can find an element $x_2 \in X_{n_2}$ such that

$$\| x_2 - x_1 \| < 2 \| X_{n_2} - X_{n_1} \|;$$

furthermore, for this x_2 we can find an element $x_3 \in X_{n_3}$ such that

$$\| x_3 - x_2 \| < 2 \| x_{n_3} - X_{n_2} \|;$$

and so forth. Obviously, $\{x_n\}$ is a fundamental sequence and therefore converges to some $x \in R$. But then the sequence $\{X_{n_k}\}$, and hence the whole sequence $\{X_n\}$, converges to the class X containing x. //

Note 1: The homomorphic mapping of the ring R into the ring of residue classes R/I with respect to a closed ideal that is obtained by assigning to each element $x \in R$ the class X containing it is an open continuous mapping.[3] For let $U \subset R$ be an open sphere with center at the origin,

$$U = \{x \in R : \| x \| < \delta\},$$

and let U' be the image of U in R/I. By the definition of the norm in the residue-class ring, this image consists of precisely those classes $X \in R/I$ for which $\| X \| < \delta$; therefore U' is an open set in R/I. In exactly the same way we can see that the image of every open sphere in R is an open set in R/I. Since the open spheres form a defining system of neighborhoods in R, it follows from this that every open set of R has an open image in R/I. On the other hand, suppose that $F' \subset R/I$ is closed. Let F be the complete inverse image of F', and let $x_1 (\in X_1), x_2 (\in X_2), \ldots, x_n (\in X_n), \ldots$ be a fundamental sequence in F and $x \in R$ its limit $(x \in X)$. Since

$$\| X - X_n \| < \| x - x_n \|,$$

we have $X = \lim_{n \to \infty} X_n$, and therefore X belongs to F'; but then $x \in F$ and therefore F is closed.

Note 2: There is a one-to-one correspondence between the closed ideals J of the ring R containing the closed ideal I and the closed ideals J' of the ring R/I under which every ideal J corresponds to its image J' in R/I.

For by the continuity of the mapping $R \to R/I$, the complete inverse image J of every closed ideal J' of R/I is a closed ideal of R, and from $J_1' \neq J_2'$ it follows that $J_1 \neq J_2$. Conversely, the image J' of every ideal J containing I is an ideal in R/I; furthermore, J is the complete inverse image of J', since if J contains x it contains all the elements that are congruent to x modulo I; and inasmuch as $R \to R/I$ is an open mapping and,

[3] That is, the image of every open set of R is an open set in R/I (the mapping is open) and the complete inverse image of every closed set of R/I is a closed set R (the mapping is continuous).

for an open mapping, the fact that the *complete* inverse image is closed implies that the image is closed, we have that the image J' of every closed ideal J containing I is a closed ideal in R/I.

Obviously, the proper ideals of R/I are the images of proper ideals of R. In particular, the maximal ideals of R/I are the images of the maximal ideals of R containing I.

THEOREM 4: *The ring of residue classes R/M of a normed ring R with respect to a maximal ideal M is a field.*

Proof: By Theorems 1 and 2, it is sufficient to verify that R/M does not contain any non-zero proper ideal. But if there were such an ideal J in R/I, then its inverse image in R would be a proper ideal containing M and not coinciding with it, in contradiction to the maximality of M. //

Note that, by Theorem 3, R/M is a normed ring, because—as we have seen above—a maximal ideal is always closed.

It is easy to see that the converse of Theorem 4 is also true:

THEOREM 5: *If the residue-class ring R/I of R with respect to a proper ideal I is a field, then I is a maximal ideal. It need not be assumed here that I is closed.*

Proof: If R were to contain a proper ideal J containing I and not coinciding with it, then its image in R/I would be a non-zero proper ideal; and this is impossible since, by assumption, R/I is a field. //

***5.** Let us consider the residue-class ring of C with respect to a maximal ideal M. Since M consists of all functions $x(t) \in C$ that vanish at some point τ (see § 2.2), every residue class X consists of all the functions $x(t) \in C$ that assume the same value λ_X at this point. Furthermore $\lambda_{X+Y} = \lambda_X + \lambda_Y$, $\lambda_{XY} = \lambda_X \lambda_Y$, and $\lambda_{\mu X} = \mu \lambda_X$. Moreover, $\| X \| = | \lambda_X |$; for if $x(t) \in X$, then $\| x \| \geqq | x(\tau) | = | \lambda_X |$ and, on the other hand, the function $x(t) \equiv \lambda_X$ belongs to the class X. Thus, C/M is isomorphic to the field of complex numbers.

We shall see later that all commutative normed rings have this property. In proving this, we shall make use of methods of the theory of analytic functions.

§ 3. Abstract Analytic Functions

1. DEFINITION 1: A function $x(\lambda)$ that is defined in some domain D of the complex plane and that takes on values in a normed ring R will be called *analytic* in D if for all $\lambda \in D$ it is strongly differentiable, i.e., if the ratio

$$[x(\lambda + h) - x(\lambda)]/h \tag{1}$$

converges in the norm to a limit $x'(\lambda)$ when $h \to 0$.

Thus, if the inverse element $x(\lambda)=(z-\lambda e)^{-1}$ exists for $\lambda=\lambda_0$ (and so, by the lemma of § 2.1, exists for all λ sufficiently near to λ_0), then it is an analytic function of λ in some neighborhood of λ_0. For

$$\frac{(z-(\lambda+h)e)^{-1}-(z-\lambda e)^{-1}}{h}=(z-(\lambda+h)e)^{-1}(z-\lambda e)^{-1}$$

and by the lemma of § 2.1 the product on the right-hand side converges to $(z-\lambda e)^{-2}$ when $h\to 0$.

If $x(\lambda)$ is an analytic function in D and f an arbitrary linear functional defined on R, then $f\{x(\lambda)\}$ is an ordinary analytic function in D. For the strong convergence of the ratio (1) also implies convergence of the ratio

$$\frac{f\{x(\lambda+h)\}-f\{x(\lambda)\}}{h}=f\left\{\frac{x(\lambda+h)-x(\lambda)}{h}\right\},$$

i.e., the differentiability of the function $f\{x(\lambda)\}$.

The fundamental results of the theory of ordinary analytic functions—most particularly, Cauchy's Theorem and Cauchy's Formula—can be extended to our 'abstract' analytic functions (with values in R), but for this purpose we have to define contour integration of abstract functions.

Let Γ be an orientated arc of a rectifiable curve in the λ-plane and $x(\lambda)$ a function with values in R defined and continuous in norm on Γ. We define the integral $\int_\Gamma x(\lambda)d\lambda$ in the usual way:

$$\int_\Gamma x(\lambda)\,d\lambda=\lim_{\max|\lambda_{k+1}-\lambda_k|\to 0}\sum_{k=0}^{n-1}x(\lambda_k')(\lambda_{k+1}-\lambda_k),$$

where $\lambda_0, \lambda_1, \ldots, \lambda_{n-1}, \lambda_n$ is a subdivision of Γ by a sequence of points, where λ_k' is an arbitrary point on Γ between λ_k and λ_{k+1}, and where the limit is understood in the sense of strong convergence. The existence and uniqueness of this limit follows from the fact that Γ is rectifiable and that $x(\lambda)$ is uniformly continuous on Γ and can be proved by the usual methods. From the definition of the integral it is also clear that

$$f\left\{\int_\Gamma x(\lambda)\,d\lambda\right\}=\int_\Gamma f\{x(\lambda)\}\,d\lambda \tag{2}$$

for every linear functional f.

CAUCHY'S THEOREM: *If a function $x(\lambda)$ with values in a normed ring R is analytic in a closed domain bounded by a simple rectifiable curve Γ, then*

$$\int_\Gamma x(\lambda)d\lambda=0.$$

Proof: We put $\int_\Gamma x(\lambda)d\lambda=y$. By (2) we have for every linear functional

f that $f\{y\} = \int_\Gamma f\{x(\lambda)\} d\lambda$. Therefore, by Cauchy's Theorem for ordinary functions, $f\{y\} = 0$ for every f. But then $y = 0$ as well, because, by the Hahn-Banach Theorem, for every $y \neq 0$ there exists a linear functional f such that $f\{y\} \neq 0$. //

Cauchy's Integral Formula is proved similarly.

CAUCHY'S INTEGRAL FORMULA: *If a function $x(\lambda)$ with values in a normed ring R is analytic in a closed domain bounded by a rectifiable curve Γ, then for all interior points λ of this domain it can be represented in the form*

$$x(\lambda) = \frac{1}{2\pi i} \int_\Gamma \frac{x(\zeta)\, d\zeta}{\zeta - \lambda}. \tag{3}$$

From formula (3) it follows in the usual way that *an analytic function $x(\lambda)$ with values in a normed ring R is infinitely differentiable and can be expanded in the neighborhood of every point of regularity $\lambda = \lambda_0$ in an absolutely convergent Taylor series*

$$x(\lambda) = x(\lambda_0) + x'(\lambda_0)(\lambda - \lambda_0) + \frac{x''(\lambda_0)}{2!}(\lambda - \lambda_0)^2 + \dots,$$

and the radius of convergence of this series is equal to the distance from λ_0 to the nearest singularity of $x(\lambda)$.

2. As an example (which will be required later) let us determine the radius of the largest circle with center at $\lambda = 0$ inside which $(e - \lambda x)^{-1}$ exists. This function is differentiable—i.e., analytic—in its entire domain of existence. Therefore its Taylor series

$$\sum_{n=0}^{\infty} \lambda^n x^n \tag{4}$$

must converge absolutely in the interior of the required circle. Conversely, the function $(e - \lambda x)^{-1}$ obviously exists in the interior of the circle of absolute convergence of (4) and coincides with the sum of this series. But the circle of absolute convergence of (4) is[4]

---

[4] From the inequality (4) of § 1 it follows readily that $\lim_{n \to \infty} \sqrt[n]{\| z^n \|}$ exists for every $z \in R$. In fact, upon setting $\| z^n \| = \alpha_n$, we have by this lemma $\alpha_n = \alpha_{mk+l} \leqq \alpha_k^m \alpha_l$ $(0 \leqq l < k)$, and hence $\alpha_n^{\frac{1}{n}} \leqq \alpha_k^{\frac{1}{k} - \frac{l}{kn}} \alpha_l^{\frac{1}{n}}$. Fixing k and taking $n \to \infty$, we obtain $\overline{\lim}\, \alpha_n^{\frac{1}{n}} \leqq \alpha_k^{\frac{1}{k}}$. Now taking $k \to \infty$, we find $\overline{\lim}\, \alpha_n^{\frac{1}{n}} \leqq \underline{\lim}\, \alpha_n^{\frac{1}{n}} = \lim \alpha_n^{\frac{1}{n}}$. At the same time, we see that $\sqrt[k]{\| z^k \|} \geqq \lim_{n \to \infty} \sqrt[n]{\| z^n \|}$ for all k. Note that we have only used the property of the sequence $\| z^n \| = \alpha_n$ $(\geqq 0)$ that $\alpha_{m+n} \leqq \alpha_m \alpha_n$.

$$|\lambda| < 1/\lim_{n\to\infty} \sqrt[n]{\|x^n\|}.$$

Thus, the required radius is $1/\lim_{n\to\infty} \sqrt[n]{\|x^n\|}$.

***3.** We can verify similarly that the radius ϱ_x of the largest spherical neighborhood of the point $x \in O$ that is entirely contained in O is equal to $1/\lim_{n\to\infty} \sqrt[n]{\|x^{-n}\|}$.

§ 4. Functions on Maximal Ideals. The Radical of a Ring

1. On the basis of the results of the preceding section we can now complete our study of the residue-class ring of a commutative normed ring with respect to a maximal ideal that was begun in § 2.

DEFINITION 1: The *spectrum* of an element x of a normed ring R is the set of all complex numbers λ for which $(x-\lambda e)^{-1}$ does not exist.

THEOREM 1: *Every element x of a normed ring R has a non-empty spectrum.*

Proof: Suppose that, in contradiction to the statement of the theorem, an element $x \in R$ has an empty spectrum, i.e., $(x-\lambda e)^{-1}$ exists for all λ. By taking, in particular, $\lambda = 0$, we see that under the present supposition x^{-1} exists. Moreover, x^{-1} also has an empty spectrum, i.e., $(x^{-1}-\lambda e)^{-1}$ also exists for all λ. For this is clear for $\lambda = 0$; and for $\lambda \neq 0$, we have:

$$(x^{-1}-\lambda e)^{-1} = -\lambda^{-1}x(x-\lambda^{-1}e)^{-1}.$$

Thus, since $(x-\lambda e)^{-1}$ and $(x^{-1}-\lambda e)^{-1}$ are entire functions, by § 3 their Taylor series $\sum_{n=0}^{\infty} x^{-n-1}\lambda^n$ and $\sum_{n=0}^{\infty} x^{n+1}\lambda^n$ converge absolutely in the entire plane, and in particular, for $\lambda = 1$. But then the limit relations $x^{-n} \to 0$ and $x^n \to 0$ must hold simultaneously; and this is impossible, since $x^n x^{-n} = e$. //

***2.** *Note*: Theorem 1 can also be proved by means of Liouville's Theorem, which for abstract analytic functions reads as follows:

If a function $x(\lambda)$ with values in a normed ring R is regular in the entire λ plane and uniformly bounded in norm, then $x(\lambda) \equiv x_0$, where x_0 is some constant element of R.

The proof of this theorem can be carried out by the same method as the proof, given in § 3.1, of Cauchy's Theorem and Cauchy's Formula. By Liouville's Theorem for ordinary analytic functions we have for every linear functional $f: f\{x(\lambda)\} \equiv \text{const.}$ But then we also have $x(\lambda) \equiv \text{const.}$, because if $x(\lambda)$ assumes two distinct values $x(\lambda_1)$ and $x(\lambda_2)$, then by the Hahn-Banach Theorem there exists a linear functional f such that

$$f\{x(\lambda_1)\} \neq f\{x(\lambda_2)\}.$$

Suppose now that $(x-\lambda e)^{-1}$ exists for all λ. We have:

$$\| (x-\lambda e)^{-1} \| = | \lambda^{-1} | \, \| (e-\lambda^{-1}x)^{-1} \|.$$

By the lemma of § 2, the second factor on the right-hand side tends to 1 for $|\lambda| \to \infty$, and therefore $\| (x-\lambda e)^{-1} \| \to 0$. Hence it follows that the function $(x-\lambda e)^{-1}$ is bounded in the entire plane; hence, by Liouville's Theorem, $(x-\lambda e)^{-1} \equiv$ const. Since $(x-\lambda e)^{-1} \to 0$ for $|\lambda| \to \infty$, we conclude that $(x-\lambda e)^{-1} \equiv 0$. But this is impossible. //

3. THEOREM 2: *A normed field R is isomorphic to the field of complex numbers.*

Proof: By Theorem 1, for every $x \in R$ there exists a λ for which the element $x-\lambda e$ does not have an inverse in R. Since R is a field, this means that $x-\lambda e = 0$, i.e., $x = \lambda e$. The correspondence

$$\lambda e \to \lambda \tag{1}$$

is then an isomorphism between R and the field of complex numbers. //

The isomorphism (1) will be called *canonical.*

From Theorem 4 of § 2 and Theorem 2 we now deduce immediately the following important theorem:

THEOREM 3: *The residue-class ring R/M of a commutative normed ring with respect to a maximal ideal M is canonically isomorphic to the field of complex numbers.*

Thus, the maximal ideal M defines a canonical homomorphic mapping of R into the field of complex numbers under which all the elements of the same class of R/M go over into the complex number that corresponds to this class under the canonical isomorphism between R/M and the field of complex numbers.

At the same time Theorem 5 of § 2 shows that, conversely, *every nontrivial algebraic homomorphic mapping of R into the field of complex numbers generates a maximal ideal.* For the kernel of this homomorphic mapping forms an ideal in R whose residue-class ring is isomorphic to the field of complex numbers; by Theorem 5 of § 2, this ideal (which consists of all the elements of the ring that go over into zero) is maximal.

4. We denote by $x(M)$ the number corresponding to the element $x \in R$ under the canonical homomorphic mapping of R into the field of complex numbers defined by the maximal ideal M. By varying M we obtain for every fixed x a function $x(M)$ defined on the set $\mathfrak{M} = \mathfrak{M}(R)$ of all maximal ideals of R. These functions obviously have the following properties:

a) If $x = x_1 + x_2$, then $x(M) = x_1(M) + x_2(M)$.
b) If $x = x_1 x_2$, then $x(M) = x_1(M) x_2(M)$.
c) If $x_2 = \lambda x_1$, then $x_2(M) = \lambda x_1(M)$.
d) $e(M) \equiv 1$.
e) If $x \in M_0$, then $x(M_0) = 0$ and, conversely, if $x(M_0) = 0$, then $x \in M_0$.
f) If $M_1 \neq M_2$, then there exists an $x \in R$ such that $x(M_1) \neq x(M_2)$.

Moreover,

g) $| x(M) | \leqq \| x \|$.

For $x(M)$ is that number λ_X which under the canonical homomorphism between R/M and the field of complex numbers corresponds to the class X containing x. Since $X = \lambda_X E$, we have $\| X \| = | \lambda_X | \, \| E \| = | \lambda_X |$. Bearing in mind the definition of the norm of residue classes we obtain:

$$| x(M) | = | \lambda_X | = \inf_{z \in X} \| z \| \leqq \| x \|.$$

The properties a)-d) show that the functions $x(M)$ form a ring $\hat{R}$ with unit element and that

$$x \rightarrow x(M) \tag{2}$$

is a homomorphic mapping of R onto this ring $\hat{R}$. We shall call (2) *the canonical homomorphic mapping* of R onto $\hat{R}$.

Furthermore the properties a), c), d), and g) show that every maximal ideal M of R generates on R a linear functional with the norm 1 defined by the equation

$$M(x) = x(M)$$

with given M and variable x. For by these properties we have

$$M(x_1 + x_2) = M(x_1) + M(x_2), \quad M(\lambda x) = \lambda M(x),$$
$$| M | \leqq \| x \| \quad \text{and} \quad M(e) = 1.$$

Furthermore, by the separability condition f), distinct maximal ideals generate distinct linear functionals. In the set of all linear functionals on R they are distinguished by the property of 'multiplicativity'

$$M(x_1 x_2) = M(x_1) M(x_2)$$

(which follows from b)); on this account they are called *multiplicative linear functionals.*

By e), we can now formulate Theorem 2 of § 2.3 in the following way:

THEOREM 3′. *For an element x of R to have an inverse it is necessary and sufficient that $x(M)$ should not vanish anywhere.*

***5.** As an illustration of these results let us determine the maximal ideals of the rings W (Example 3, § 1.2) and A (Example 6, § 1.2).

1. Suppose that under the canonical homomorphic mapping with respect to the maximal ideal M the element $\exp(it)$ of W goes over into the number a, so that $\exp(-it)$ goes over into a^{-1}. By g) we have that

$$|a| \leqq \| \exp(it) \| = 1 \quad \text{and} \quad |a^{-1}| \leqq \| \exp(-it) \| = 1,$$

and therefore $a = \exp(it_0)$ $(0 \leqq t_0 < 2\pi)$. Thus, $\exp(it)$ goes over into $\exp(it_0)$. But then $\sum_{n=-\infty}^{\infty} c_n \exp(int) \in W$ goes over into $\sum_{n=-\infty}^{\infty} c_n \exp(int_0)$. Here M consists of all the series that go over into zero, i.e., all the functions $\sum_{n=-\infty}^{\infty} c_n \exp(int)$ that vanish at the point t_0.

On the basis of this result we can now prove the Theorem of Wiener mentioned in § 2.2. In fact, suppose that the sum of the absolutely convergent series $\sum_{n=-\infty}^{\infty} c_n \exp(int)$ vanishes nowhere. By what we have proved, this means that as an element of W it does not belong to any maximal ideal and then, by Theorem 2 of § 2.3, there exists an inverse element in W, i.e., $1/\left\{\sum_{n=-\infty}^{\infty} c_n \exp(int)\right\}$ can also be expanded in an absolutely convergent trigonometrical series.

2. Just as in the discussion of the ring C (§ 2.2), we can satisfy ourselves that the set of all functions of A that vanish at some point ζ_0 of the circle $|\zeta| \leqq 1$ form a maximal ideal in A. Let us show that the converse is also true. Let M_0 be a maximal ideal of A and ζ_0 the number into which the function $x(\zeta) \equiv \zeta \in A$ goes over under the canonical homomorphic mapping with respect to this maximal ideal. The function ζ is a *generator* of the ring A: all the functions of A are limits of uniformly convergent sequences of polomials.[5] Hence it follows that we have for every element $x(\zeta) \in A$ that

[5] For suppose that $x \in A$. $x(\zeta)$ can be uniformly approximated in the circle $|\zeta| \leqq 1$ by the functions $x(\zeta/(1+\epsilon))$ $(\epsilon > 0)$; and the latter, being analytic in the corresponding circle $|\zeta| < 1 + \epsilon$, can be uniformly approximated by polynomials in ζ in the interior of the corresponding circle $|\zeta| \leqq 1$.

$x(M_0) = x(\zeta_0)$, and hence M_0 coincides with the set of all $x(\zeta) \in A$ for which $x(\zeta_0) = 0$. We note that A does not satisfy the condition a) of § 2.2.

***6.** *Note*: Our investigation of the maximal ideals of the rings C, D_n, W, and A has shown that there exists a one-to-one correspondence $t \rightleftarrows M_t$ between the points t of the domain of definition of the functions $x(t)$ in this ring and the maximal ideals M under which $x(t) \equiv x(M_t)$. In these rings we shall therefore identify the maximal ideals with their corresponding points.

7. There is a simple connection between the functions $x(M)$ and the spectrum of the element x (Definition 1):

THEOREM 4: *The spectrum of the element x coincides with the set of values assumed by the functions $x(M)$.*

Proof: If $x(M_0) = \lambda_0$, then $(x - \lambda_0 e)(M_0) = 0$, hence $x - \lambda_0 e \in M_0$, and therefore $(x - \lambda_0 e)^{-1}$ does not exist. Conversely, if $(x - \lambda_0 e)^{-1}$ does not exist, then it means that $(x - \lambda_0 e)(M)$ vanishes on some maximal ideal M_0, i.e., $x(M_0) = \lambda_0$.

8. The kernel of the canonical homomorphic mapping of R onto the ring $\hat{R}$ formed by the functions $x(M)$ is the set of all elements $x \in R$ for which $x(M) \equiv 0$, i.e., that are contained in all the maximal ideals of the ring. Let x be such an element. By Theorem 4, $(x - \lambda e)^{-1}$ exists for all λ except $\lambda = 0$ and therefore the function $(e - \lambda x)^{-1}$ is an entire function. But this means that the radius of convergence of its Taylor series $\sum_{n=0}^{\infty} \lambda^n x^n$, which is equal to $\lim_{n \to \infty} 1/\sqrt[n]{\| x^n \|}$ (see § 2.2) is infinite, i.e.,

$$\lim_{n \to \infty} \sqrt[n]{\| x^n \|} = 0. \tag{3}$$

Conversely, if the relation (3) is satisfied for an element x of R, then the Taylor series of $(e - \lambda x)^{-1}$ is absolutely convergent for all λ; therefore $(x - \lambda e)^{-1}$ exists for all λ except $\lambda = 0$; and therefore, by Theorem 4, $x(M) \equiv 0$.

Thus, the kernel of the canonical homomorphic mapping of R onto the ring $\hat{R}$ of the functions $x(M)$ consists of precisely those elements $x \in R$ for which the relation (3) holds. //

DEFINITION 2: An element $x \in R$ for which the relation (3) holds is called *generalized nilpotent*[6] and the set of all generalized nilpotent elements is called the *radical* of the ring. A ring in which there are no non-zero

[6] Obviously, ordinary nilpotent elements, i.e., elements x satisfying the condition $x^n = 0$ for some n, are also generalized nilpotent elements. The converse, however, is not true in general.

generalized nilpotent elements is called, for short, a *ring without radical.* Our result can now be formulated in the following way:

THEOREM 5: *The intersection of all maximal ideals of a commutative normed ring coincides with the radical of this ring.*

***9.** A trivial example of a ring with radical, i.e., a ring in which there are non-zero generalized nilpotent elements, is the ring $I^{(n)}$ of Example 4 of § 1.2. In this ring $D^{n+1}=0$, and therefore all the elements that do not contain a constant term are nilpotent; and hence they are also generalized nilpotent and form the unique maximal ideal of this ring.

A non-trivial example of a ring with radical is the ring I of Example 5 of § 1.2. It has as a generating function $x_0 \equiv 1$.[7] For $x_0{}^n = t^{n-1}/(n-1)!$, and therefore x_0 generates all the polynomials; and the latter are everywhere dense in norm in the space $L^1(0, 1)$. But

$$\| x_0^n \| = \frac{1}{(n-1)!} \int_0^1 t^{n-1}\, dt = \frac{1}{n!},$$

so that $\sqrt[n]{\| x_0{}^n \|} \to 0$ for $n \to \infty$. Thus, x_0 is a generalized nilpotent element. Since it is a generator, all the elements of I that do not contain the adjoint unit element are generalized nilpotent. They constitute the unique maximal ideal of I.

Hence it follows that every element $\lambda e + x(t) \in I$ in which $\lambda \neq 0$ has an inverse element $\mu e + u(t)$ in I. By choosing $x = -\lambda^2 k$ here we obtain that the equation

$$u(t) = k(t) + \lambda \int_0^t k(t-\tau)\, u(\tau)\, d\tau,$$

where $k(t) \in L^1(0, 1)$, can be solved in $L^1(0, 1)$ for every λ.

Examples of rings without radical are the rings C, D_n, W, and A examined above. For in these rings we have established a one-to-one correspondence between the maximal ideals and the points of the set on which the functions in the ring are given; furthermore, the value of the function $x(M)$ on the maximal ideal M proved to coincide with the value of the function $x(t)$ at the corresponding point t. Thus, in these rings the only element that vanishes on all the maximal ideals is identically zero.

***10.** Let us now investigate the ring generated by a Hermitian kernel $k(s, t)$, where $\overline{k(t, s)} \equiv k(s, t)$, defined and continuous in the square $0 \leqq s, t \leqq 1$ with the norm $\| h \| = \max\limits_{0 \leqq s, t \leqq 1} | h(s, t) |$ and the multiplication

[7] The unit element of the ring is not counted among the generators.

$$g(s, t) * h(s, t) = \int_0^1 g(s, \tau)\, h(\tau, t)\, d\tau,$$

and with a formally adjoined unit element. The absence of non-zero generalized nilpotent elements is here equivalent to the theorem that a Hermitian kernel $h(s, t)$ has at least one eigenvalue. For if $g(s, t)$ is a generalized nilpotent element, then the Hermitian kernel

$$g(s, t) * \overline{g(t, s)} = \int_0^1 g(s, \tau)\, \overline{g(t, \tau)}\, d\tau$$

is also generalized nilpotent and is equal to zero only when $g(s, t) \equiv 0$. But it is well known that the eigenvalues of the Hermitian kernel $h(s, t)$ are the singular points of the Neumann series formed for this kernel, i.e., the singular points of the function $(e - \lambda h)^{-1}$. Now if $h(s, t) \not\equiv 0$ were a generalized nilpotent element, then this function could not have singularities, i.e., h could not have any eigenvalue, in contradiction to the well-known result in the theory of integral equations.

11. The relation (3) which characterizes generalized nilpotent elements is a special case of the following general relation:

THEOREM 6: *For every* $x \in R$

$$\sup_{M \in \mathfrak{M}} |x(M)| = \lim_{n \to \infty} \sqrt[n]{\|x^n\|}, \tag{4}$$

where $\mathfrak{M}$, *on the left-hand side, denotes the set of all maximal ideals of* R.

Proof: We put $\sup\limits_{M \in \mathfrak{M}} |x(M)| = a$. By Theorem 4 the element $x - \mu e$ has an inverse for all μ with $|\mu| > a$, and therefore the function $(e - \lambda x)^{-1}$, where $\lambda = 1/\mu$, is analytic in the circle $|\lambda| < 1/a$. Hence it follows that $1/a$ does not exceed the radius of convergence

$$\frac{1}{\overline{\lim\limits_{n \to \infty}} \sqrt[n]{\|x^n\|}}$$

of the Taylor series of this function, or

$$\overline{\lim_{n \to \infty}} \sqrt[n]{\|x^n\|} \leqq a. \tag{5}$$

On the other hand, since $\sup\limits_{M \in \mathfrak{M}} |x^n(M)| = a^n$ for all n, we have $\sqrt[n]{\|x^n\|} \geqq a$ and hence

$$\lim_{n \to \infty} \sqrt[n]{\|x^n\|} \geqq a. \tag{6}$$

A comparison of the inequalities (5) and (6) leads to the relation (4). The existence of $\lim_{n \to \infty} \sqrt[n]{\| x^n \|}$ is obtained here as an incidental result. //

***12.** As an illustration of Theorem 6 let us consider the ring W of Example 3 of § 1.2. We have found above (§ 4.5) that there exists a one-to-one correspondence $t \rightleftarrows M_t$ between the maximal ideals M of this ring and the points t of the half-closed interval $0 \leqq t < 2\pi$ under which $x(M_t) \equiv x(t)$ for all $x \in W$. Therefore $\sup_{M \in \mathfrak{M}} | x(M) | = \max | x(t) |$. Thus, Theorem 6 applied to this case shows that *the maximum modulus of the sum of an absolutely convergent trigonometrical series is equal to the limit of the n-th root of the sums of the absolute values of the coefficient of the series obtained by raising the given series to the n-th power.* When Theorem 6 is applied to the ring discussed in § 4.10, it shows that *the modulus of the first eigenvalue of a continuous Hermitian kernel is equal to the limit of the reciprocal values of the n-th power of the maximum moduli of the n-th iterated kernel.*

§ 5. The Space of Maximal Ideals

1. As we wish to make the functions $x(M)$ continuous we naturally introduce the following topology in the set $\mathfrak{M}(R)$ of all maximal ideals of R.

DEFINITION 1: Let $x_1, \ldots, x_n$ be arbitrary elements of R and let ε be an arbitrary positive number. The *neighborhood* $[M_0; x_1, \ldots, x_n; \varepsilon]$ of the maximal ideal M_0 is defined as the set of all maximal ideals of M for which the inequalities

$$| x_1(M) - x_1(M_0) | < \varepsilon, \ \ldots, \ | x_n(M) - x_n(M_0) | < \varepsilon$$

hold.

With this definition of neighborhood $\mathfrak{M}(R)$ becomes a Hausdorff space. For, first of all, every maximal ideal obviously has neighborhoods and is contained in each of its neighborhoods. Further, the intersection of the neighborhoods

$$[M_0; x_1, \ldots, x_n; \varepsilon_1] \quad \text{and} \quad [M_0; x_{n+1}, \ldots, x_{n+m}; \varepsilon_2]$$

of the maximal ideal M_0 contains the neighborhood

$$[M_0; x_1, \ldots, x_n, x_{n+1}, \ldots, x_{n+m}; \min(\varepsilon_1, \varepsilon_2)].$$

Moreover, if $M_1 \in [M_0; x_1, \ldots, x_n; \varepsilon]$, then we also have

$$[M_1;\ x_1,\ \ldots,\ x_n;\ \delta]\subset[M_0;\ x_1,\ \ldots,\ x_n;\ \varepsilon],$$

where

$$(0<)\,\delta<\min\{\varepsilon-|x_1(M_1)-x_1(M_0)|,\ \ldots,\ \varepsilon-|x_n(M_1)-x_n(M_0)|\}.$$

Finally, if $M'\neq M$, then by the separability condition f) (§ 4.4) we can find an element $x\in R$ for which $x(M')\neq x(M)$ and the neighborhoods $[M\,;x\,;\varepsilon/2]$ and $[M'\,;x\,;\varepsilon/2]$ do not intersect if we take $\varepsilon<|x(M')-x(M)|$.

The continuity of the functions $x(M)$ now follows immediately from the definition of neighborhood; in order to find a neighborhood of a maximal ideal M in which the values of the function $x(M')$ differ from its value in M by less than ε it is sufficient to take the set $[M\,;x;\,\varepsilon]$, which is a neighborhood by definition.

***2.** Let us verify that in the rings C, D_n, W, and A the topology just assigned to $\mathfrak{M}$ coincides with the topology of the domain of definition of the functions that occur in the corresponding ring (i.e., with the topology of the interval for the rings C and D_n, of the circle for the ring W, and of the disc for the ring A). To do this, we have to show that every old neighborhood is contained in a new one and that, conversely, every new neighborhood is contained in an old one. The first follows from the fact that the functions $x(t)$ that occur in the ring in question are continuous and therefore the sets $[t_0;\ x_1,\ \ldots,\ x_n;\ \varepsilon]$ are open. For the proof of the converse statement we note that in each of the rings under consideration there exists a function that vanishes only at the given point t_0. Let U be an arbitrary neighborhood of t_0 and $x(t)$ a function having the property just mentioned. Now if for every ε the set $[t_0\,;\ x\,;\ \varepsilon]$ were to contain points that are not in U, it would mean that the minimum of the function $|x(t)|$ is zero on the complement of U. But since $x(t)$ is continuous and the domain of definition of the functions in all these rings is compact, $|x(t)|$ would assume this minimum, i.e., $x(t)$ would be zero at some point other than t_0, contrary to assumption. Thus, for some ε the new neighborhood $[t_0;\ x\,;\ \varepsilon]$ is entirely contained in the old neighborhood U. //

3. THEOREM 1: *The space $\mathfrak{M}(R)$ of maximal ideals of a commutative normed ring R, topologized in accordance with Definition* 1, *is compact.*

Proof: With every element x of R we associate the circle Q_x in the complex plane with radius $\|x\|$ and center at the origin. Let Q be the topological product of all these circles, i.e., the space whose points are arbitrary sets (λ_x) of numbers $\lambda_x\in Q_x$ (where x ranges over the whole of R) and in which a fundamental system of neighborhoods of the point $\{\lambda_x^{(0)}\}$ is constituted by the set

$$\left\{(\lambda_x)\colon\ |\lambda_{x_1}-\lambda_{x_1}^{(0)}|<\varepsilon,\ \ldots,\ |\lambda_{x_n}-\lambda_{x_n}^{(0)}|<\varepsilon\right\},$$

given by arbitrary finite collections of elements $x_1, \ldots, x_n \in R$ and arbitrary positive numbers ε. Since all the Q_x are compact, Q is also compact, by Tychonov's Theorem.[8]

In virtue of the inequality $|x(M)| \leqq \|x\|$, every maximal ideal M corresponds to a point $M' = \{\mu_x\} \in Q$, where $\mu_x = x(M)$. Furthermore, by the separability condition f) (§ 4.4) two distinct maximal ideals correspond to points that differ from each other in at least one coordinate and are therefore distinct. Thus, $\mathfrak{M} = \mathfrak{M}(R)$ is mapped one to one onto a part $\mathfrak{M}'$ of the space Q. A comparison of the definitions of the topologies in $\mathfrak{M}$ and in Q now shows that this mapping is homeomorphic. To satisfy ourselves of the compactness of the space $\mathfrak{M}$ it now remains to show that $\mathfrak{M}'$ is closed in Q.

Let $\Lambda = \{\lambda_x\}$ be an adherence point of $\mathfrak{M}'$. Let us show that $\Lambda \in \mathfrak{M}'$, i.e., that there exists a maximal ideal M_0 such that $\lambda_x = x(M_0)$ for all $x \in R$. To this end, we show that $\lambda_{x+y} = \lambda_x + \lambda_y$, $\mu\lambda_x = \lambda_{\mu x}$, and $\lambda_{xy} = \lambda_x\lambda_y$. We confine ourselves to proving the last of these relations; the proof of the first two proceeds along similar lines. Let us consider the neighborhood of Λ determined by the elements e, x, y and xy and an arbitrary positive number ε. Since Λ is an adherence point of $\mathfrak{M}'$, we can find a point $M' \in \mathfrak{M}'$ belonging to this neighborhood, i.e., for some $M \in \mathfrak{M}$ we have

$$|\lambda_e - e(M)| = |\lambda_e - 1| < \varepsilon,$$
$$|\lambda_x - x(M)| < \varepsilon, \quad |\lambda_y - y(M)| < \varepsilon,$$
$$|\lambda_{xy} - (xy)(M)| = |\lambda_{xy} - x(M)\,y(M)| < \varepsilon.$$

But then

$$\begin{aligned} |\lambda_{xy} - \lambda_x\lambda_y| &\leqq |\lambda_{xy} - x(M)\,y(M)| + |x(M)\,[y(M) - \lambda_y]| + \\ &+ |\lambda_y\,[x(M) - \lambda_x]| \leqq |\lambda_{xy} - x(M)\,y(M)| + \|x\|\,|y(M) - \lambda_y| + \\ &+ |\lambda_y|\,|x(M) - \lambda_x| < \varepsilon(1 + \|x\| + |\lambda_y|), \end{aligned}$$

and since ε is arbitrary, it follows that

$$\lambda_{xy} = \lambda_x\lambda_y, \quad \lambda_e = 1.$$

Thus, the correspondence $x \to \lambda_x$ is a homomorphic mapping of R into the field of complex numbers; the kernel of this mapping is then a maximal ideal M_0 such that $x(M_0) = \lambda_x$ for all $x \in R$. //

---

[8] See P. S. Aleksandrov, *Introduction to the General Theory of Sets and Functions*, Moscow-Leningrad, 1948, p. 394 [in Russian; English translation in prep.]. (Or see, for example, J. L. Kelley, *General Topology*, p. 143; Royden, *Real Analysis*, p. 144.)

Note: We have arrived at the definition of the topology in the set $\mathfrak{M}$ of maximal ideals as a result of having set out to make the functions $x(M)$ continuous on $\mathfrak{M}$. The following proposition shows that this requirement uniquely determines the topology in $\mathfrak{M}$ for which $\mathfrak{M}$ becomes compact.

THEOREM 1′: *Let $\mathfrak{M}'$ be the space of all maximal ideals of a commutative normed ring, topologized in any way so that* 1) *$\mathfrak{M}'$ is compact and* 2) *the functions $x(M)$ are continuous on $\mathfrak{M}'$. $\mathfrak{M}'$ is then homeomorphic to $\mathfrak{M}$, where $\mathfrak{M}$ is the space of these maximal ideals, topologized in accordance with Definition* 1.

Proof: The mapping of $\mathfrak{M}'$ onto $\mathfrak{M}$ that associates every maximal ideal $M \in \mathfrak{M}'$ with itself in $\mathfrak{M}$ is one to one. In consequence of condition 2) the inverse image in $\mathfrak{M}'$ of every neighborhood, and therefore also of every open set, of $\mathfrak{M}$ is an open set in $\mathfrak{M}'$. Thus, this mapping of $\mathfrak{M}'$ onto $\mathfrak{M}$ is continuous. But since $\mathfrak{M}$ and $\mathfrak{M}'$ are compact (condition 1)) and $\mathfrak{M}$ is even a Hausdorff space, the inverse mapping is also continuous by a well known theorem of topology, i.e., $\mathfrak{M}'$ is homeomorphic to $\mathfrak{M}$. //

***4.** This theorem contains as special cases facts proved above (§ 5.2) that in the rings C, D_n, W, and A the topologies of the sets of maximal ideals coincide with the original topologies of the domain of definition of the corresponding ring. From this it also follows that the space of maximal ideals of the ring $C(S)$ of all continuous complex functions on the compact Hausdorff space S that was studied in § 2 is homeomorphic to S.

5. In conjunction with results obtained in § 4, Theorem 1 yields Theorem 2.

THEOREM 2: *Every commutative normed ring can be mapped homomorphically onto a suitable ring of continuous functions on a compact Hausdorff space and the kernel of this homomorphic mapping is the radical of the ring. Thus, if the radical of the ring consists of* 0 *only, this mapping is an isomorphism.*

***6.** Does the result just formulated perhaps give something new when applied to rings of functions with the ordinary algebraic operations? We answer this question as follows: By finding the sets of all maximal ideals we establish a natural domain of definition of the functions that form the ring.

First of all, it is clear that every point t_0 of the originally given domain of definition of the functions of the ring generates a maximal ideal: the mapping $x(t) \to x(t_0)$ is a homomorphic mapping of the ring onto the field of complex numbers; and in virtue of the definition, $x(M_0)$, where M_0 is the maximal ideal generated by the point t_0, is here equal to $x(t_0)$. But if two distinct points t_1 and t_2 are to generate two distinct maximal ideals, it is necessary that these points be 'separable' in the ring in question, i.e., that there exist a function $x(t)$ in this ring for which $x(t_1) \neq x(t_2)$. But if the points t_1 and t_2 are not separable in the given ring, then if we identify all the points t with the maximal ideals corresponding to them they are 'fused,' coalesced into one maximal ideal. Thus, the points that differ by an integer multiple of 2π are fused when the points of the line are identified with the corresponding maximal ideals of W; the line is then converted into the circumference of a circle of radius 1—the natural domain of definition of the functions of the ring W, inasmuch as they all have the period 2π.

But this is not the only change that the originally given domain of definition of the functions forming the ring can undergo when the transition to maximal ideals is made: it may also be expanded. Let us consider, for example, the ring C, taking the functions occurring in it as given only at the rational points of $[0, 1]$. Since the rational points are everywhere dense in $[0, 1]$, the functions of C, being continuous, are completely determined by their values at these points. However, the set of maximal ideals of the ring C as thus defined is not exhausted by the set of rational points: maximal ideals are, as before, all the points of $[0, 1]$. The transition to the ring of functions in question as given on the set of all maximal ideals means here the extending of the domain of definition of the functions to its natural limits; in topological terms, a transition from the space of rational points (in the topology induced on $[0, 1]$) to its *compact extension*.[9]

In this connection let us consider the ring B formed by the continuous almost periodic functions (Bohr) on a line, with the norm

$$\| x \| = \sup_{-\infty < t < +\infty} | x(t) |$$

and with the usual algebraic operations. The points of the line are separable in this ring, so that no fusion occurs on transition to the maximal ideals. However, the set of all maximal ideals is not exhausted by the points of the line. For if that were the case, then the fact that an almost periodic function does not vanish would imply that its reciprocal is also an almost periodic function; but this is refuted by the example of the function $2 - \sin x - \sin \lambda x$ for λ irrational. It does not vanish anywhere; however, the greatest lower bound of its values is zero and its reciprocal is therefore not bounded and so does not belong to B.

***7.** In the last two examples the points of the original domain of definition of the functions of the ring were everywhere dense in the compact space of maximal ideals, and finding all the maximal ideals has the character

[9] See § 43.

of a topological completion of the original domain.[1] Extensions of another kind, however, are also possible. Consider, for example, the ring A. Since the functions belonging to this ring are completely determined by their values on the boundary of the circle $|\zeta| \leqq 1$, we could regard A as a ring of continuous functions given a priori not on the whole circle, but only on the circumference $|\zeta| = 1$. In introducing the maximum modulus on the circumference as the norm, we would be preserving the old norm, because the maximum modulus of a function regular in the interior of the circle $|\zeta| \leqq 1$ is assumed on the boundary of this circle. However, the set of maximal ideals of A is not exhausted by the points of the circumference $|\zeta| = 1$; moreover, these points are not even located everywhere dense in the compact space of maximal ideals, which, as we know, coincides with the whole circle. The transition from the points of the domain of definition, given a priori, of the functions of the ring to the space of maximal ideals has the character here not of a topological, but of an 'analytical' extension of this domain.

To sum up, we can say that on transition from the original domain of values of the ring of functions to the space of maximal ideals the following occur: 1) a 'fusion' of points that are inseparable functions of the ring; 2) a closure of the set of points so obtained; and possibly, 3) its 'analytical' extension, when further maximal ideals (maximal in the topological sense) are added, where these maximal ideals are, however, completely determined by giving the ring on the original domain.

8. In order to give the topology in the set of maximal ideals it is not necessary to make use of all the elements of the ring.

Definition 2: A set K of elements of a normed ring R is called a *system of generators* of this ring if the smallest closed subring with unit element containing K is the entire ring R.

Note: The unit element is not included among the generators.

Theorem 3: *The set $\{U\}$ of neighborhoods of the form $[M; x_1, \ldots, x_n; \varepsilon]$, where x_i ranges over all the elements of a system K of generators of R, is a fundamental system of neighborhoods in M.*

Proof: We have to show that every neighborhood $[M; y_1, \ldots, y_n; \varepsilon]$ contains a neighborhood of $\{U\}$. Let

$$P_1(x_{11}, \ldots, x_{1k_1}), \ldots, P_n(x_{n1}, \ldots, x_{nk_n})$$

[1] In the second example, the topology on the line $-\infty < t < +\infty$ induced by the compact space $\mathfrak{M}$ is different from the usual topology of the line; see §§ 29.8 and 43.2. The space $\mathfrak{M}$ of maximal ideals of the ring B can be found in § 29.

be polynomials in elements $x_{ik} \in K$ $(1 \leqq i \leqq n,\ 1 \leqq k \leqq k_i)$, differing in norm from the corresponding elements $y_1, \ldots, y_n$ by less than $\varepsilon/3$ and let δ be such that $[M; x_{11}, \ldots, x_{nk_n}; \delta]$ is contained in $[M; P_1, \ldots, P_n; \varepsilon/3]$. We claim that the neighborhood $[M; x_{11}, \ldots, x_{nk_n}; \delta]$ (which belongs to $\{U\}$) is contained in $[M; y_1, \ldots, y_n; \varepsilon]$. Indeed, we have

$$\begin{aligned}|y_i(M') - y_i(M)| &\leqq |y_i(M') - P_i\{x_{ik}(M')\}| + \\ &+ |P_i\{x_{ik}(M')\} - P_i\{x_{ik}(M)\}| + |P_i\{x_{ik}(M)\} - y_i(M)| \leqq \\ &\leqq 2\|P_i - y_i\| + |P_i\{x_{ik}(M')\} - P_i\{x_{ik}(M)\}| \leqq 2\frac{\varepsilon}{3} + \frac{\varepsilon}{3} = \varepsilon\end{aligned}$$

for all M' of this neighborhood and all $i = 1, \ldots, n$. //

***9.** *Note: If R admits a countable system of neighborhoods $x_1, \ldots, x_n, \ldots$ or, what is equivalent, if R is separable, then $\mathfrak{M}$ satisfies the second axiom of countability* (and hence, by the theorem of Uryson,[2] $\mathfrak{M}$ is *metrizable*). For let $r_1, \ldots, r_n, \ldots$ be a sequence of all the complex rational numbers (i.e., the numbers of the form $r + si$, where r and s are real rational numbers). We claim that a fundamental system of neighborhoods in $\mathfrak{M}$ is the (countable) collection of all open sets of the form

$$\{M \in \mathfrak{M} : |x_{n_k}(M) - r_{m_k}| < 1/p \ \ (k = 1, \ldots, l;\ p \text{ an integer})\}. \tag{1}$$

By Theorem 3, it is sufficient to show that every neighborhood $[M_0; x_{n_1}, \ldots, x_{n_l}; \varepsilon]$ contains such a set. But for this purpose we only have to take in (1) $p > 2/\varepsilon$ and $|r_{m_k} - x_{n_k}(M_0)| < 1/p$ $(k = 1, \ldots, l)$.

10. Of special interest is the case where the ring has a finite number of generators.

THEOREM 4: *If the ring R has a finite number of generators $x_1, \ldots, x_n$, then $\mathfrak{M} = \mathfrak{M}(R)$ is homeomorphic to a closed bounded subset of complex n-dimensional space.*

Proof: $M \to (x_1(M), \ldots, x_n(M))$ is a continuous single-valued mapping of $\mathfrak{M}$ onto some closed bounded subset $\mathfrak{M}'$ of complex n-dimensional space. We shall show that this mapping is one to one; since $\mathfrak{M}$ and $\mathfrak{M}'$ are compact Hausdorff spaces, it is then continuous both ways, i.e., $\mathfrak{M}$ is homeomorphic to $\mathfrak{M}'$.

---

[2] See P. S. Aleksandrov, *Introduction to the General Theory of Sets and Functions*, Moscow-Leningrad, 1948, p. 388 [in Russian; English translation in prep.]. (Or see, for example, W. Sierpiński, *General Topology*, Toronto, 1952, p. 128; J. L. Kelley, *General Topology*, p. 125; P. Alexandroff and H. Hopf, *Topologie, I*, Berlin, 1935, pp. 81 ff.)

Suppose that two points M_1 and M_2 are mapped into the same point of $\mathfrak{M}'$. This means that $x_1(M_1)=x_1(M_2), \ldots, x_n(M_1)=x_n(M_2)$. Then for every polynomial P formed from $x_1, \ldots, x_n$ we have $P(M_1)=P(M_2)$ and since $x_1, \ldots, x_n$ are generators of the ring, the equality $x(M_1)=x(M_2)$ is satisfied for all $x \in R$. But in that case, by the separability condition f) of § 4.4, we have $M_1=M_2$. //

Theorem 4 implies, in particular, that *if R is a ring with a single generator, then $\mathfrak{M}$ is homeomorphic to a closed bounded set of points of the complex plane.*

***11.** The question arises: how can we characterize the sets that are homeomorphic to the space of maximal ideals of rings with n generators among all the closed bounded subsets of the complex n-dimensional space $\mathbf{C}^n$? The answer to this question is as follows: *A set $F \subset \mathbf{C}^n$ is homeomorphic to the space of maximal ideals of a normed ring with n generators if and only if it is bounded, closed, and convex with respect to polynomials.* By convex with respect to polynomials we mean the following: For every point $\zeta^{(0)}=(\zeta_1^{(0)}, \ldots, \zeta_n^{(0)})$ not belonging to F we can find a polynomial $P(\zeta_1, \ldots, \zeta_n)$ that assumes the value 1 at $\zeta^{(0)}$ and values whose absolute value is less than 1 at every point $\zeta \in F$.

To prove this we argue as follows. Let R be a normed ring with n generators $z_1, \ldots, z_n$, $\mathfrak{M}$ the space of its maximal ideals, and F the image of $\mathfrak{M}$ under the mapping $M \to \zeta=(z_1(M), \ldots, z_n(M))$. F is a closed and bounded set in $\mathbf{C}^n$; we have to show that it is convex with respect to polynomials.

Suppose that there is a point $\zeta^{(0)}=(\zeta_1^{(0)}, \ldots, \zeta_n^{(0)}) \in \mathbf{C}^n$ not belonging to F. This means that there exists no homomorphism of R into the field of complex numbers that carries each generator z_j into $\zeta_j^{(0)}$ or, what is the same, there exists no maximal ideal of R containing all the differences $z_j-\zeta_j^{(0)}e$, where e is the unit element of the ring. But the set of all finite sums of the form

$$\sum_{j=1}^{n} (z_j-\zeta_j^{(0)}e)\, q_j, \tag{2}$$

where q_j are arbitrary elements of R, obviously forms an ideal in R. Since this ideal does not belong to any maximal ideal, it must coincide with the whole ring R; in particular, for a suitable choice of the elements q_j the sum (2) gives the unit element of the ring:

$$e=\sum_{j=1}^{n} (z_j-\zeta_j^{(0)}e)\, q_j. \tag{3}$$

The elements q_j that occur in (3) can be approximated in norm with arbitrary accuracy by polynomials P_j in the generators $z_1, \ldots, z_n$. In particular, these polynomials can be chosen so that the norm of the difference

$$e - \sum_{j=1}^{n} (z_j - \zeta_j^{(0)} e) P_j \tag{4}$$

becomes less than 1. The expression (4) represents a certain polynomial $P(z_1, \ldots, z_n)$ in the generators $z_1, \ldots, z_n$. Let us consider the polynomial corresponding to it

$$P(\zeta_1, \ldots, \zeta_n) = 1 - \sum_{j=1}^{n} (\zeta_j - \zeta_j^{(0)}) P_j(\zeta_1, \ldots, \zeta_n).$$

By construction, this polynomial

a) assumes the value 1 at the point $\zeta^{(0)}$;

b) assumes values of absolute value less than 1 at every point $\zeta \in F$.

Thus, F is convex with respect to polynomials.

Suppose, conversely, that F is a closed bounded set in $\mathbf{C}^n$, convex with respect to polynomials. In the ring R' of all possible polynomials $P(\zeta_1, \ldots, \zeta_n)$ we introduce a norm given by the formula

$$\| P \| = \max_{\zeta \in F} | P(\zeta_1, \ldots, \zeta_n) | ; \tag{5}$$

and we let R be the completion of R' in this norm.

R is a ring of continuous functions on F having the functions $z_j(\zeta) = \zeta_j$ $(j = 1, \ldots, n)$ as generators. Obviously, every point $\zeta^{(0)} \in F$ determines a maximal ideal of R corresponding to the homomorphism $z_j \to \zeta_j^{(0)}$ $(j = 1, \ldots, n)$ and distinct points determine distinct maximal ideals. Since R is a ring with the generators $z_1, \ldots, z_n$, the space $\mathfrak{M}(R)$ of its maximal ideals can be regarded as a subset of the space $\mathbf{C}^n$ containing F. Suppose that the point $\zeta^{(1)} \in \mathbf{C}^n$ does not belong to F. Then, by the condition on F, there is a polynomial $P(\zeta_1, \ldots, \zeta_n)$ whose values on F are of absolute value less than 1 and at the point $\zeta^{(1)}$ is equal to 1. Since F is a closed bounded set, the norm of this polynomial, in accordance with the definition (5), is less than 1; therefore the polynomial $P(z_1, \ldots, z_n) \in R$ assumes values less than 1 in absolute value on every maximal ideal and therefore the point $\zeta^{(1)}$ cannot be in $\mathfrak{M}(R)$.

Thus, F coincides with $\mathfrak{M}(R)$ and our statement is completely proved. //

For $n = 1$, the set $\mathfrak{M}$ admits a simpler characterization, one that can be given, moreover, in purely topological terms: $\mathfrak{M}$ is a closed bounded set in the complex plane $\mathbf{C}^1$ that does not split the plane (i.e., has a connected complement). A proof of this statement will be given in § 10. Note that even for a ring with two generators there cannot be a purely topological characterization of the set of maximal ideals as a subset of the complex two-dimensional space $\mathbf{C}^2$. For example, the circle $|z_1|=1$, $|z_2|=0$ cannot be the set of maximal ideals of any normed ring with the generators z_1, z_2, while the circle $z_1=\exp(i\varphi)$, $z_2=\exp(-i\varphi)$ $(0 \leqq \varphi < 2\pi)$ can be identified with the space of maximal ideals of the ring of all continuous functions on this circle whose generators are z_1 and z_2.

§ 6. Analytic Functions of an Element of a Ring

1. If a normed ring R contains the element x, it also contains all the polynomials in x and, more generally, all the 'entire' functions of x, i.e., all the elements of the form $\sum_{n=0}^{\infty} c_n x^n$, where the series $\sum_{n=0}^{\infty} c_n \zeta^n$ represents an entire analytic function of the complex variable ζ. For in this case the series $\sum_{n=0}^{\infty} c_n x^n$ is majorized by the convergent series $\sum_{n=0}^{\infty} |c_n| \, \|x\|^n$. Thus, to every entire analytic function $f(\zeta) = \sum_{n=0}^{\infty} c_n \zeta^n$ there corresponds an 'abstract' analytic function $\tilde{f}(x) = \sum_{n=0}^{\infty} c_n x^n$ defined on all the elements of the ring. Moreover,

$$\tilde{f}(x)(M) = \sum_{n=0}^{\infty} c_n x^n(M) = f(x(M)),$$

since whenever the ring $\hat{R}$ contains the function $x(M)$ it also contains the function $f(x(M))$, whatever the entire analytic function $f(\zeta)$ of the complex variable ζ. Furthermore, we have seen that the ring contains the inverse element x^{-1} of each of its elements x for which $x(M)$ does not vanish on any maximal ideal. Thus, to the analytic function $1/\zeta$ with a pole at $\zeta=0$ there corresponds the abstract 'analytic' function x^{-1}, which is defined for all the elements x whose spectrum does not contain the point 0, i.e., is completely contained in the domain of regularity of $1/\zeta$; moreover,

$$(x^{-1})(M) = 1/x(M).$$

In just the same way, more generally, there corresponds to every rational function $R(\zeta) = P(\zeta)/Q(\zeta)$ of the complex variable ζ an abstract 'rational' function $\tilde{R}(x) = \tilde{P}(x)[\tilde{Q}(x)]^{-1}$ which is defined for all the elements x whose spectrum does not contain any of the zeros of the polynomial $Q(\zeta)$, i.e., is completely contained in the domain of regularity of $R(\zeta)$; moreover, $R(x(M)) = P(x(M))/Q(x(M))$.

Naturally, the question arises whether this correspondence, which has been established for entire and for rational functions, can be extended to arbitrary analytic functions. This question is answered in the affirmative by the following theorem:

THEOREM 1: *Let $f(\zeta)$ be an analytic function of the complex variable ζ and x an arbitrary element of R whose spectrum is contained in the domain of regularity of this function. The integral*

$$\frac{1}{2\pi i}\int_{\Gamma}(\lambda e - x)^{-1} f(\lambda)\, d\lambda, \tag{1}$$

where Γ is an arbitrary rectifiable contour contained in the domain of regularity of $f(\zeta)$ and such that the spectrum S_x of x is contained in the interior of the domain bounded by it, exists and does not depend on the choice of Γ, subject only to the conditions stated. Under the canonical homomorphic mapping of R into the ring $\hat{R}$ of functions $x(M)$ the element $\tilde{f}(x)$ that is represented by the integral (1) *goes over into the function $f(x(M))$:*

$$\tilde{f}(x)(M) = f(x(M)),$$

so that whenever $\hat{R}$ contains $x(M)$ it also contains $f(x(M))$, no matter what the analytic function $f(\zeta)$ of the complex variable ζ may be, as long as it is regular on the set of values of $x(M)$.

Proof: The function $z(\lambda) = (\lambda e - x)^{-1} f(\lambda)$ of the variable λ that occurs under the integral sign in (1) exists and is continuous in norm along the contour Γ, in virtue of the choice of this contour. Therefore, in accordance with what we have said in § 3.1, the integral (1) actually exists in the sense of convergence in norm. Moreover, it does not depend on the choice of Γ, subject of course to the conditions of the theorem, since $z(\zeta)$ is an abstract analytic function of ζ regular outside S_x and, as we have seen in § 3.1, Cauchy's Theorem holds for such functions. Finally, since for fixed M and x variable $x(M)$ is a linear functional of x (see § 4.4), we have by the formula (2) of § 3 and Cauchy's Integral Formula

$$\tilde{f}(x)(M) = \frac{1}{2\pi i}\int_{\Gamma}(\lambda e - x)^{-1}(M) f(\lambda)\, d\lambda =$$

$$= \frac{1}{2\pi i}\int_{\Gamma}\frac{f(\lambda)\, d\lambda}{\lambda - x(M)} = f(x(M)).$$

The proof of the theorem is now complete. //

*__2.__ This theorem contains as a special case the following generalization of Wiener's Theorem, mentioned in §§ 2.2 and 4.5, which is due to P. Levy [34] :[3]

Suppose that the Fourier series of the periodic function $x(t)$ is absolutely convergent and that the values of $x(t)$ lie in the circle $|\zeta - \zeta^0| < \varrho$. If $f(\zeta)$ is a function of the complex variable ζ, regular for all points of the circle, then the Fourier series of $f(x(t))$ is absolutely convergent.

3. An extension of Theorem 1 to the case of (in general, many-valued) analytic functions of several elements of the ring will be given in § 13.

*__4.__ We conclude the present section by establishing the algebraic and topological properties that characterize the mapping

$$f(\zeta) \to \frac{1}{2\pi i} \int_{\Gamma} (\lambda e - x)^{-1} f(\lambda)\, d\lambda. \tag{1'}$$

Let D be a non-empty closed bounded set in the plane of the complex variable ζ; A_D, the ring of all analytic functions regular on D (with the usual operations of addition and multiplication); $\mathfrak{D}$, the set of all elements of a normed ring R whose spectrum is entirely contained[4] in D; and $R_{\mathfrak{D}}$, the ring of all functions defined on $\mathfrak{D}$ with values in R.

THEOREM 2: *The mapping of the ring A_D into $R_{\mathfrak{D}}$ given by formula* (1′) *has the following properties, which determine it uniquely*: a) *it is an algebraic isomorphism under which* b) $f(\zeta) \equiv 1$ *goes over into* $\tilde{f}(x) \equiv e$, c) $f(\zeta) \equiv \zeta$ *goes over into* $\tilde{f}(x) \equiv x$, *and* d) *a sequence of functions* $f_n(\zeta)$ *that is uniformly convergent in some closed domain containing D goes over into a sequence of functions* $\tilde{f}_n(x)$ *that converges in norm for every* $x \in \mathfrak{D}$.

Proof: To begin with, we show that if a mapping having all these properties exists, then it is expressed by the formula (1′), and then we shall satisfy ourselves of the fact that the mapping (1′) does in fact have all the properties.

Suppose, then, that the mapping $f(\zeta) \to \tilde{f}(x)$ in question exists. By the conditions a) and b) the function $1/f(\zeta)$, if it is contained in A_D together with $f(\zeta)$, goes over into $\tilde{f}^{-1}(x) = [\tilde{f}(x)]^{-1}$. Now let ζ_0 be any point not belonging to D. Then the function $(\zeta - \zeta_0)^{-1}$ is contained in A_D; and by what has already been shown, it must go over into the function $(x - \zeta_0 e)^{-1}$, because $\zeta - \zeta_0$ goves over into $x - \zeta_0 e$ in virtue of the conditions a), b), and c). Since $x \in \mathfrak{D}$, by assumption, and $\zeta_0 \notin D$, the latter function actually exists.

[3] See also A. Zygmund, *Trigonometric Series,* Vol. I, Cambridge, 1959, pp. 245-247.
[4] $\mathfrak{D}$ is not empty : if $\lambda_0 \in D$, then in any case $\lambda_0 e \in \mathfrak{D}$.

We now consider an arbitrary function $f(\zeta) \in A_D$. Since $f(\zeta)$ is regular on the closed bounded set D, there exists a rectifiable contour Γ which lies entirely in the domain of regularity of $f(\zeta)$ and which encloses D and has a positive distance from D. By Cauchy's Formula we have, for all points $\zeta \in D$,

$$f(\zeta) = \frac{1}{2\pi i} \int_\Gamma \frac{f(\lambda)\, d\lambda}{\lambda - \zeta}.$$

The right-hand side of this equation is the limit of a sequence of sums of the form

$$f_n(\zeta) = \frac{1}{2\pi i} \sum_{k=0}^{n} \frac{f(\lambda_k)(\lambda_{k+1} - \lambda_k)}{\lambda_k - \zeta} \qquad (\lambda_k \in \Gamma),$$

which converges uniformly on the whole set D. In consequence of what we have shown above and of condition a), there corresponds to each such sum an abstract function

$$\tilde{f}_n(x) = \frac{1}{2\pi i} \sum_{k=0}^{n} f(\lambda_k)(\lambda_k e - x)^{-1}(\lambda_{k+1} - \lambda_k).$$

By the condition d) these functions must converge in norm to a limit for every $x \in \mathfrak{D}$. But they are approximating sums for the integral (1), which, as we have seen in the proof of Theorem 1, actually exists in the sense of convergence in norm and does not depend on the choice of the contour Γ lying in the domain of regularity of $f(\zeta)$ and enclosing the spectrum of x.

Thus, we have come to the conclusion that to every function $f(\zeta) \in A_D$ there corresponds an abstract function

$$\tilde{f}(x) = \frac{1}{2\pi i} \int_\Gamma (\lambda e - x)^{-1} f(\lambda)\, d\lambda, \tag{1''}$$

and this function actually exists and is uniquely determined for all $x \in \mathfrak{D}$.

We shall now show that the mapping $f(\zeta) \to \tilde{f}(x)$ given by formula (1′) has, in fact, the properties stated in the theorem.

Let us show that this mapping is homomorphic. The proof requires only that products go over into products. In other words, we have to prove the equation

$$\frac{1}{2\pi i} \int_\Gamma (\lambda e - x)^{-1} f(\lambda) g(\lambda)\, d\lambda =$$
$$= \frac{1}{2\pi i} \int_\Gamma (\lambda e - x)^{-1} f(\lambda)\, d\lambda \cdot \frac{1}{2\pi i} \int_\Gamma (\lambda e - x)^{-1} g(\lambda)\, d\lambda, \tag{2}$$

where Γ is a rectifiable contour completely contained in the common domain of regularity of the functions $f(\zeta)$ and $g(\zeta)$ and containing the spectrum of x in the interior of the domain bounded by it. In the second integral on the right-hand side, upon replacing the contour Γ by a contour Γ_1 that contains it, has a positive distance from Γ, and is still completely contained in the domain of regularity of the functions $f(\zeta)$ and $g(\zeta)$, we obtain

$$\frac{1}{2\pi i}\int_{\Gamma}(\lambda e-x)^{-1}f(\lambda)\,d\lambda\cdot\frac{1}{2\pi i}\int_{\Gamma_1}(\lambda e-x)^{-1}g(\lambda)\,d\lambda=$$

$$=-\frac{1}{4\pi^2}\int_{\Gamma}\int_{\Gamma_1}(\lambda e-x)^{-1}(\mu e-x)^{-1}f(\lambda)\,g(\mu)\,d\lambda\,d\mu=$$

$$=-\frac{1}{4\pi^2}\int_{\Gamma}\int_{\Gamma_1}\frac{f(\lambda)\,g(\mu)}{\mu-\lambda}[(\lambda e-x)^{-1}-(\mu e-x)^{-1}]\,d\lambda\,d\mu=$$

$$=\frac{1}{2\pi i}\int_{\Gamma}(\lambda e-x)^{-1}f(\lambda)\left(\frac{1}{2\pi i}\int_{\Gamma_1}\frac{g(\mu)\,d\mu}{\mu-\lambda}\right)d\lambda+$$

$$+\frac{1}{4\pi^2}\int_{\Gamma_1}(\mu e-x)^{-1}g(\mu)\left(\int_{\Gamma}\frac{f(\lambda)\,d\lambda}{\mu-\lambda}\right)d\mu.$$

Since the function $f(\lambda)/(\mu-\lambda)$ is regular for all $\mu\in\Gamma_1$ in the closed domain of the λ-plane bounded by Γ, the second summand on the right-hand side is zero by Cauchy's Theorem. Further, since λ is an interior point of the domain of the μ-plane bounded by Γ_1, the inner integral in the first summand is equal to $g(\lambda)$ by Cauchy's Formula. We thus obtain equation (2).

Let us now show that ζ goes over into x, i.e., that the equation

$$\frac{1}{2\pi i}\int_{\Gamma}(\zeta e-x)^{-1}\zeta\,d\zeta=x$$

holds. Since the function $f(\zeta)\equiv\zeta$ is regular in the whole plane, we can take for our Γ a circle of arbitrarily large radius with center at 0. We choose this circle so that the series

$$\zeta(\zeta e-x)^{-1}=e+\zeta^{-1}x+\zeta^{-2}x^2+\ \dots$$

converges absolutely on it. Upon integrating term by term, we obtain

$$\frac{1}{2\pi i}\int_{\Gamma}(\zeta e-x)^{-1}\zeta\,d\zeta=$$

$$=\frac{1}{2\pi i}\int_{\Gamma}(e+\zeta^{-1}x+\zeta^{-2}x^2+\ \dots)\,d\zeta=x\cdot\frac{1}{2\pi i}\int_{\Gamma}\zeta^{-1}\,d\zeta=x.$$

Similarly, we can establish that 1 goes over into e.

Suppose now that $f_n(\zeta) \to f(\zeta)$ uniformly in a closed domain D' containing D. By taking for Γ a contour lying inside D', containing D, and having a positive distance from D, we have

$$\|\tilde{f}(x) - \tilde{f}_n(x)\| = \left\| \frac{1}{2\pi i} \int_\Gamma (\lambda e - x)^{-1} [f(\lambda) - f_n(\lambda)] \, d\lambda \right\| \leqq$$

$$\leqq \max_{\lambda \in \Gamma} |f(\lambda) - f_n(\lambda)| \cdot \frac{1}{2\pi} \int_\Gamma \|(\lambda e - x)^{-1}\| \, |d\lambda| \to 0 \qquad (n \to \infty).$$

It remains to show that the homomorphic mapping $f(\zeta) \to \tilde{f}(x)$ is an isomorphism, i.e., that $\tilde{f}(x)$ is identically zero only when $f(\zeta) \equiv 0$.

But if $f(\zeta) \not\equiv 0$, then there exists a closed set $D_1 < D$ such that $f(\zeta) \neq 0$ at all the points of D_1. Then $1/f(\zeta) \in A_{D_1}$, and the function $\tilde{f}^{-1}(x)$ is defined for all x whose spectrum is contained in D_1. By what we have shown above, we have $\tilde{f}(x)\tilde{f}^{-1}(x) = e$; and hence $\tilde{f}(x) \neq 0$ for all these x. //

§ 7. The Ring $\hat{R}$ of Functions $x(M)$

1. In § 4 it was shown that to every element x of a commutative normed ring R there corresponds in a natural manner a function $x(M)$ defined on the set $\mathfrak{M}$ of all maximal ideals of R, that these functions form a ring $\hat{R}$ with unit element under the ordinary algebraic operations, and that $x \to x(M)$ is a homomorphic mapping of R onto $\hat{R}$ whose kernel is the radical of R. It was thus shown, in particular, that every commutative normed ring R without a radical admits a canonical representation in the form of a ring of functions $\hat{R}$. The supply of functions $x(M)$ is large enough to 'separate' any two distinct points of $\mathfrak{M}$; and in § 5 we used these functions to define a topology in $\mathfrak{M}$ under which $\mathfrak{M}$ becomes a compact Hausdorff space and the $x(M)$ turn out to be continuous functions on this space $\mathfrak{M}$ of maximal ideals.

The question naturally suggests itself: Is the supply of functions $x(M)$ large enough so that all the continuous functions on $\mathfrak{M}$ can be approximated by the $x(M)$?

***2.** In the ring A this problem has a negative answer: As we have seen, the space of maximal ideals of this ring is the circle $|\zeta| \leqq 1$; and since A is closed under passage to the limit, uniformly on this circle, a non-analytic function that is continuous in $|\zeta| \leqq 1$ cannot be uniformly approximated by functions of A.

3. DEFINITION 1: Let R be a commutative normed ring. The ring $\hat{R}$ of functions $x(M)$ defined by R will be called *symmetric* if the complex conjugate of every function in R is also contained in R, i.e., if for every $x \in R$

there exists a $y \in R$ such that $y(M) = \overline{x(M)}$ on the set $\mathfrak{M}$ of all maximal ideals M of R.

***4.** The ring A (which, as we have seen, can be regarded as coinciding with $\hat{A}$) is unsymmetric: $x(\zeta)$ and $\overline{x(\zeta)}$ can only be simultaneously analytic in the circle $|\zeta| < 1$ if $x(\zeta) = \text{const.}$ By way of contrast, the rings C, D_n, and W (regarded as rings of functions on the sets of their maximal ideals) are symmetric: for C and D_n, this is obvious; and in the ring W, the conjugate of $\sum_{n=-\infty}^{\infty} c_n \exp(int)$ is $\sum_{n=-\infty}^{\infty} \bar{c}_{-n} \exp(int)$.

5. THEOREM 1: *If $\hat{R}$ is a symmetric ring, then every function $f(M)$ continuous on $\mathfrak{M}$ is the limit of a uniformly convergent sequence of functions $x(M) \in \hat{R}$.*

Let us note, first of all, that by the symmetry of $\hat{R}$ the real and imaginary parts of each function $x(M) \in \hat{R}$ also belong to $\hat{R}$. Moreover, if

$$x(M_1) \neq x(M_2),$$

then at least one of the corresponding inequalities for $\mathfrak{R}x(M)$ or $\mathfrak{I}x(M)$ must be satisfied, so that distinct maximal ideals can already be separated by real functions of $\hat{R}$. Furthermore, when $x(M)$ is close to $f(M)$, then $\mathfrak{R}x(M)$ is close to $\mathfrak{R}f(M)$ and $\mathfrak{I}x(M)$ to $\mathfrak{I}f(M)$, and vice versa. Thus, Theorem 1 reduces to the following theorem.

6. THEOREM 1′: *On the compact space $\mathfrak{M}$ let a set K of continuous real functions be given that contains all the constants and is an algebraic ring with the usual operations of addition and multiplication and is such that for any two points $M_1 \neq M_2$ there exists a function $\varphi(M) \in K$ for which $\varphi(M_1) \neq \varphi(M_2)$. Then every continuous real function given on $\mathfrak{M}$ is the limit of a uniformly convergent sequence of functions of K.*

Proof of Theorem 1′: We denote by $C^R(\mathfrak{M})$ the space of all continuous real functions on $\mathfrak{M}$ with the norm $\| f(M) \| = \max |f(M)|$ and by $\bar{K}$ the closure of K in $C^R(\mathfrak{M})$. We have to show that $\bar{K} \equiv C^R(\mathfrak{M})$. We break the proof up into several successive steps.

1. *If $f(M) \in \bar{K}$, then $|f(M)| \in \bar{K}$.*

For $f(M)$ as a continuous function given on a compact space is bounded: $|f(M)| \leq a$. We have

$$|f(M)| = \sqrt{a^2 - [a^2 - f^2(M)]} = a \sqrt{1 - \left(1 - \frac{f^2(M)}{a^2}\right)} =$$

$$= a \left\{ 1 - \sum_{n=1}^{\infty} \frac{1 \cdot 1 \cdot 3 \ldots (2n-3)}{2 \cdot 4 \cdot 6 \ldots 2n} \left(1 - \frac{f^2(M)}{a^2}\right)^n \right\},$$

where the series converges uniformly, since $0 \leqq 1 - f^2(M)/a^2 \leqq 1$. Thus, $|f(M)|$ is the limit of a uniformly convergent sequence of functions of $\overline{K}$ (namely, the partial sums of this series), so that it also belongs to $\overline{K}$.

2. *If $f(M)$, $g(M)$, ..., $l(M)$ belong to $\overline{K}$, then*

$$\max [f(M), g(M), \ldots, l(M)]$$

and

$$\min [f(M), g(M), \ldots, l(M)]$$

also belong to $\overline{K}$.

It is obviously sufficient to verify this for the case of two functions. But

$$\max [f(M), g(M)] = [f(M) + g(M) + |f(M) - g(M)|]/2$$

and

$$\min [f(M), g(M)] = [f(M) + g(M) - |f(M) - g(M)|]/2,$$

so that the truth of our statement follows from the result obtained in the preceding step.

3. *For any two points $M_1 \neq M_2$ there exists in $\overline{K}$ a non-negative function that does not exceed 1 and is equal to 1 in M_1 and to 0 in some neighborhood of M_2.*

By assumption there exists a function $\varphi(M) \in K$ such that $\varphi(M_1) \neq \varphi(M_2)$. We put $\psi(M) = [\varphi(M) - \varphi(M_2)]/[\varphi(M_1) - \varphi(M_2)]$. Obviously,

$$\psi(M_1) = 1 \text{ and } \psi(M_2) = 0.$$

By continuity, $\psi(M) < \varepsilon < 1$ in some neighborhood of M_2. The function $\chi(M) = \max [\psi(M) - \varepsilon, 0]/(1 - \varepsilon)$ is then equal to 1 in M_1 and to 0 in the entire neighborhood just mentioned of M_2 and is non-negative; hence, $\omega(M) = \min [\chi(M), 1]$ satisfies all our conditions.

4. *Let F be a closed set in $\mathfrak{M}$ and M_1 a point not belonging to it. Then there exists in $\overline{K}$ a non-negative function that does not exceed 1 and is equal to 1 in M_1 and to 0 on the whole of F.*

For to every point $M_2 \in F$ there corresponds some function with the properties listed in 3. above and, in particular, having some neighborhood in which the function is equal to zero. Since F is compact, there exists a finite number n of such neighborhoods covering the entire set F. Let $\omega_1(M)$, ..., $\omega_n(M)$ be the corresponding functions. Then the function $\omega(M) = \min [\omega_1(M), \ldots, \omega_n(M)]$ obviously satisfies all our conditions.

5. *Let F_1 and F_2 be two non-intersecting closed sets. Then there exists in $\overline{K}$ a non-negative function that does not exceed 1 and is equal to 1 on F_1 and to 0 on F_2.*

For by what we have shown in the preceding stage of the proof, to every point $M_1 \in F_1$ there corresponds a certain non-negative function $\varphi(M) \in \bar{K}$, that does not exceed 1 and is equal to 1 at that point and to 0 on F_2. By continuity, $\varphi(M) > 1 - \varepsilon > 0$ in some neighborhood of M_1. Then the function $\psi(M) = \min [\varphi(M), 1 - \varepsilon]/(1 - \varepsilon)$ is equal to 0 on F_2 and to 1 on the whole of this neighborhood of M_1. By letting M_1 range over the whole of F_1, we obtain a covering of the compact set F_1 by neighborhoods. From this we choose a finite covering $U_1, \ldots, U_n$; let $\psi_1(M), \ldots, \psi_n(M)$ be the corresponding functions. Then the function $\psi(M) = \max [\psi_1(M), \ldots, \psi_n(M)]$ satisfies all our conditions.

6. $\bar{K} \equiv C^{\mathrm{R}}(\mathfrak{M})$.

For let $f(M)$ be an arbitrary continuous real function on $\mathfrak{M}$. Without loss of generality we may assume that $\min f(M) = 0$ and $\max f(M) = 1$. Let n be a natural number. We denote by P_k the set of points M at which $f(M) \leqq k/n$, and by Q_k the set of points M at which $f(M) \geqq (k+1)/n$ $(k = 0, 1, \ldots, n-1)$. By the continuity of $f(M)$, all these sets are closed. Here $P_0 \subset P_1 \subset \ldots \subset P_{n-1}$, $Q_0 \supset Q_1 \supset \ldots \supset Q_{n-1}$, and P_k does not intersect Q_k. Let $\psi_k(M)$ be a function of $\bar{K}$ constructed for Q_k and P_k as described in the preceding part of the proof, i.e., a function that is equal to 1 on Q_k and to 0 on P_k and has 0 and 1 as its lower and upper bounds. We put $\psi(M) = \frac{1}{n} \sum_{k=0}^{n} \psi_k(M)$. We claim that $|f(M) - \psi(M)| \leqq 1/n$. For suppose that at the point M the value $f(M)$ lies between k/n and $(k+1)/n$, so that M is contained in Q_{k-1} and P_{k+1} Then $\psi_0(M) = \psi_1(M) = \ldots = \psi_{k-1}(M) = 1$, $\psi_{k+1}(M) = \psi_{k+2}(M) = \ldots = \psi_{n-1}(M) = 0$; and since $0 \leqq \psi_k(M) \leqq 1$, we have $k/n \leqq \psi(M) \leqq (k+1)/n$; consequently $f(M)$ and $\psi(M)$ lie in the same interval of length $1/n$, and hence

$$|f(M) - \psi(M)| \leqq 1/n.$$

Now, since n is arbitrary and $\bar{K}$ is closed, we have $f(M) \in \bar{K}$. Thus, $\bar{K} \equiv C^{\mathrm{R}}(\mathfrak{M})$, as asserted in Theorem 1′. //

Theorem 1 is an immediate consequence of Theorem 1′. //

*__7.__ Obviously, Theorem 1′ contains as a special case the well-known approximation theorem of Weierstrass (for algebraic and trigonometric polynomials in an arbitrary number of variables). As another illustration of Theorem 1′, we consider an algebraic ring K defined on the topological product $S \times T$ of the compact Hausdorff spaces S and T and generated by the collection of all functions of the form $\varphi(s)\psi(t)$, where $\varphi(s)$ and $\psi(t)$ are continuous real functions defined on S and T, respectively. The elements of this ring are obviously continuous on the compact space $S \times T$. Further, since S is completely regular, for two arbitrary points $s_1 \neq s_2$ of S there exists a continuous function $\varphi(s)$ such that $\varphi(s_1) \neq \varphi(s_2)$, and the same is true of T. Therefore we can also find, for two arbitrary points $(s_1, t_1) \neq (s_2, t_2)$ of $S \times T$, a function in K that assumes distinct values at these points. Thus, the set K satisfies all the conditions of Theorem 1′ and we have the following result: *Every continuous real kernel $k(s, t)$ given on $S \times T$ can be uniformly approximated by 'degenerate' kernels of the form $\sum_{k=1}^{n} \varphi_k(s)\psi_k(t)$, where $\varphi_k(s)$ and $\psi_k(t)$ are real continuous functions given on S and T, respectively.*

Note: The condition that the set K of Theorem 1′ contains the products of its elements was used only in part 1 of the proof of this theorem. Thus, we have in fact proved the following proposition:

Let a set K of real continuous functions given on a compact space $\mathfrak{M}$ satisfy the following conditions: 1) *It contains the function $\varphi(M) \equiv 1$;* 2) *it contains the sum of every pair of functions belonging to K;* 3) *for every $\varphi(M)$ in K it contains $\lambda\varphi(M)$, where λ is an arbitrary real number;* 4) *if $\varphi(M)$ is in K, then $|\varphi(M)|$ is also in K; and* 5) *for every pair of points $M_1 \neq M_2$, it contains a function $\varphi(M)$ such that $\varphi(M_1) \neq \varphi(M_2)$. Then every continuous real function given on $\mathfrak{M}$ is the limit of a uniformly convergent sequence of functions of K.*

8. From Theorem 1 we immediately deduce the following theorem.

THEOREM 2: *If $\hat{R}$ is symmetric and if the uniform convergence of the functions $x_n(M) \in \hat{R}$ implies the convergence in norm of the elements x_n of R, then R is isomorphic to the ring $C(\mathfrak{M})$ of all continuous complex functions on $\mathfrak{M}$ with the norm $\| f \| = \max | f(M) |$.*

COROLLARY: *If the equation $\| x^2 \| = \| x \|^2$ is satisfied for every element x of R and if $\hat{R}$ is symmetric, then R is isometrically isomorphic to the ring $C(\mathfrak{M})$ of all continuous complex functions on $\mathfrak{M}$.*

For from this equality and from Theorem 6 of §4, it follows that

$$\max | x(M) | = \lim_{n \to \infty} \sqrt[2^n]{\| x^{2^n} \|} = \lim_{n \to \infty} \sqrt[2^n]{\| x \|^{2^n}} = \| x \|,$$

and hence the uniform convergence of the functions $x(M)$ implies the convergence in norm of the element x.

§ 8. Rings With an Involution

1. DEFINITION 1: An *involution* in a ring R is an operation $x \to x^*$ that assigns to every element $x \in R$ a uniquely defined element $x^* \in R$ such that the following conditions are identically satisfied:

a) $(x^*)^* = x$;

b) $(\lambda x + \mu y)^* = \bar{\lambda} x^* + \bar{\mu} y^*$;

c) $(xy)^* = y^* x^*$.

A ring R in which an involution is given will be called a *ring with an involution,* and the element x^* will be called *conjugate* to x.

2. Let us give some examples of commutative normed rings with an involution.

1. Let R be the normed ring of bounded functions on an arbitrary set S with the usual operations and the 'uniform' norm $\| x \| = \sup_{t \in S} | x(t) |$. We shall assume, moreover, that the complex conjugate of every function belonging to R is also contained in R. Then $x(t) \to \overline{x(t)}$ is an involution in R.

***3.** 2. Let Q_0 be an arbitrary set of pairwise permutable bounded Hermitian operators in a Hilbert space H. We denote by Q the smallest set of bounded linear operators in H containing Q_0 and the unit operator E and closed with respect to the operations of multiplication by complex numbers, addition, multiplication, taking the conjugate operator, and passing to a limit in the operator norm. Q is a commutative normed ring under the usual operations on operators and the operator norm, and the transition to the conjugate operator is an involution in Q.

3. The ring A of Example 6 of § 1.2 has the involution

$$x(\zeta) = \sum_{n=0}^{\infty} c_n \zeta^n \to \overline{x(\bar{\zeta})} = \sum_{n=0}^{\infty} \bar{c}_n \zeta^n.$$

4. The ring I of Example 5 of § 1.2 has the involution

$$\lambda e + x(t) \to \bar{\lambda} e + \overline{x(t)}.$$

4. In a ring with an involution $x \to x^*$, the elements for which $x^* = x$ will be called *self-conjugate.*

The unit element e of a ring with an involution is a self-conjugate element. For by the conditions a) and c) of Definition 1, we have:

$$e = (e^*)^* = (e^* e)^* = e^*(e^*)^* = e^* e = e^*.$$

In a ring R with an involution, xx^* is a self-conjugate element for every $x \in R$. For by the same conditions, we have $(xx^*)^* = (x^*)^* x^* = xx^*$.

In particular, 0 ($=00^*$) is a self-conjugate element, i.e., $0^*=0$.

Every element x of a ring with an involution can be represented uniquely in the form $y+iz$, where y and z are self-conjugate elements For if $x=y+iz$ is such a representation, then by condition b) of Definition 1, $x^*=y-iz$; hence $y=(x+x^*)/2$ and $z=(x-x^*)/2i$; and the conditions a) and b) show that the y and z so defined are in fact self-conjugate elements.

Definition 2: A commutative normed ring R with an involution will be called a *symmetric ring* (and its involution will be called a *symmetric involution*) if

$$x^*(M)=\overline{x(M)}$$

for all $x\in R$ and all maximal ideals M of the ring R.

Note: Thus, if R is symmetric in the sense of this definition, then the complex conjugate function $\overline{x(M)}$ of every function $x(M)$ contained in $\hat{R}$ is itself contained in $\hat{R}$, i.e., $\hat{R}$ is symmetric in the sense of Definition 1 of § 7 (so that, in particular, Theorem 1 of § 7 is applicable to it).

***5.** Since $\hat{R}$ is algebraically isomorphic to the factor ring of R with respect to its radical, it can be regarded as a commutative normed ring with the norm

$$\| x(M) \| = \inf \{ \| x' \| : \ x' \in R, \ x'(M) \equiv x(M) \}$$

(see Theorem 3 of § 2). Then symmetry of $\hat{R}$ in the sense of Definition 1 of § 7 means the same as symmetry in the sense of the Definition 2 just given. For the set of maximal ideals of $\hat{R}$ can be identified naturally with the set $\mathfrak{M}$ of maximal ideals of R, which is the domain of definition of the functions of $\hat{R}$, and symmetry of R in the sense of the Definition 1 of § 7 is equivalent to the existence of the involution $x(M)\to\overline{x(M)}$, which is symmetric in the sense of Definition 2.

6. Definition 3: A normed ring R with an involution is called *completely regular* if in addition to satisfying conditions a) to c) of Definition 1 its involution also satisfies the condition

$$\text{d)} \quad \| xx^* \| = \| x \| \, \| x^* \| \quad \text{for all} \quad x\in R.$$

Theorem 1: *A commutative completely regular normed ring R is isometrically isomorphic to the ring $C(\mathfrak{M})$ of all continuous complex functions on the space $\mathfrak{M}$ of its maximal ideals.*

Proof: By the note to Definition 2 and the Corollary of Theorem 2 of § 7 it is sufficient to show that a) R is symmetric and b) $\| x^2 \| = \| x \|^2$ for every $x\in R$.

1. Suppose, in contradiction to a), that R is not symmetric, i.e., that there exists an $x_0 \in R$ and an $M_0 \in \mathfrak{M}$ such that $x_0^*(M_0) \neq \overline{x_0(M_0)}$, so that $\mathfrak{J}(x_0 + x_0^*)(M_0) \neq 0$. We put

$$h = \frac{x_0 + x_0^* - \mathfrak{R}(x_0 + x_0^*)(M_0) \cdot e}{\mathfrak{J}(x_0 + x_0^*)(M_0)},$$

where e is the unit element of R. Then $h^* = h$. Further, $h(M_0) = i$ and therefore the element $h - ie$ of R does not have an inverse. From condition c) of Definition 1 it follows easily that $(h - ie)^* = h + ie$ then also does not have an inverse and hence that there exists an $M_1 \in \mathfrak{M}$ such that $h(M_1) = -i$. We choose an arbitrary $N > 0$. Since

$$(h + Nie)(M_0) = (1 + N)i \quad \text{and} \quad (h - Nie)(M_1) = -(1 + N)i,$$

then

$$\| h + Nie \| \geqq 1 + N \quad \text{and} \quad \| h - Nie \| \geqq 1 + N,$$

and hence, by condition d) of Definition 3,

$$\| h^2 + N^2 e \| = \| (h + Nie)(h + Nie)^* \| = \\ = \| h + Nie \| \, \| h - Nie \| \geqq (1 + N)^2.$$

But, on the other hand, $\| h^2 + N^2 e \| \leqq \| h^2 \| + N^2$ and we obtain that $(1 + N)^2 \leqq \| h^2 \| + N^2$ for all $N > 0$, which is impossible.

2. By condition d) of Definition 3,

$$\| (xx^*)^2 \| = \| (xx^*)(xx^*)^* \| = \| xx^* \|^2 = \| x \|^2 \| x^* \|^2$$

and

$$\| x^2(x^*)^2 \| = \| x^2(x^2)^* \| = \| x^2 \| \, \| (x^*)^2 \|.$$

But since $xx^* = x^*x$, we have $x^2(x^*)^2 = (xx^*)^2$. Therefore

$$\| x \|^2 \| x^* \|^2 = \| x^2 \| \, \| (x^*)^2 \|. \tag{1}$$

But $\| x^2 \| \leqq \| x \|^2$ and $\| (x^*)^2 \| \leqq \| x^* \|^2$. Therefore, if we were to assume that $\| x^2 \| < \| x \|^2$, it would follow from equation (1) that $x^* = 0$; but then we would also have $x = (x^*)^* = 0^* = 0$ and this would contradict the inequality assumed. Hence $\| x^2 \| = \| x \|^2$ for all $x \in R$. //

***7.** Since the condition d) of Definition 3 is satisfied for linear operators in a Hilbert space, the ring Q of Example 2 is completely regular. Thus, it follows immediately from Theorem 1 that *every commutative normed ring Q of bounded linear operators in a Hilbert space which is provided with an* *operator norm and is such that the complex conjugate operator of every operator in Q is also in Q is isometrically isomorphic to the ring of all continuous complex functions on the compact space of its maximal ideals.*

From this property of Q one can easily obtain the spectral decomposition of bounded normal operators.[5]

8. The symmetry of a ring can be characterized without the use of the values of its elements on the maximal ideals.

THEOREM 2: *For a commutative normed ring R with an involution $x \to x^*$ to be symmetric it is necessary and sufficient that $(e + x^*x)^{-1}$ should exist in R for all $x \in R$.*

Proof: 1. *Necessity.* If R is symmetric, then

$$(e + x^*x)(M) = 1 + |x(M)|^2 \neq 0$$

for all maximal ideals M, so that $(e + x^*x)^{-1}$ exists.

2. *Sufficiency.* To begin with, we show that *if the condition of the theorem is satisfied, then the function $x(M)$ is real for every self-conjugate element $x \in R$.* By Theorem 4 of §4, it is sufficient for this purpose to show that $(x - (\alpha + i\beta)e)^{-1}$, where α and β are real numbers, exists for all $\beta \neq 0$. But $(x - \alpha e)^* = x^* - \bar{\alpha} e^* = x - \alpha e$. Therefore, for $\beta \neq 0$ we have

$$(x - (\alpha + i\beta)e)(x - (\alpha - i\beta)e) = (x - \alpha e)^2 + \beta^2 e = \\ = (x - \alpha e)^*(x - \alpha e) + \beta^2 e = \beta^2(e + z^*z),$$

where $z = (x - \alpha e)/\beta$. And since the right-hand side, by the condition of the theorem, has an inverse element, the element

$$(x - (\alpha + i\beta)e)^{-1} = (x - (\alpha - i\beta)e)((x - \alpha e)^2 + \beta^2 e)^{-1}$$

also exists.

Now let x be an arbitrary element of R. Representing x in the form $y + iz$, where y and z are self-conjugate elements, we have that $x(M) = y(M) + iz(M)$ and $x^*(M) = y(M) - iz(M)$; and consequently $x^*(M) = \overline{x(M)}$, because $y(M)$ and $z(M)$ (as we have proved) are real. //

9. Theorem 2 enables us in certain cases to establish the symmetry of a ring without finding its maximal ideals. Let us show, for example, that *the involution $x(t) \to \overline{x(t)}$ on the ring R of Example* 1 *is symmetric.* By Theorem 2, it is sufficient for this purpose to show that if $x(t) \in R$, then $1/(1 + |x(t)|^2) \in R$ as well.

---

[5] See, for example, [35], pp. 94-95.

Suppose that $\| x \| = a$. Then

$$\frac{1}{1+|x(t)|^2} = \frac{1}{(a^2+1)-[a^2-|x(t)|^2]} =$$

$$= \frac{1}{a^2+1} \frac{1}{1-\dfrac{a^2-|x(t)|^2}{a^2+1}} = \frac{1}{a^2+1} \sum_{n=0}^{\infty} \left(\frac{a^2-|x(t)|^2}{a^2+1} \right)^n,$$

where the series converges uniformly, because

$$\left| \frac{a^2-|x(t)|^2}{a^2+1} \right| \leqq \frac{a^2}{a^2+1} < 1.$$

But all the partial sums of this series belong to R. Since R is closed with respect to uniform passage to the limit, the sum of the series, that is, the function $1/(1+|x(t)|^2)$, also belongs to R. Hence R (and so $\hat{R}$ as well) is symmetric.

Since the condition $\| x^2 \| = \| x \|^2$ is also satisfied in R, we conclude, bearing in mind the Corollary of Theorem 2 of § 7, that the following theorem holds.

THEOREM 3: *Every normed ring R of bounded functions (with the usual operations) on an arbitrary set S that is provided with a uniform norm and is such that the complex conjugate of every function in R is also isometrically isomorphic to the ring $C(\mathfrak{M})$ of all continuous complex functions on the compact space $\mathfrak{M}$ of its maximal ideals; and complex conjugate functions of R correspond to complex conjugate functions in $C(\mathfrak{M})$.*

***10.** In particular, this is true for the ring B of continuous, almost periodic functions on a line (or on any topological group).

We also note that the symmetry of the ring Q of Example 2, which follows from Theorem 1, can also be proved on the basis of Theorem 2. Let $A \in Q$ and $\| A \| = a$. It is sufficient to prove that

$$\| (a^2E - A^*A)/(a^2+1) \| < 1$$

because we then have, exactly as in the preceding case,

$$(E+A^*A)^{-1} = \frac{1}{a^2+1} \sum_{n=0}^{\infty} \left(\frac{a^2E-A^*A}{a^2+1} \right)^n.$$

But since $(A^*Af, A^*Af) = (AA^*Af, Af) \leqq \| AA^* \| (Af, Af) = a^2(Af, Af)$, then

$$\|a^2E - A^*A\|^2 = \sup_{f\neq 0} \frac{((a^2E - A^*A)f, (a^2E - A^*A)f)}{(f, f)} =$$

$$= \sup_{f\neq 0} \frac{a^4(f, f) - 2a^2(Af, Af) + (A^*Af, A^*Af)}{(f, f)} \leqq$$

$$\leqq \sup_{f\neq 0} \frac{a^4(f, f) - a^2(Af, Af)}{(f, f)} \leqq a^4,$$

from which it follows that

$$\left\| \frac{a^2E - A^*A}{a^2+1} \right\| \leqq \frac{a^2}{a^2+1} < 1.$$

11. DEFINITION 4: A linear functional f defined on a ring R with an involution is called *positive* if $f(xx^*) \geqq 0$ for all $x \in R$ (or, what is the same, $f(x^*x) \geqq 0$ for all $x \in R$). An involution in R will be called *essential* if for every non-zero element $x_0 \in R$ there exists a positive linear functional f_0 such that $f_0(x_0x_0^*) \neq 0$.[6]

***12.** Let R be the ring of Example 1. Every point $t_0 \in S$ determines a linear functional $f_{t_0}(x) = x(t_0)$ on R which is positive with respect to the involution $x(t) \to \overline{x(t)}$, and since $(x_0x_0^*)(t) \equiv |x_0(t)|^2 \not\equiv 0$ means that $f_t(x_0x_0^*) \neq 0$ for some $t \in S$, this involution is essential. As we have seen, it is also symmetric.

Let Q be the ring of Example 2. Every element $f \in H$ determines the linear functional $f(A) = (Af, f)$ $(A \in Q)$ on Q; $f(A)$ is positive, because $f(A^*A) = (A^*Af, f) = (Af, Af) \geqq 0$. Moreover, since for every $A_0 \neq 0$ there exists an element $f_0 \in H$ such that $(A_0^*A_0f_0, f_0) = \|A_0f_0\|^2 \neq 0$, the involution $A \to A^*$ in Q is essential. As we have seen, this involution is also symmetric.

***13.** The involution $x(\zeta) \to \overline{x(\bar{\zeta})}$ in the ring A of Example 6 of § 1.2 is also essential. For every non-decreasing real function $\sigma(t)$ given on the interval $[-1, 1]$ determines on A the positive linear functional[7]

$$f_\sigma(x) = \int_{-1}^{1} x(t)d\sigma(t);$$

if $f_\sigma(xx^*) = \int_{-1}^{1} |x(t)|^2 d\sigma(t) = 0$ for all σ, then $x(t) = 0$ for all $t \in [-1, 1]$

[6] It can be shown that, in a ring with an essential involution, for every non-zero element x_0 there exists a positive linear functional f_0 such that $f_0(x_0) \neq 0$.

[7] This is the general form of a positive linear functional on A; see §§ 47 and 50.

and hence, being an analytic function, $x(\zeta) \equiv 0$. Note that the involution in question is not symmetric. For $x(\zeta) \equiv \zeta \in A$, but the function

$$1/[1 + x(\zeta)\overline{x(\bar{\zeta})}] = 1/[1 + \zeta^2]$$

has poles at the points $\zeta = \pm i$ and therefore does not belong to A. Besides, the asymmetry of the ring A, established in § 7, shows that no involution in this ring can be symmetric.

From Theorem 4 to be proved below, it follows that no involution in the ring I of Example 5 of § 1.2 can be essential. However, by associating the element $z^* = \bar{\lambda}e + x^*(t)$ with the element $z = \lambda e + x(t)$, we see that it is symmetric, since the values of z and z^* on the unique maximal ideal of I (which is formed by the elements z for which $\lambda = 0$; see § 4.9) are equal to λ and $\bar{\lambda}$, respectively.

14. Let f be a positive linear functional on the commutative normed ring R with the involution $x \to x^*$. From the fact that for all complex numbers λ the inequality

$$f((x + \lambda e)(x + \lambda e)^*) = f(xx^*) + \lambda f(x^*) + \bar{\lambda} f(x) + |\lambda|^2 f(e) \geqq 0$$

holds it follows easily that

$$f(x^*) = \overline{f(x)} \tag{2}$$

and

$$|f(x)|^2 \leqq f(e) f(xx^*). \tag{3}$$

Upon putting $xx^* = z$, applying inequality (3) repeatedly, and bearing in mind that $(z^m)^* = z^m$, we obtain

$$|f(x)| \leqq f(e)^{\frac{1}{2}} f(z)^{\frac{1}{2}} \leqq f(e)^{\frac{1}{2}+\frac{1}{4}} f(z^2)^{\frac{1}{4}} \leqq \cdots$$
$$\cdots \leqq f(e)^{\frac{1}{2}+\frac{1}{4}+\cdots+\frac{1}{2^n}} f\left(z^{2^{n-1}}\right)^{\frac{1}{2^n}}.$$

Since $|f(y)| \leqq \|f\| \, \|y\|$, it now follows that

$$|f(x)| \leqq f(e)^{1-\frac{1}{2^n}} \|f\|^{\frac{1}{2^n}} \left\|z^{2^{n-1}}\right\|^{\frac{1}{2^n}}.$$

Passing to the limit $n \to \infty$ and bearing Theorem 6 of § 4 in mind, we obtain

$$|f(x)| \leqq f(e) \max_{M \in \mathfrak{M}} |(xx^*)(M)|^{\frac{1}{2}}. \tag{4}$$

This inequality has an important consequence.

THEOREM 4: *A commutative normed ring R with an essential involution is a ring without radical.*

Proof: When we replace x by xx^* in (4), we obtain

$$|f(xx^*)| \leqq f(e) \max_{M \in \mathfrak{M}} |(xx^*)(M)|.$$

This inequality shows that *if x is a generalized nilpotent element, i.e., $x(M) \equiv 0$, then $f(xx^*) = 0$ for all positive linear functionals f.* But if the involution in R is essential, then it follows from this that $x = 0$, i.e., that R is a ring without radical. //

In Chapter IV we shall give applications of Theorem 4 to harmonic analysis on groups.

THEOREM 5: *Every positive linear functional f on a commutative normed ring R with a symmetric involution $x \to x^*$ defines a positive linear functional $\hat{f}(x(M)) = f(x)$ on the space $\hat{R}$ of functions $x(M)$ that can be extended uniquely to a positive linear functional on the space $C(\mathfrak{M})$ of all continuous complex functions on $\mathfrak{M}$, where $\mathfrak{M}$ is the space of the maximal ideals of R. Positiveness of a functional on the space of functions is understood here in the usual way.*

Proof: Since R is a ring with a symmetric involution, $x^*(M) = \overline{x(M)}$, and therefore the inequality (4) assumes the form

$$|f(x)| \leqq f(e) \max_{M \in \mathfrak{M}} |x(M)|. \tag{4'}$$

This shows that $\hat{f}$ is a uniquely defined linear functional on R provided with a uniform norm $\|x\|_0 = \max_{M \in \mathfrak{M}} |x(M)|$. Let us show that this functional is positive, i.e., that $\hat{f}(x(M)) \geqq 0$ if $x(M) \geqq 0$ for all $M \in \mathfrak{M}$. We note, first of all, that if $x(M)$ is real for all $M \in \mathfrak{M}$, then $f(x)$ is real. For since $x^*(M) = x(M)$ in that case, x^* can differ from x only by a generalized nilpotent element; and on such an element, f vanishes; therefore $f(x^*) = f(x)$, i.e., by (2), $\overline{f(x)} = f(x)$. Suppose now that $x(M) \geqq 0$ for all $M \in \mathfrak{M}$. We put $\|x\|_0 = a$. Since

$$0 \leqq a - x(M) = (ae - x)(M) \leqq a$$

and, by what we have just proved, $f(ae - x)$ is real, on applying the inequality (4′), we obtain

$$af(e) - f(x) = f(ae - x) \leqq f(e) \|ae - x\|_0 \leqq af(e).$$

But this shows that $\hat{f}(x(M)) = f(x) \geqq 0$.

By the symmetry of $\hat{R}$ and Theorem 1 of § 7, the completion of $\hat{R}$ with respect to the uniform norm coincides with $C(\mathfrak{M})$. Hence it follows that the functional $\hat{f}$ can be extended uniquely to a linear functional on $C(\mathfrak{M})$. From Theorem 1 of § 7 it also follows easily that every non-negative continuous real function given on $\mathfrak{M}$ can be uniformly approximated by non-negative real functions of $\hat{R}$. In conjunction with the preceding, it follows from this that $\hat{f}$ remains positive on the entire space $C(\mathfrak{M})$. //

***15.** As Markov [33] has shown, every positive linear functional φ on the space $C(\mathrm{K})$ of all continuous complex functions given on a compactum K can be uniquely represented in the form

$$\varphi(z)=\int_{\mathrm{K}} z(t)d\Phi(t),$$

where Φ is a non-negative real completely additive regular[8] set function on the field of Borel subsets of K. Therefore Theorem 5 immediately implies the following theorem.

THEOREM 5′: *Every positive linear functional f defined on a commutative normed ring R with a symmetric involution is representable uniquely in the form*

$$f(x)=\int_{\mathfrak{M}} x(M)d\Phi(M),$$

where Φ is a non-negative real completely additive regular set function on the field of Borel sets of the space $\mathfrak{M}$ of maximal ideals of R.

In Chapter IV this theorem will be applied to yield a proof of the Theorem of Bochner on positive-definite functions on a group.

16. Obviously, *the linear functional $M(x)=x(M)$ on a commutative normed ring with a symmetric involution that generates the maximal ideal M is positive.* For $M(x^*)=\overline{M(x)}$; and consequently

$$M(xx^*)=|M(x)|^2\geqq 0$$

for all $x\in R$.

Incidentally, it follows from this that for symmetric rings the converse of Theorem 4 is also true, i.e., *if R is a symmetric ring without a radical, then*

[8] A function defined on the field of Borel sets is called *regular* if the value of its total variation on every Borel set B is equal to the lower bound of the values of its total variation on the open sets containing B and to the upper bound of such values on the compact sets contained in B.

the involution $x \to x^*$ *defined in* R *is essential.* Indeed, for every non-zero element $x_0 \in R$ there exists a maximal ideal M_0, and hence also the positive linear functional $M_0(x)$ generated by it, such that

$$M_0(x_0 x_0^*) = |\, x_0(M_0) \,|^2 \neq 0.$$

The reader can find further properties of rings with an involution in the paper by Gelfand and Naimark, "Normed rings with an involution, and their representations," which constitutes Chapter VIII of this book.

CHAPTER II

THE GENERAL THEORY OF COMMUTATIVE NORMED RINGS (*continued*)

§ 9. The Connection between Algebraic and Topological Isomorphisms

1. If R is a ring without radical, then, by Theorem 2 of § 5, R is isomorphic to the ring $\hat{R}$ of functions $x(M)$ on the set of maximal ideals, generated by R. Therefore, rings without radical will also be called *rings of functions.*

In this section, we shall study the relation between the topological and algebraic properties of rings of functions.

THEOREM 1: *Let R_1 and R_2 be two normed rings of complex functions, defined on the same set which serves as the set of maximal ideals for each of them, and let R_1 be contained in R_2. Then every sequence of functions $x_1, \ldots, x_n, \ldots$ of R_1 that converges in the norm of R_1 also converges in the norm of R_2.*

Proof: We denote the norm of the function $x \in R_1$ in R_1 by $\| x \|_1$ and that in R_2 by $\| x \|_2$. We put

$$\| x \| = \max \{ \| x \|_1, \| x \|_2 \}. \tag{1}$$

It is easy to verify that all the conditions imposed on the norm are then satisfied. Let us show that R_1 is still a complete space in this new norm. Let $x_1, \ldots, x_n, \ldots$ be a sequence of functions of R_1 that is a fundamental sequence in the sense of the norm (1). Obviously, it is a fundamental sequence with respect to the two previous norms. Let x and x' be its limits in these norms. Since the sequence of functions $x_n(M)$ converges uniformly to $x(M)$ and to $x'(M)$, we have $x(M) \equiv x'(M)$, and therefore $x = x'$. Hence

$$\| x - x_n \| = \max \{ \| x - x_n \|_1, \| x - x_n \|_2 \} \to 0 \qquad (n \to \infty).$$

Therefore x is also the limit of the sequence $x_1, \dots, x_n, \dots$ in the norm (1) as well. R_1 is thus complete in this norm.

Obviously, convergence in R_1 in the norm (1) implies convergence in the original norm of R_1. Since R_1 is complete in both norms, we can appeal to Banach's Theorem,[1] which states that a one-to-one mapping of complete spaces that is continuous in one direction is also continuous in the other, and deduce from this that convergence in the original norm of R_1 implies convergence in the norm (1) and hence also in the norm of R_2. //

***2.** As an illustration of Theorem 1 we prove the following proposition.

Let R be a normed ring of infinitely differentiable functions $x(t)$ on the interval $0 \leqq t \leqq 1$ which serves as set of maximal ideals for R. There exists a sequence of positive numbers $m_0, m_1, \dots, m_n, \dots$ such that for every function $x(t) \in R$

$$\max_{0 \leqq t \leqq 1} |x^{(n)}(t)| < C m_n, \tag{2}$$

where the constant C depends only on the function and not on n.

Proof: It is sufficient to produce, for every n, a positive number m_n such that it follows from $x(t) \in R$, $\| x \| \leqq 1$ that $\max_{0 \leqq t \leqq 1} |x^{(n)}(t)| \leqq m_n$. Suppose that there is an n for which such a number m_n cannot be found; then R must contain a sequence of functions $x_1(t), \dots, x_k(t), \dots$ such that $\| x_k \| \leqq 1$ for all k and $\max_{0 \leqq t \leqq 1} |x_k^{(n)}(t)| > k$. The functions $y_k(t) = (1/k) x_k(t)$ converge to 0 in norm. Since R is obviously contained in D_n (Example 2 of § 1.2; see also § 2.2), Theorem 1 shows that the $y_k(t)$ also converge to zero in the sense of convergence in D_n; but this is impossible, since

$$\max_{0 \leqq t \leqq 1} |y_k^{(n)}(t)| > 1. \quad //$$

COROLLARY: *The ring D_∞ of all infinitely differentiable functions on the interval $0 \leqq t \leqq 1$ cannot be normed.*

Proof: Suppose that D_∞ is normed and that $m_0, m_1, \dots, m_n, \dots$ are the numbers that occur in the proposition proved above. It is known that an infinitely differentiable function $y(t)$ can be constructed such that

$$\max_{0 \leqq t \leqq 1} |y^{(n)}(t)| > n m_n \qquad (n = 1, 2, \dots).^2$$

---

[1] See footnote 6 (§ 1.4).

[2] Let $1 = k_1 < k_2 < \dots < k_n < \dots$ be a sequence of natural numbers. We put $a_k = 1/(2\pi k)^n$ for $k_n \leqq k < k_{n+1}$ $(n = 1, 2, \dots)$. Since a_k tends to zero faster than

But this function cannot satisfy the inequality (2), in contradiction to what we have shown. //

3. From Theorem 1 we obtain immediately the following theorem.

THEOREM 2: *Algebraically isomorphic commutative normed rings without radical are also topologically isomorphic, i.e., the convergence of elements in one of these rings is equivalent to the convergence of the corresponding elements in the other.*

The meaning and the significance of this proposition consists in the following. In order to specify a normed ring one has to give, first of all, a list of its elements with their algebraic properties and, second, a norm. Theorem 2 shows that for rings of functions the norm is uniquely determined to within equivalence by merely giving the list of elements with their algebraic properties.

***4.** From Theorem 2 it follows, in particular, that the sufficient condition, stated in Theorem 2 of § 7, for the isomorphism of a symmetric ring R with the ring of all continuous functions on the set of its maximal ideals is also necessary. Thus, the algebraic ring of all continuous functions on the compact Hausdorff space S can be normed in a unique way to within equivalence: all possible norms of this ring are equivalent to the norm $\|x\| = \max_{t \in S} |x(t)|$.

5. THEOREM 3: *Every automorphism of a commutative normed ring R without radical is continuous.*

Proof: Let an automorphism of R be given, i.e., a one-to-one mapping of this ring onto itself with the following properties: If x goes over into x', then λx goes over into $\lambda x'$ for all complex numbers λ; if x, y go over into x' and y', respectively, then $x + y$ goes over into $x' + y'$ and xy into $x'y'$. We may regard this automorphism as an algebraic isomorphism between the ring R of functions x and the ring R' of the corresponding functions x'. By Theorem 2, the rings R and R' are also topologically isomorphic, i.e., convergence of a sequence $x_1, \ldots, x_n, \ldots$ is equivalent to convergence of the corresponding sequence $x_1', \ldots, x_n', \ldots$. But this shows that the automorphism is continuous. //

---

any power of $1/k$, the function $y(t) = \sum_{k=1}^{\infty} a_k \exp(2\pi ikt)$ is infinitely differentiable. Moreover,

$$|y^{(n)}(0)| = \sum_{k=1}^{\infty} a_k (2\pi k)^n > \sum_{k=k_n}^{k_{n+1}-1} a_k (2\pi k)^n = k_{n+1} - k_n.$$

By choosing the sequence $\{k_n\}$ so that $k_{n+1} - k_n > nm_n$ $(n = 1, 2, \ldots)$, we obtain a function $y(t)$ with the required properties.

THEOREM 4: *In a symmetric ring without radical the involution is continuous.*

Proof: Let R be a symmetric ring without radical. In that case the element x^* conjugate to x is completely determined. The mapping $x \to x^*$ of R onto itself differs from an automorphism only in that λx goes over not into λx^*, but into $\bar{\lambda} x^*$. Theorem 1 shows that this mapping is continuous, i.e., that the convergence of x_n to x implies the convergence of x_n^* to x^*.[3] To satisfy ourselves of this, it is sufficient to introduce in R, apart from the original norm $\| x \|$, the new norm $| x | = \| x^* \|$. //

§ 10. Generalized Divisors of Zero

1. In algebra, a divisor of zero of a ring R is an element x of the ring for which the product with some element $y \neq 0$ is zero.

***2.** In the ring C, for example, the function $x(t)$ that is equal to zero on the interval $\Delta = [0, 1/2]$ is a divisor of zero, because its product with a function $y(t) \in C$ equal to zero everywhere outside Δ but not identically zero is equal to zero.

3. This concept can be generalized to normed rings in a very natural way, as follows:

DEFINITION 1: An element x of a normed ring R is called a *generalized divisor of zero* if there exists a sequence $\{y_n\} \subset R$ such that

1) $\inf_n \| y_n \| > 0$;

2) $\lim_{n \to \infty} \| x y_n \| = 0$.

Obviously, ordinary divisors of zero are also generalized divisors of zero. In a finite-dimensional ring,[4] where a convergent sequence can be selected from every bounded sequence, the converse is also true: Every generalized divisor of zero is a divisor of zero in the ordinary sense. But generalized divisors of zero need not in general be ordinary divisors of zero.

***4.** Let us consider a function $x(t)$ in C that is equal to zero at the point $t = 0$ only; it is not an (ordinary) divisor of zero. Further, let $x_n(t)$ be a positive function that is equal to zero on $[1/n, 1]$ and assumes its maximal

[3] The converse statement—the convergence of x_n^* to x^* implies the convergence of x_n to x—also holds here, because for every $z \in R$ we have $z^{**} = z$.

[4] I.e., in every ring whose elements form a finite-dimensional space.

value on $[0, 1/n]$, with $\| x_n(t) \| = 1$. Obviously, $xx_n \to 0$; therefore $x(t)$ is a generalized divisor of zero.

5. *An element x can only be a generalized divisor of zero if it does not have an inverse.* For suppose that x^{-1} exists. If $\lim xy_n = 0$, then $\lim y_n = \lim x^{-1}xy_n = 0$ as well; hence the conditions 1) and 2) of Definition 1 are incompatible, so that x is not a generalized divisor of zero.

***6.** It is easy to see that in C *every* element that does not have an inverse is a generalized divisor of zero (see above). However, not all rings have this property.

Let us consider, for example, the element $x(\zeta) = \zeta$ of the ring A (§ 1.2, Example 6), which obviously does not have an inverse. Let a sequence $\{y_n(\zeta)\} \subset A$ be such that $\| \zeta y_n(\zeta) \| \to 0$. On the boundary of the unit circle $|\zeta| = 1$ we therefore have $| y_n(\zeta) | \to 0$. By the maximum-modulus principle, $y_n(\zeta)$ converges uniformly to zero on the entire circle, i.e., $\| y_n \| \to 0$. Therefore $x(\zeta) = \zeta$ is not a generalized divisor of zero in A.

One can arrive at the concept of a generalized divisor of zero from the following arguments. Let us regard the multiplication by x as a linear operator in the space R, giving a single-valued and continuous mapping of R onto part of R. The following cases are possible:

1. The mapping $R \to R$ is not one-to-one, i.e., there exist distinct elements y and z such that $xy = xz$. In that case, x is a divisor of zero, since $x(y - z) = 0$.

2. The mapping $R \to R$ is one-to-one and the image fills out the whole space R. In that case, x has an inverse, because there exists a $y \in R$ such that $xy = e$.

3. The mapping $R \to R$ is one-to-one, but the image fills out not the whole space R but only a certain complete subspace $R' \subset R$. Then by Banach's Theorem the inverse mapping $R' \to R$ is also continuous, i.e., $xy_n \to 0$ implies $y_n \to 0$. This indicates that x is not a generalized divisor of zero in R.[5]

4. The mapping $R \to R$ is one-to-one, but the image fills out an incomplete subspace $R' \subset R$. In that case, the inverse mapping $R' \to R$ is not continuous (for otherwise R' would be homeomorphic to R and consequently complete), i.e., there exists a sequence $\{y_n\} \subset R$ such that $xy_n \to 0$ holds, but not $y_n \to 0$. In that case, x is a generalized divisor of zero.

7. THEOREM 1: *Let x be an arbitrary element of a commutative normed ring R and λ_0 a boundary point of its spectrum, i.e., of the set S_x of values $x(M)$; then $x - \lambda_0 e$ is a generalized divisor of zero.*

[5] As Arens [2] has shown (and before him, in certain special cases, Shilov [60]), such elements x (and such elements only) have an inverse in an extension of R.

Proof: We consider a sequence of numbers λ_n ($n = 1, 2, \ldots$) that do not belong to S_x and converge to λ_0. The elements $(x - \lambda_n e)^{-1}$ exist in R. We put $y_n = (x - \lambda_n e)^{-1} / \| (x - \lambda_n e)^{-1} \|$. We have $\| y_n \| = 1$. On the other hand, since

$$\|(x - \lambda_n e)^{-1}\| \geqq \max_{M \in \mathfrak{M}} | x(M) - \lambda_n |^{-1} \geqq | \lambda_0 - \lambda_n |^{-1} \to \infty,$$

we have

$$(x - \lambda_0 e) y_n = (x - \lambda_n e) y_n + (\lambda_n - \lambda_0) y_n =$$
$$= \frac{e}{\|(x - \lambda_n e)^{-1}\|} + (\lambda_n - \lambda_0) y_n \to 0,$$

and this is what was to be proved. //

COROLLARY 1: *If 0 is the only generalized divisor of zero of a commutative ring R, then R is isomorphic to the field of complex numbers.*

Proof: In virtue of our assumption and of Theorem 1, for every $x \in R$ there exists a λ_0 such that $x - \lambda_0 e = 0$, i.e., $x = \lambda_0 e$. //

***8.** This result obviously contains the following theorem of Mazur [38]:

If $\| xy \| = \| x \| \| y \|$ for arbitrary x and y of R, then R is the field of complex numbers.

9. COROLLARY 2: *Every element x of a symmetric ring R without radical that does not have an inverse is a generalized divisor of zero.*

Proof: If x does not have an inverse, then 0 is a boundary point of the set of values of the (non-negative real) function $(xx^*)(M)$ and, by Theorem 1, xx^* is a generalized divisor of zero: there exists a sequence $\{z_n\} \subset R$ such that $\inf_n \| z_n \| > 0$ and $xx^* z_n = x(x^* z_n) \to 0$ $(n \to \infty)$. Now if $\inf_n \| x^* z_n \| > 0$, then the theorem is proved; otherwise, there exists a subsequence z_{n_k} for which $x^* z_{n_k} \to 0$; but then, by Theorem 4 of § 9, we also have $x z_{n_k}^* = (x^* z_{n_k})^* \to 0$, and so we again find that x is a generalized divisor of zero, because, again by Theorem 4 of § 9, $\inf_n \| z_n \| > 0$ implies that $\inf_n \| z_n^* \| > 0$. //

***10.** In § 5 it was proved that the space of maximal ideals of a ring with one generator, considered as a set of values of the function $z(M)$, where z is the generator, is a closed bounded set in the complex plane.

With the help of Theorem 1 we can now give a complete topological description of this set:

THEOREM 2: *A point set S in the complex plane is the set of values of a function $z(M)$ for a ring with one generator z if and only if it is closed and bounded and does not divide the plane* (i.e., its complement is connected).

Proof: Suppose that S satisfies these assumptions. We construct a ring R as follows: for polynomials $P(z)$ we put

$$\| P(z) \| = \max_{\zeta \in S} | P(\zeta) | ;$$

we identify polynomials that coincide on S (if S is finite) and then we take the completion in the norm just introduced. Every element of R can be regarded as a continuous function $f(\zeta)$ defined on S and provided with the norm

$$\| f(\zeta) \| = \max_{\zeta \in S} | f(\zeta) | .$$

We denote by S' the set of values of $z(M)$, where M ranges over the set of all maximal ideals of R. If $\zeta_0 \in S$, then there exists a homomorphism $f(\zeta) \to f(\zeta_0)$ of R into the field of complex numbers that carries z into ζ_0. Hence $\zeta_0 \in S'$ and hence $S' \supset S$. Suppose that there is a point $\zeta_1 \in S' \setminus S$. By assumption, we can draw a continuous line from this point leading to infinity and not having points in common with S; let ζ_2 be the last point of S' on this line. Then ζ_2 is a boundary value of the function $z(M)$, so that $z - \zeta_2 e$, by Theorem 1, is a generalized divisor of zero, i.e., there exists a sequence $y_n \in R$ such that $\| y_n(z - \zeta_2 e) \| = \max_{\zeta \in S} | y_n(\zeta)(\zeta - \zeta_2) | \to 0$; moreover, $\inf_n \| y_n \| > 0$. On S we have $| \zeta - \zeta_2 | \geqq \varrho(\zeta_2, S) > 0$ (where, as usual, $\varrho(\zeta, S)$ denotes the distance of ζ from S); therefore the sequence of functions $y_n(\zeta)$ must converge uniformly to zero on S and consequently it also converges to zero in norm. This contradiction shows that $S' = S$.

Suppose, conversely, that R is a ring with one generator z and S is the set of all values of the function $z(M)$. We have already seen in § 5 that S is closed and bounded. If its complement were to contain a bounded connected component G, then every fundamental sequence of polynomials in R, being uniformly convergent on S, would also converge uniformly on G by the maximum modulus principle. By taking an arbitrary point $\zeta_0 \in G$ and assigning to every element $y = \lim_{n \to \infty} P_n(z) \in R$ the number $\lim_{n \to \infty} P_n(\zeta_0)$, we would obtain a homomorphism of R into the field of complex numbers carrying the generator z into ζ_0; thus, S would not exhaust all the values of the function $z(M)$. //

By making use of the general characteristic of the set of maximal ideals of a commutative normed ring with a finite number of generators that was given in § 5.11, we now arrive at the following result:

COROLLARY: *For every point* ζ_0 *that does not belong to a set* S *satisfying the conditions of Theorem 2 we can find a polynomial* $P(z)$ *such that* $|P(\zeta_0)| > \max_{\zeta \in S} |P(\zeta)|$.

§ 11. The Boundary of the Space of Maximal Ideals

1. In the ring A of functions of a complex variable that are continuous in the closed circle $|\zeta| \leqq 1$ and analytic in $|\zeta| < 1$ the bounding contour $|\zeta| = 1$ plays a special role: the modulus of every function $x(\zeta) \in A$ assumes its maximum on it. A similar situation, it turns out, obtains in every commutative normed ring.

DEFINITION 1: Let R be a commutative normed ring and $\mathfrak{M}$ the space of its maximal ideals. A closed set $F \subset \mathfrak{M}$ shall be called a *defining* set if the absolute value of every function $x(M)$ $(x \in R)$ assumes its maximum value on F.

Obviously, at least one defining set exists: namely, $\mathfrak{M}$ itself.

DEFINITION 2: A minimal defining set (i.e., a defining set every proper subset of which is no longer defining) shall be called the *boundary* of $\mathfrak{M}$ (or the *ring boundary*).

THEOREM 1: *The space* $\mathfrak{M}$ *of maximal ideals of every commutative normed ring* R *has a uniquely defined boundary.*

Proof of the existence of a boundary: If the defining set $F = F_1$ is not minimal, then there exists a defining set $F_2 \subset F_1$; if F_2 is not minimal, then there exists a defining set $F_3 \subset F_2$, and so forth. Suppose that we have obtained in this way a countable chain of defining sets $F_1 \supset F_2 \supset F_3 \supset \ldots$, each containing the next; their intersection F_ω, which is not empty because $\mathfrak{M}$ is compact, is also a defining set: for let $m = \max_{M \in \mathfrak{M}} |x(M)|$ and let $M_n \in F_n$ be such that $|x(M_n)| = m$; since $|x(M)|$ is continuous, we have, for every limit point M_ω of M_n, $|x(M_\omega)| = m$; but $M_\omega \in F_\omega$; and hence F_ω is a defining set. If F_ω is not minimal, then the construction of sequences of defining sets each containing the next can be continued; in this way, we obtain a transfinite decreasing sequence of closed sets; the intersection of all the sets of this sequence is not empty, because of the compactness of $\mathfrak{M}$, and gives us the required minimal set.

Proof of the uniqueness of the boundary: Let us assume that $\mathfrak{M}$ has two boundaries, Γ_1 and Γ_2. Let M_1 be an arbitrary point of Γ_1; we shall show that every neighborhood of M_1 contains points of Γ_2. Since Γ_2 is closed, this then shows that $M_1 \in \Gamma_2$ and, since M_1 is an arbitrary point, that $\Gamma_1 \subset \Gamma_2$; but then $\Gamma_1 = \Gamma_2$, because the inclusion $\Gamma_2 \subset \Gamma_1$ holds equally well.

Suppose, then, that U is an arbitrary neighborhood of M_1; it contains a neighborhood U' of M_1 defined by n inequalities $|x_i(M)| < \varepsilon$ $(i = 1, \ldots, n)$, where $x_1, \ldots, x_n$ are arbitrary elements of the ring belonging to M_1. Since Γ_1 is a minimal defining set, there exists a function $y(M)$ $(y \in R)$ whose absolute value assumes its maximum m on Γ_1 (this is equal, by assumption, to the maximum on the whole of $\mathfrak{M}$) within U' and remains less than m on Γ_1 outside this neighborhood; for otherwise the closed set $\Gamma_1 \setminus U' \subset \Gamma_1$ would also be a defining set and consequently Γ_1 would not be minimal. Without loss of generality, we can assume that $m = 1$; then by replacing y, if necessary, by some power of y we can assume that at the points of Γ_1 outside U' we have $|y(M)| < \varepsilon / \max_i \| x_i \|$. Then the products $x_i y$ $(i = 1, \ldots, n)$ remain less than ε in absolute value everywhere on Γ_1 and consequently on the whole of $\mathfrak{M}$. But there must exist a point M_2 on Γ_2 for which $|y(M_2)| = 1$; since $|(x_i y)(M_2)| < \varepsilon$ on M_2, we have

$$|x_i(M_2)| < \varepsilon \qquad (i = 1, \ldots, n).$$

But this means that $M_2 \in U'$ $(\subset U)$. We have thus established the uniqueness of the boundary Γ of $\mathfrak{M}$. //

THEOREM 2: *A point $M_0 \in \mathfrak{M}$ belongs to the boundary Γ if and only if for every neighborhood $U(M_0)$ of M_0 there exists a function $y(M)$ $(y \in R)$ whose absolute value assumes its maximum within $U(M_0)$ and is less than this maximum outside of $U(M_0)$.*

Proof: The necessity of the condition of Theorem 2 follows from the argument in the second part of the proof of Theorem 1; the sufficiency, from the fact that when this condition is satisfied, every neighborhood of M_0 contains points of the boundary. //

***2.** *For rings R in which $\| x \| = \max |x(M)|$, the generalized divisors of zero are those elements $x \in R$ that vanish at at least one point $M_0 \in \Gamma$.*

For R can be regarded as a ring of continuous functions on Γ. Let R' be the ring of all continuous complex functions on Γ. If $x \in R$ and $x(M)$ does not vanish at any point $M \in \Gamma$, then x has an inverse element x^{-1} in R', and therefore x cannot be a generalized divisor of zero in the subring R of R' Conversely, if $x \in R$ and $x(M_0) = 0$ at the point $M_0 \in \Gamma$, then for every $\varepsilon > 0$ the inequality $|x(M)| < \varepsilon$ is satisfied in some neighborhood $U(M_0)$ of M_0 in $\mathfrak{M}(R)$; on the other hand, since $M_0 \in \Gamma$, by what we have shown above there exists an element $h \in R$ such that $|h(M)|$ assumes its greatest value 1 in $U(M_0)$ and does not exceed $\varepsilon / \| x \|$ outside of $U(M_0)$. Then

$$\| xh \| = \max | x(M)h(M) | \leqq \varepsilon,$$

and since $\| h \| = 1$, x is a generalized divisor of zero.

The boundary Γ can now be defined as *the smallest closed subset of* $\mathfrak{M}$ *on which every generalized divisor of zero vanishes.*

For the proof it is sufficient to verify, taking into account what we have shown above, that for every neighborhood $U(M_0)$ of $M_0 \in \Gamma$ there is a generalized divisor of zero that vanishes only in this neighborhood. Let us consider an element $h \in R$ for which $| h(M) |$ assumes the maximum value 1 within $U(M_0)$, say at the point $M_1 \in \Gamma$, and is less than, say, 1/2 outside $U(M_0)$. Then $h - h(M_1)e$ is a generalized divisor of zero (because the function $h(M) - h(M_1)$ vanishes at the point M_1 of the boundary) and is necessarily different from zero outside $U(M_0)$. //

***3.** Let us find the position of the boundary in a few particular cases.

1. Let R be a ring with one generator. Then, by Theorem 4 of § 5, $\mathfrak{M}(R)$ can be regarded as a closed bounded set of points of the complex plane. In this case Γ coincides with the ordinary topological boundary of $\mathfrak{M}(R)$. For by the maximum-modulus principle the conditions of Theorem 2 cannot possibly be satisfied for interior points (in the usual sense) of $\mathfrak{M}$, and so such points cannot belong to Γ; but if λ_0 is a boundary point (in the usual sense) of $\mathfrak{M}$, then there exists a sequence $\lambda_n \to \lambda_0$ such that $\lambda_n \notin \mathfrak{M}$; the functions $(x - \lambda_n e)^{-1}$, where x is the generator, belong to R and for n greater than a sufficiently large N satisfy the conditions of Theorem 2; hence $\lambda_0 \in \Gamma$.

2. Let R be the ring of functions obtained from the class of all polynomials of two complex variables z_1 and z_2 by uniform passage to the limit in the domain $| z_1 | \leqq 1$, $| z_2 | \leqq 1$. It is easy to see that $\mathfrak{M}(R)$ coincides with the set of all points of this domain. The topological boundary of $\mathfrak{M}(R)$ consists of all the points at which at least one coordinate is equal to 1 in absolute value. The ring boundary is formed by all the points at which both coordinates are equal to 1 in absolute value. For the set S of all these points, as is easy to see, is a defining set, so that $\Gamma \subset S$; on the other hand, for every point $(\exp(i\varphi_1), \exp(i\varphi_2))$ of S the function

$$[z_1 + \exp(i\varphi_1)][z_2 + \exp(i\varphi_2)]$$

satisfies the conditions of Theorem 2, and hence $S \subset \Gamma$.

This example shows that the ring boundary Γ need not coincide with the topological boundary of $\mathfrak{M}$ (which is assumed to be contained in the complex space naturally connected with it; see Theorem 4 of § 5).

3. Let R be the ring of functions obtained from the class of all polynomials in two independent variables z_1 and z_2 by uniform passage to the

limit in the domain $|z_1|+|z_2|\leqq 1$. Here $\mathfrak{M}(R)$ is the set of all points of this domain and the topological boundary and the ring boundary coincide. (The proof proceeds along the same lines as in Example 4 below.)

Every function $f(M)$ for which $1/f(M)$ does not exist vanishes on the boundary. Suppose, to begin with, that $f(M)=f(z_1, z_2)$ is a polynomial; since $1/f(z_1, z_2)$ does not exist, we have $f(z_1{}^0, z_2{}^0)=0$, where

$$|z_1{}^0|+|z_2{}^0|\leqq 1.$$

Suppose that $|z_1{}^0|+|z_2{}^0|<1$. The polynomial $f(z_1{}^0, z_2)$ in the argument z_2 has the root $z_2{}^0$; when the parameter $z_1{}^0$ changes continuously, then the root $z_2{}^0$ also changes continuously; therefore if we increase $|z_1{}^0|$, then the time will come when $|z_1{}^0|+|z_2{}^0|=1$. Thus, for polynomials our statement is true. Suppose now that $f(M)$ is an arbitrary function of R for which $1/f(M)$ does not exist, so that $f(M_0)=0$ at a certain point $M_0\in\mathfrak{M}(R)$. $f(M)$ can be approximated with arbitrary accuracy by polynomials that also vanish at M_0 and consequently vanish somewhere on the boundary as well. Therefore $\min|f(M)|$ cannot be positive on the boundary.

COROLLARY: *Every element of the ring R that does not have an inverse is a generalized divisor of zero.*

4. Let us consider the set S_a of points $\{z\}=\{z_1, \ldots, z_n, \ldots\}$ of an infinite-dimensional complex space satisfying the inequality $\sum_{j=1}^{\infty}|z_j|\leqq a$. We form a ring R_a from the limits of sequences of polynomials that converge uniformly on S_a, where every polynomial depends only on a finite number of variables. Let us find the sets $\mathfrak{M}$ and $\Gamma(\mathfrak{M})$ of this ring. In S_a we introduce the usual Tychonov topology, in which it becomes a compact space.

Let us determine $\mathfrak{M}(R_a)$. Just as for rings with a finite number of generators, every maximal ideal M is represented by a point

$$\{z(M)\}=\{z_1(M), \ldots, z_n(M), \ldots\}$$

of the space in question. We know that every point of S_a gives a maximal ideal of R_a, i.e., $S_a\subset\mathfrak{M}(R_a)$. Now let M_0 be an arbitrary maximal ideal and let $\varrho_k \exp(i\varphi_k)=z_k(M_0)$. We consider the function

$$F(z)=\sum_{k=1}^{\infty} z_k \exp(-i\varphi_k).$$

This series converges absolutely on S_a, so that $F(z)\in R_a$. The norm of $F(z)$, i.e., the maximum absolute value of $F(z)$ on S_a, does not exceed a, so that $|F(M_0)| \leqq a$. But $F(M_0) = \sum_{k=1}^{\infty} z_k(M_0) \exp(-i\varphi_k) = \sum_{k=1}^{\infty} \varrho_k$; therefore $\sum_{k=1}^{\infty} \varrho_k \leqq a$, and thus $M_0 \in S_a$. Hence $\mathfrak{M}(R_a)$ coincides with S_a.

We now consider the polynomial $P(z) = \prod_{k=1}^{n} (z_k - z_k^0)$, where $z_k^0 \neq 0$ and the absolute values $r_k = |z_k^0|$ have the property that the largest of them differs from the average $c = (1/n) \sum_{k=1}^{n} r_k$ by not more than a/n. Let us show that such a polynomial assumes its maximum absolute value on S_a at a unique point. Since S_a is compact, at least *one* point exists at which $|P(z)|$ assumes a maximum; let this be $\zeta = \{\zeta_k\} = \{\exp(i\varphi_k)\varrho_k\}$. Then $\varphi_k = \pi + \arg z_k^0$, since otherwise we could move ζ_k along a circle of radius ϱ_k and increase $|P(z)|$ without leaving S_a. Therefore

$$|P(\zeta)| = \prod_{k=1}^{n} (r_k + \varrho_k).$$

At a point at which this product assumes its maximum, we must obviously have the condition $\sum_{k=1}^{n} \varrho_k = a$, or $\sum_{k=1}^{n} (r_k + \varrho_k) = a + \sum_{k=1}^{n} r_k$. But by these conditions the factors $r_k + \varrho_k$ for which the product assumes the maximum are uniquely determined by these conditions: each of them is equal to

$$r_k + \varrho_k = \frac{a + \sum_{k=1}^{n} r_k}{n} = \frac{a}{n} + c,$$

so that we find $\varrho_k = a/n + c - r_k \geqq 0$. Thus, the point at which $|P(z)|$ assumes its maximum is unique. Let us now show that for every point $\{\zeta\} = \{\zeta_1, \ldots, \zeta_n, 0, 0, \ldots\}$ with $\sum_{k=1}^{n} |\zeta_k| = a$ we can find a polynomial $P(z)$ that assumes its maximum absolute value on S_a precisely at this point $\{\zeta\}$. Let $\zeta_k = \varrho_k \exp(i\varphi_k)$. We define numbers r_k by the equations

$$r_k = c + a/n - \varrho_k \qquad (k = 1, \ldots, n),$$

where c is an arbitrary number greater than $\varrho_k - a/n$ $(k = 1, \ldots, n)$. It is easy to see that c is the arithmetic mean of the numbers r_k and that the greatest of these numbers differs from it by not more than a/n. We put $z_k^0 = r_k \exp((\pi + \varphi_k)i)$ and consider the polynomial $P(z) = \prod_{k=1}^{n} (z_k - z_k^0)$; by what we have proved, it assumes its maximum at a unique point, namely at $\{\zeta\}$.

By Theorem 2, every point $\{\zeta\} = \{\zeta_1, \ldots, \zeta_n, 0, 0, \ldots\}$ belongs to $\Gamma(\mathfrak{M})$. But such points form an everywhere-dense set in S_a; and since Γ is closed, we have $\Gamma = S_a = \mathfrak{M}$. Here we have an example of an unsymmetric ring with a uniform norm in which $\Gamma = \mathfrak{M}$. It is an open problem whether a ring with this property can have a finite number of generators.

§ 12. Extension of Maximal Ideals

1. In the theory of normed spaces an important role is played by the Hahn-Banach Theorem, which states that every linear functional can be extended from a given normed space to any larger space with preservation of the norm. In normed rings the analogous question can be asked, whether it is possible to extend multiplicative linear functionals, i.e., linear functionals $f(x)$ satisfying the additional condition $f(xy) = f(x)f(y)$ for all x, y (see §4.4.). In general, the answer to this question is in the negative. For example, in the ring A of Example 6 of § 1 the multiplicative linear functional $f(x) = x(0)$ cannot be extended with preservation of multiplication to the ring of all continuous functions on the circle $|\zeta| = 1$, which contains A as a closed subring. (It can, however, be extended as a linear functional—for example, by the formula $f(x) = (2\pi i)^{-1} \int_{|\lambda|=1} \lambda^{-1} x(\lambda) d\lambda$.) But it does turn out that in every commutative normed ring R_1 there is a set of multiplicative linear functionals that can be extended with preservation of multiplication to every commutative normed ring R containing R_1 as a closed subring. These functionals are precisely the multiplicative linear functionals that correspond to the maximal ideals forming the boundary of the space $\mathfrak{M}(R_1)$ of maximal ideals of R_1.

Before proceeding to a proof of this statement, let us note the following. Suppose that a multiplicative linear functional $f(x)$ is extended from a ring R_1 to a larger ring R. Then, in particular, the set M_1 of those $x_1 \in R_1$ for which $f(x_1) = 0$ becomes part of the set M of those $x \in R$ for which $f(x) = 0$. But M_1 is a maximal ideal of R_1 and M a maximal ideal of R. Thus, the extension of the multiplicative functional $f(x_1)$ goes hand in hand with an extension of the maximal ideal M_1 of R_1 to a maximal ideal M of R. Conversely, if the maximal ideal M_1 of R_1 is extended to the maximal ideal M of R, then the multiplicative linear functional $M(x) = x(M)$ on R is an extension of the multiplicative linear functional $M_1(x)$ given on R_1. For let $x_0 \in R_1$ and $M_1(x_0) = \lambda_0$. Then $M_1(x_0 - \lambda_0 e) = 0$, so that

$$x_0 - \lambda_0 e \in M_1 \subset M,$$

and hence we also have that $M(x_0) = \lambda_0$. Consequently, the problem of extending a multiplicative linear functional is equivalent to that of extending the corresponding maximal ideal.

Theorem 1: *In an arbitrary normed ring R containing R_1 as a closed subring, every maximal ideal of the boundary Γ_1 of the space $\mathfrak{M}(R_1)$ can be extended to a maximal ideal of R.*

Proof: First of all, observe that if $x \in R_1$, then

$$\max_{M \subset R} | x(M) | = \max_{M_1 \subset R_1} | x(M_1) |.$$

This follows immediately from the formula

$$\max | x(M) | = \lim_{n \to \infty} \sqrt[n]{\| x^n \|}$$

when we bear in mind that all x^n are contained simultaneously in R_1 and in R and that their norms in the two rings coincide. Let us assume now that there is a maximal ideal $M_1 \in \Gamma_1$ not contained in any maximal ideal of R. This means that there exists no proper ideal in R containing all the elements of M_1, i.e., that the totality of all sums of the form

$$\sum_{i=1}^{n} x_i z_i \qquad (x_i \in M_1,\ z_i \in R)$$

coincides with the whole ring R; in particular, one of these sums yields the unit element of the ring R:

$$e = \sum_{i=1}^{n} x_i z_i.$$

We may assume here without loss of generality that $\max | x_i(M) | \leqq 1$.

Suppose that $\mu > \max_i \{\max_M | z_i(M) | \}$; we consider the neighborhood of M_1 defined by the inequalities

$$| x_i(M) | < 1/2n\mu \qquad (i = 1, \ldots, n;\ M \subset R_1),$$

and a function $y(M)$ $(y \in R_1)$ whose absolute value assumes its maximum 1 within this neighborhood and does not exceed $1/2n\mu$ outside it (see Theorem 2 of § 11). By the remark made above, the product $y \cdot \sum_{i=1}^{n} x_i z_i = ye = y$ also assumes the absolute value 1 on the maximal ideals of R. But on the other hand,

$$\max_{M\subset R}\left|\left(y\cdot\sum_{i=1}^{n}x_i z_i\right)(M)\right|\leqq\sum_{i=1}^{n}\max|y(M)x_i(M)|\max|z_i(M)|\leqq$$

$$\leqq\sum_{i=1}^{n}1\cdot\frac{1}{2n\mu}\cdot\mu=\frac{1}{2}.$$

This contradiction proves the theorem. //

COROLLARY: *Every maximal ideal of a symmetric ring R_1 can be extended to a maximal ideal of every larger ring R.*

Proof: In virtue of the theorem just proved, it is sufficient to show that for a symmetric ring, $\Gamma=\mathfrak{M}$. Let M_0 be any point of $\mathfrak{M}$ and $U(M_0)$ a neighborhood of M_0. By part 4 of the proof of Theorem 1′ of § 7, there exists an element $x\in R$ such that $\max|x(M)-f(M)|<1/2$, where $f(M)$ is a continuous function that is equal to 0 outside of $U(M_0)$ and equal to 1 at M_0. In this case the maximal value of $|x(M)|$ must be assumed within $U(M_0)$. But this means that the condition of Theorem 2 of § 11 is satisfied, and $M_0\in\Gamma$. //

***2.** If $\|x\|=\max|x(M)|$ in R_1, then we can state that there exists a ring $R\supset R_1$ for which only those $M\subset R_1$ can be extended to maximal ideals that belong to the boundary Γ of the space $\mathfrak{M}(R_1)$. This ring R can be chosen as the ring of all continuous functions defined on Γ; it contains R_1 as a subring, and since Γ is a compact space, the set of all maximal ideals of R coincides with Γ.

§ 13. Locally Analytic Operations on Certain Elements of a Ring

1. We have seen in § 6 that every analytic function $f(\zeta)$, single-valued on the spectrum of the element x of a normed ring R, has an element $y\in R$ corresponding to it such that $y(M)=f(x(M))$ for every maximal ideal of R.

This theorem as it stands is inadequate for many important applications. For example, it permits us to establish the existence of the elements $\sqrt{x}$ or $\log x$ only when the values of the functions $\sqrt{\zeta}$ or $\log\zeta$ on the spectrum of x lie on a single sheet of the corresponding Riemann surface. This condition is a rather artificial one in our problem.

Let us consider, for example, the case of the ring $C(\mathrm{K})$ of all continuous functions on the circle K $(0\leqq t<2\pi)$. The question as to what elements have square roots in the ring has a simple answer in the present case: if $x(t)$ does not vanish (i.e., if the function $\sqrt{\zeta}$ does not have a branch point on the spectrum of x), then $\sqrt{x(t)}$ exists in $C(\mathrm{K})$ if and only if the increment of the argument of $x(t)$, when the independent variable ranges over the whole circumference of K, is equal to $4\pi n$, where n is an integer.

When we go over from $C(\mathrm{K})$ to the ring W, say, of all functions with an absolutely convergent Fourier series, then the question as to which elements have square roots in this ring becomes non-trivial. However, it turns out that the condition given above—the complete increment of the argument of $x(t)$ is equal to an integer multiple of 4π—is necessary and sufficient even in this case, and not only for the ring W, but for any ring R having the circle as its space of maximal ideals.

2. For a rigorous formulation of the general result that holds in all normed rings it is convenient to introduce the concept of a locally analytic operation.

We shall use the term *locally analytic operation on the element* z of the normed ring R for the transition from the function $x(M)$ to a function $f(M)$ that admits in a neighborhood of every point M_0 a representation in the form of a convergent series

$$f(M) = \sum_{n=0}^{\infty} a_n [z(M) - z(M_0)]^n.$$

It is obvious that $\sqrt{z}$ and $\log z$ are locally analytic operations on the element z if the functions $\sqrt{\zeta}$ and $\log \zeta$ do not have singularities (branch points) on the spectrum of z, although the values of these functions need not lie, as before, on the same sheet of the Riemann surface of the corresponding function.

We shall establish the following result: *If a locally analytic operation on the element z carries the function $z(M)$ into a single-valued function $f(M)$ on the whole space $\mathfrak{M}$, then this function $f(M)$ is generated by an element of R.*

Now this will solve the problem of the existence, in the general case, of the functions $\sqrt{z(M)}$ and $\log z(M)$ in a ring R without radical: namely, these functions exist in R whenever $z(M)$ does, provided they are single-valued on $\mathfrak{M}$. The condition above on the complete increment of the argument is, for the circle, precisely the condition that the function $\sqrt{z(M)}$ should be single-valued. In the case of an arbitrary space $\mathfrak{M}$, the role of the condition on the complete increment of the argument of $z(M)$ will be taken each time by a condition of a topological nature; we shall not go into any of the details.

3. We cannot give a proof of this theorem entirely within the framework of the theory of functions of a single variable. Our proof rests essentially on the possibility of going into the realm of functions of several complex variables. It is not worthwhile, therefore, to confine ourselves in the statement of the theorem to functions of a single variable. Let us make the following general definition:

DEFINITION 1: The term *locally analytic operation with respect to elements of a normed ring R* will be used for the transition from functions $z(M)$ corresponding to the elements of this ring to a function $f(M)$ that in a neighborhood of every point $M_0 \in \mathfrak{M}$ admits a representation

$$f(M) = \sum_{k_1, \ldots, k_m = 0}^{\infty} a_{k_1 \ldots k_m} [z_1(M) - z_1(M_0)]^{k_1} \cdots$$
$$\cdots [z_m(M) - z_m(M_0)]^{k_m}, \qquad (1)$$

where the power series

$$\sum_{k_1, \ldots, k_m = 0}^{\infty} a_{k_1 \ldots k_m} \zeta_1^{k_1} \cdots \zeta_m^{k_m}$$

converges in a neighborhood of the origin in an m-dimensional complex space and where the elements $z_1, \ldots, z_m$ in the equation (1), as well as their number m, may depend on the choice of the point M_0.

Furthermore, for the sake of brevity we shall call the function $f(M)$ itself *locally analytic*, having in mind, of course, that this term refers to its dependence not on M but on the corresponding functions $z(M)$.

4. We shall shortly prove the following theorem.

THEOREM 1: *To every locally analytic operation with respect to elements of a normed ring R there corresponds an element $x \in R$ such that $x(M) = f(M)$ for all $M \in \mathfrak{M}(R)$, where $f(M)$ is the corresponding locally analytic function.*

This theorem obviously contains Theorem 1 of § 6 as a special case. For the function $f(\zeta)$ that occurs in Theorem 1 of § 6 can be associated with the function $f_1(M) = f(x(M))$ defined on the space $\mathfrak{M} = \mathfrak{M}(R)$. Since, by assumption, $f(\zeta)$ is analytic on the spectrum of x, for every point $M_0 \in \mathfrak{M}$ we can exhibit a neighborhood of the form $\{M \in \mathfrak{M} : |x(M) - x(M_0)| < \varepsilon\}$ in which $f_1(M)$ can be expanded in a Taylor series in powers of $x(M) - x(M_0)$. Therefore, $x(M) \to f_1(M)$ is a locally analytic operation; but then, by the theorem that has been stated just above, there exists an element $x_1 \in R$ such that $x_1(M) \equiv f_1(M)$, and this is what was required.

The theorem of the present section differs from the one in § 6 first of all by the fact that it permits the use of multiple series instead of simple Taylor series, where the elements that occur in these multiple series may change from place to place. But even in the simplest case, where the given locally analytic function can be expressed in the neighborhood of each point $M_0 \in \mathfrak{M}$ as a power series in $x(M) - x(M_0)$ with one and the same x, the theorem stated here does not reduce to Theorem 1 of § 6, since $f(\zeta)$ need not be a single-valued function on the spectrum of x.

5. *Proof*: The proof of Theorem 1 is not easy; we carry it out in several steps.

A) We begin by discussing the case where R is a ring with a finite number of generators $x_1, \ldots, x_n$ and the elements $z_1, \ldots, z_m$ in all the expansions (1) are chosen from among these generators. As we know (§ 5), the space $\mathfrak{M} = \mathfrak{M}(R)$ can be identified with a certain closed bounded subset F of an n-dimensional complex space $\mathbf{C}^n$, by assigning to every maximal ideal M the point $\zeta = \{x_1(M), \ldots, x_n(M)\} \in \mathbf{C}^n$. Since this correspondence is one to one, the single-valued function $f(M)$ on $\mathfrak{M}$ is also single-valued on the set F and is an analytic function in the variables $\zeta_1, \ldots, \zeta_n$ on this set, so that it is single-valued and analytic in a certain domain G containing the set $F = \mathfrak{M}$.

It was shown at the end of § 5 that $\mathfrak{M}$ is convex with respect to polynomials and is therefore the intersection of a certain (infinite) number of domains of the form

$$\{\zeta \in \mathbf{C}^n : |P(\zeta_1, \ldots, \zeta_n)| < 1\}, \tag{2}$$

where $\zeta = (\zeta_1, \ldots, \zeta_n)$ and P is the symbol for a polynomial in the variables $\zeta_1, \ldots, \zeta_n$.

6. We define a *Weil domain* as an arbitrary domain that is the intersection of a *finite* number of domains of the form (2)—say, of the domains

$$G_k = \{\zeta \in \mathbf{C}^n: \ |P_k(\zeta_1, \ldots, \zeta_n)| < 1\} \qquad (k = 1, \ldots, N),$$

where every system of n manifolds

$$\Gamma_k = \{\zeta \in \mathbf{C}^n: \ |P_k(\zeta_1, \ldots, \zeta_n)| = 1\}$$

has an intersection of (real) dimension not exceeding n.

Let us show that there exists a Weil domain containing $\mathfrak{M}$ and contained in G.

Let $\mathfrak{M} = \bigcap_\nu G_\nu$, where

$$G_\nu = \{\zeta \in \mathbf{C}^n: \ |P_\nu(\zeta_1, \ldots, \zeta_n)| < 1\}.$$

Since $\mathfrak{M}$ is bounded, n domains of the form

$$Q_j = \left\{\zeta \in \mathbf{C}^n: \ \left|\frac{\zeta_j}{c_j}\right| < 1\right\} \qquad (j = 1, \ldots, n),$$

where c_j are positive constants, can be added to the domains G_ν. Let

$$m_\nu = \max_{\zeta \in \mathfrak{M}} |P_\nu(\zeta_1, \ldots, \zeta_n)|.$$

Obviously, $m_\nu < 1$. We consider the closed sets

$$F_\nu = \{\zeta \in \mathbf{C}^n: \quad |P_\nu(\zeta_1, \ldots, \zeta_n)| \leqq \theta_\nu\}$$

and

$$S_j = \{\zeta \in \mathbf{C}^n: \quad |\zeta_j| \leqq a_j\},$$

where $m_\nu < \theta_\nu < 1$ and

$$\max_{M \in \mathfrak{M}} |z_j(M)| = \max_{\zeta \in \mathfrak{M}} |\zeta_j| < a_j < c_j \qquad (j = 1, \ldots, n).$$

Since $\mathfrak{M} \subset F_\nu \subset G_\nu$ and $\mathfrak{M} \subset S_j \subset Q_j$, the intersection of all the sets F_ν and S_j also coincides with $\mathfrak{M}$. But the intersection of all the sets F_ν and S_j with the closed set CG complementary to G is empty; since $S = \bigcap_{j=1}^{n} S_j$ is a closed bounded set in $\mathbf{C}^n$, we can therefore choose a finite number of indices $\nu_1, \ldots, \nu_m$ such that intersection of the sets $F_{\nu_1}, \ldots, F_{\nu_m}, S$ with CG is empty; but this means that

$$F_{\nu_1} \cap \cdots \cap F_{\nu_m} \cap S_1 \cap \cdots \cap S_n \subset G.$$

The intersection of all the domains

$$\left\{\zeta \in \mathbf{C}^n: \ \left|\frac{1}{\theta_{\nu_i}} P_{\nu_i}(\zeta_1, \ldots, \zeta_n)\right| < 1\right\} \qquad (i = 1, \ldots, m)$$

and

$$\left\{\zeta \in \mathbf{C}^n: \ \left|\frac{\zeta_j}{a_j}\right| < 1\right\} \qquad (j = 1, \ldots, n)$$

is a fortiori contained in G. But this intersection is a domain contained in G and containing $\mathfrak{M}$.

This domain can always be made to satisfy the last condition defining a Weil domain; this is so in virtue of the following argument, which shows that, apart from the polynomial $P_\nu = P_\nu(\zeta_1, \ldots, \zeta_n)$, every polynomial $\tilde{P}_\nu$ sufficiently near to P_ν is suitable for the construction of the required domain.

We put $\tilde{P}_\nu = P_\nu/\theta_\nu + R_\nu$, where $R_\nu(\zeta)$ is an arbitrary polynomial whose absolute value does not exceed a given number $\varepsilon > 0$ in the domain

$$Q = \{\zeta \in \mathbf{C}^n : |\zeta_j| \leqq c_j \ \ (j = 1, \ldots, n)\}.$$

Since $\mathfrak{M} \subset Q$, we have $\max_{\zeta \in \mathfrak{M}} |\tilde{P}_\nu(\zeta)| \leqq \frac{m_\nu}{\theta_\nu} + \varepsilon$. On the other hand, in Q it follows from $|\tilde{P}_\nu(\zeta)| < 1$ that $|P_\nu(\zeta)| < \theta_\nu(1 + \varepsilon)$. Thus, for sufficiently small ε the intersection of the domain $\{\zeta \in \mathbf{C}^n : |\tilde{P}_\nu(\zeta)| < 1\}$ with Q contains $\mathfrak{M}$ and is contained in $\{\zeta \in \mathbf{C}^n : |P_\nu(\zeta)| < 1\}$, and this means that in the construction of the Weil domain the polynomial $P_\nu(\zeta)$ can be replaced by $\tilde{P}_\nu(\zeta)$. By selecting the latter polynomials in such a way that no system of n functions $|\tilde{P}_\nu(\zeta)|$ is functionally dependent, we then obtain the required Weil domain.

7. Now we can proceed to the completion of the proof of our theorem in the case A under consideration. For rings with uniform convergence the truth of this statement follows directly from a theorem of Weil, which states that every analytic function in a Weil domain G can be represented in the form of a series of polynomials uniformly convergent inside G (i.e., on every closed bounded subset of G).[6]

In the general case, we consider the so-called *integral representation of Weil* (which he constructed for the explicit purpose of proving the theorem of the present section). It is obtained as follows.

The Weil domain constructed above is defined by inequalities of the form

$$|P_i(\zeta_1, \ldots, \zeta_n)| < 1 \qquad (i = 1, \ldots, N),$$

where P_i are certain polynomials in $\zeta_1, \ldots, \zeta_n$. Let

$$s_i = \{\zeta \in \mathbf{C}^n : |P_i(\zeta_1, \ldots, \zeta_n) = 1|\}.$$

We denote by $\sigma_{i_1 \ldots i_n}$ the intersection of the manifolds $s_{i_1}, \ldots, s_{i_n}$; it is orientated in a definite way, but there is no need for us to go into the details. For every point $(\tau_1, \ldots, \tau_n) \in \mathfrak{M}$ we can write down the expansion

$$P_i(\zeta_1, \ldots, \zeta_n) - P_i(\tau_1, \ldots, \tau_n) = \sum_{j=1}^{n} (\zeta_j - \tau_j) Q_{ij}(\zeta_1, \ldots, \zeta_n),$$

where Q_{ij} are certain polynomials in $\zeta_1, \ldots, \zeta_n$ whose coefficients depend on $\tau_1, \ldots, \tau_n$. We put

$$D_{i_1 \ldots i_n} = \det \| Q_{i_\alpha j} \| \qquad (j, \alpha = 1, \ldots, n).$$

Then *every analytic function* $f(\zeta_1, \ldots, \zeta_n)$ *in* G *can be represented on* $\mathfrak{M}$ *in the form of an integral*

$$f(\tau_1, \ldots, \tau_n) = \\ = \frac{1}{(2\pi i)^n} \sum_{(i_1, \ldots, i_n)} \int_{\sigma_{i_1 \ldots i_n}} \frac{D_{i_1 \ldots i_n} f(\zeta_1, \ldots, \zeta_n)\, d\zeta_1 \ldots d\zeta_n}{\prod_{\nu=1}^{n} [P_{i_\nu}(\zeta_1, \ldots, \zeta_n) - P_{i_\nu}(\tau_1, \ldots, \tau_n)]}, \qquad (3)$$

[6] B. A. Fuks, *Theory of Analytic Functions of Several Complex Variables*, Moscow-Leningrad, 1948, p. 329 [in Russian]. See also B. A. Fuks *Introduction to the Theory of Analytic Functions of Several Complex Variables*, Providence, 1963, p. 300.

where the summation extends over all combinations of the indices $i_1 < i_2 < \ldots < i_n$ ranging from 1 to N.[7]

Since the polynomial $P_{i_\nu}(\zeta_1, \ldots, \zeta_n) - P_{i_\nu}(\tau_1, \ldots, \tau_n)$ does not vanish when $(\zeta_1, \ldots, \zeta_n)$ ranges over $\sigma_{i_1 \ldots i_n}$ and $(\tau_1, \ldots, \tau_n)$ ranges over $\mathfrak{M}$, the integral (3) can be interpreted within the ring. Indeed, first of all, the inverse element

$$[P_{i_\nu}(\zeta_1, \ldots, \zeta_n)e - P_{i_\nu}(x_1, \ldots, x_n)]^{-1}$$

exists in R (where e is the unit element of R) and obviously depends continuously on the parameters $\zeta_1, \ldots, \zeta_n$. Furthermore, since $D_{i_1 \ldots i_n}$ as a polynomial also depends continuously on the parameters $\zeta_1, \ldots, \zeta_n$, we can integrate the ring element

$$D_{i_1 \ldots i_n} f(\zeta_1, \ldots, \zeta_n) \prod_{\nu=1}^{n} [P_{i_\nu}(\zeta_1, \ldots, \zeta_n)e - P_{i_\nu}(x_1, \ldots, x_n)]^{-1}$$

with respect to the parameters $\zeta_1, \ldots, \zeta_n$ ranging over $\sigma_{i_1 \ldots i_n}$. When we sum these results over the indices $i_1, \ldots, i_n$ and multiply by $1/(2\pi i)^n$, we again obtain an element of R. In virtue of the continuity of the canonical homomorphism $R \to R/M$, the values of the element so obtained on the maximal ideals of R coincide with the values of the function $f(\tau_1, \ldots, \tau_n)$ at the corresponding points of $\mathfrak{M}$. Theorem 1 is thus proved in the case A.

8. B) For what follows we introduce the important concept of the *joint spectrum* of n elements $x_1, \ldots, x_n$ of a normed ring R. This is the name for the set $S = S_R(x_1, \ldots, x_n)$ of all points $\{x_1(M), \ldots, x_n(M)\}$ of the n-dimensional complex space $\mathbf{C}^n$, where M ranges over the space $\mathfrak{M}$ of maximal ideals of R. Obviously S is a closed bounded set that lies in the domain

$$Z = \{\zeta \in \mathbf{C}^n : |\zeta_j| \leqq \|x_j\| \ (j = 1, \ldots, n)\}.$$

Note that $S_R(x_1, \ldots, x_n)$ does not, in general, coincide with the joint spectrum $S_{R_0}(x_1, \ldots, x_n)$ of the elements $x_1, \ldots, x_n$ in the subring $R_0 \subset R$ generated by them, i.e., by Theorem 4 of § 5, with the space of maximal ideals of this subring R_0; in general, we only have the inclusion

$$S_R(x_1, \ldots, x_n) \subset S_{R_0}(x_1, \ldots, x_n)$$

(which follows in an obvious way from the fact that every maximal ideal of R yields, by intersection with R_0, a maximal ideal of R_0).

***9.** For example, let R be the ring of all continuous functions on the circle $|\zeta| = 1$ of the complex plane. The spectrum of the element $z(\zeta) = \zeta$

[7] *Loc. cit.*, p. 306.

is the set of the points of the circle $|\zeta|=1$. The subring R_0 generated by these elements is the ring A of all functions analytic in $|\zeta|<1$ and continuous on $|\zeta|\leqq 1$; the set of its maximal ideals is the disc $|\zeta|\leqq 1$, which is larger than the spectrum of the element ζ in R.

10. We shall now establish the following important proposition:

LEMMA: *For every (single-valued) function $f(\zeta)=f(\zeta_1, \ldots, \zeta_n)$ analytic on the joint spectrum S of the elements $x_1, \ldots, x_n$ of R there exists an element $x\in R$ such that for every $M\in\mathfrak{M}$ we have the equality*

$$f(x_1(M), \ldots, x_n(M))=x(M).$$

Proof: In the special case where $x_1, \ldots, x_n$ are generators of R, their joint spectrum S coincides with $\mathfrak{M}$ and the truth of the lemma follows from what has been proved under A. Now it turns out that in the general case we can add to $x_1, \ldots, x_n$ a certain number of elements $y_1, \ldots, y_p$ of R so that the projection $S_{R'}(x_1, \ldots, x_n)$ onto $\mathbf{C}^n$ of the space $\mathfrak{M}'$ of maximal ideals of the subring $R'\subset R$ generated by the elements $x_1, \ldots, x_n, y_1, \ldots, y_p$, considered as a subset of the $(n+p)$-dimensional complex space $\mathbf{C}^{n+p}$, is contained in a preassigned open neighborhood U of S—for example, one in which the function $f(\zeta)$ remains analytic and single-valued. Then $f(\zeta)$ can be continued as a constant with respect to the remaining p arguments to $\mathfrak{M}'$, where it is again a single-valued analytic function and, by what has been proved under A, corresponds to a certain element $x\in R'$; this completes the proof.

Let us now show how to select the required elements $y_1, \ldots, y_p$. If the point $\zeta^0=\{\zeta_1{}^0, \ldots, \zeta_n{}^0\}\in\mathbf{C}^n$ does not belong to $S_R(x_1, \ldots, x_n)$, this means that there is no maximal ideal in R that contains all the differences $x_j-\zeta_j{}^0e$ $(j=1, \ldots, n)$. But then the ideal generated by these differences is the whole ring R and consequently for a certain choice of the elements $y_1, \ldots, y_n$ we have

$$\sum_{j=1}^{n}(x_j-\zeta_j{}^0e)y_j=e.$$

In a certain neighborhood of ζ^0 the sum $\sum_{j=1}^{n}(x_j-\zeta_je)y_j$ differs in norm from e by less than 1 and therefore has an inverse element in R. Moreover, this inverse element is contained in the subring R_1 generated by $x_1, \ldots, x_n, y_1, \ldots, y_n$. Thus, even in R_1 there is no maximal ideal containing all the differences $x_j-\zeta_je$, i.e., $(\zeta_1, \ldots, \zeta_n)\notin S_{R_1}(x_1, \ldots, x_n)$, and a fortiori $(\zeta_1, \ldots, \zeta_n)\notin S_{R'}(x_1, \ldots, x_n)$ for every $R'\supset R_1$. Now we observe that the set-theoretical difference $Z\setminus U$ of the domain Z with U is compact and therefore can be covered by a finite number of the neighborhoods in question. We consider the subring R' generated by all the corresponding elements $x_1, \ldots, x_n, y_1, \ldots, y_n, \ldots, y_p$. Its space of maximal ideals $\mathfrak{M}'$ considered as a subset of the $(n+p)$-dimensional complex space $\mathbf{C}^{n+p}$ projects in the space $\mathbf{C}^n$ of the first n coordinates onto the set $S_{R'}(x_1, \ldots, x_n)$ contained in Z. Furthermore, by what has been proved, no point of $Z \setminus U$ can fall into the projection $S_{R'}(x_1, \ldots, x_n)$, which is therefore entirely contained in U. The lemma is now proved.

11. C) We now proceed to the proof of the theorem in the general case. By assumption, with every point $M_0 \in \mathfrak{M}$ there are associated its neighborhood $U(M_0)$ and n arbitrary elements $z_1, \ldots, z_n$. Since $\mathfrak{M}$ is compact, it can be covered by a finite number of these neighborhoods, say, by $U_1, \ldots, U_m$, and we may assume that U_j is given by inequalities of the form

$$|y_{1j}(M) - y_{1j}(M_j)| < \varepsilon_j, \ \ldots, \ |y_{kj}(M) - y_{kj}(M_j)| < \varepsilon_j. \tag{4}$$

The elements involved in the expansions of the form (1) for the neighborhoods $U_1, \ldots, U_m$ and also in the inequalities (4) describing these neighborhoods will be denoted by $x_1, \ldots, x_q$. Let $S = S_R(x_1, \ldots, x_q)$ be the joint spectrum of the elements $x_1, \ldots, x_q$, *i.e.*, the set of points

$$\zeta = \{x_1(M), \ldots, x_q(M)\}$$

of the q-dimensional complex space $\mathbf{C}^q$, where M ranges over $\mathfrak{M}$.

If the points M' and M'' do not belong to the same one of the neighborhoods $U_1, \ldots, U_m$, then by construction the corresponding points ζ' and ζ'' are distinct. Therefore the complete inverse image $\mathfrak{M}_\zeta \subset \mathfrak{M}$ of every point $\zeta \in S$ lies entirely in one of the neighborhoods $U_1, \ldots, U_m$. On the whole of this inverse image a single expansion of the form (1) is operative. Since the functions $z_j(M)$ involved in the expansions (1) are constant on $\mathfrak{M}_\zeta$, the function $f(M)$ has the same value on the whole of $\mathfrak{M}_\zeta$. Thus, we can define a single-valued function $f(\zeta)$ on S by putting it equal to $f(M)$ for every $M \in \mathfrak{M}_\zeta$. All the functions $x_1(M), \ldots, x_q(M)$ also intersect on the set S, which is formed simply by the cartesian coordinates of its points in $\mathbf{C}^q$. The expansion (1) shows that $f(\zeta)$ is an analytic function in a neighborhood of every point $\zeta \in S$, i.e., is analytic on the whole of S. But then, by the lemma proved under B (§ 13.10), there exists an element $x \in R$ such that the function $x(M)$ assumes the value $f(\zeta)$, where $\zeta = \{x_1(M), \ldots, x_q(M)\}$, for every $M \in \mathfrak{M}$ and therefore coincides with $f(M)$. The proof of the theorem is now complete. //

***12.** As our first example of an application of Theorem 1, let us prove the following proposition, which gives a criterion for our ring of functions to contain all the functions that are analytic on a given set.

THEOREM 2: *Let R be a ring of functions of a complex variable ζ having as its set of maximal ideals a closed bounded set S in the ζ-plane. If for every point $\zeta_0 \in S$ there exists a function $\varphi_0(\zeta) \in R$ that is analytic in a neighborhood V_0 of ζ_0 and has a non-zero derivative in V_0, then R contains every function that is analytic on S.*

Proof: Every function $f(\zeta)$ analytic on V_0 is analytic with respect to $\varphi_0(\zeta)$, i.e., can be expanded in a series of powers of $\varphi_0(\zeta)$ (because $\varphi_0(\zeta)$ effects a conformal transformation onto a domain in its plane and $f(\zeta)$ can be regarded as given and analytic in this domain). Therefore every function $f(\zeta)$ analytic on S is, in virtue of the condition of the theorem, locally analytic with respect to R and, by what has been proved, must belong to this ring. //

***13.** As a second example, we shall make an application of Theorem 1 to the problem of the solvability of analytic equations in a ring without radical.

We begin with the equation

$$x^2 = a, \tag{5}$$

where a is a known and x an unknown element of the normed ring R.

Clearly, such an equation is not always solvable; thus, in the ring A of functions $z(\zeta)$ analytic in the circle $|\zeta| < 1$ and continuous in the closed circle $|\zeta| \leqq 1$ the square root of the element $z(\zeta) = \zeta$ cannot be extracted. An obvious necessary condition for solvability of the equation (5) is that there exists on $\mathfrak{M}(R)$ a (single-valued) continuous function $f(M)$ for which $f^2(M) = a(M)$. In the particular example quoted, of course, this condition is not satisfied. But in the general case it is still not a sufficient condition. For example, in the ring A_0 of functions $f(\zeta)$ analytic in the circle $|\zeta| < 1$, continuous in the circle $|\zeta| \leqq 1$, and having at $\zeta = 0$ a derivative equal to zero, the function $z(\zeta) = \zeta^2$ does not have a square root, although the corresponding continuous function $f(\zeta) = \zeta$ exists on the space of maximal ideals.

Using Theorem 1 we can give the following sufficient condition for the existence of a square root of the element a: *For the solvability of the equation (5) in a ring R without radical it is sufficient that the equation $f^2(M) = a(M)$ has a solution in the class of all (single-valued) continuous functions on $\mathfrak{M}(R)$ and that a has an inverse.* For under these assumptions the continuous function $f(M)$ admits an expansion in a series of powers of $a(M) - a(M_0)$ in the neighborhood of every point M_0, and therefore, by Theorem 1, there exists an element x such that $x(M) = f(M)$, and thus $x^2 = a$.

Hence it follows, for example, that if R is the ring of functions on $\zeta = \exp(it)$ $(0 \leqq t < 2\pi)$, which has this circle as its space of maximal ideals, and if R contains the function $\exp(2it)$, then R also contains the function $\exp(it)$. This result cannot be deduced from the theorems of § 6.

The arguments used here can be extended almost without change to an analytic equation

$$F(x, a_1, \ldots, a_n) = 0 \tag{6}$$

and also to a system of such equations

$$F_i(x_1, \ldots, x_k, a_1, \ldots, a_k) = 0 \qquad (i = 1, \ldots, k); \tag{7}$$

we call an equation of the form (6) analytic here if the function F can be expressed as a power series in the arguments ζ, $a_1(M), \ldots, a_n(M)$ in the neighborhood of every system of values of these arguments.

A sufficient condition for the equation (6) to be solvable with respect to x in a ring R without radical is as follows: *This equation is solvable in the class of all continuous functions on $\mathfrak{M}(R)$ and the expression*

$$\frac{\partial F(x(M), a_1(M), \ldots, a_n(M))}{\partial x}$$

does not vanish on $\mathfrak{M}(R)$.

A sufficient condition for the system of equations (7) to be solvable with respect to $x_1, \ldots, x_k$ in a ring R without radical is as follows: *These equations have a solution in the class of all continuous functions on $M(R)$; moreover*

$$\det \left\| \frac{\partial F_i(x_1(M), \ldots, x_k(M), a_1(M), \ldots, a_n(M))}{\partial x_j} \right\|$$

does not vanish on $\mathfrak{M}(R)$.

For in these cases, by the classical theorems on implicit functions, the solutions $x(M)$ (in the case of the equation (6)) and $x_j(M)$ (in the case of the system (7)) are locally analytic functions of the elements $a_1, \ldots, a_n$ and, by Theorem 1, belong to R.

14. We conclude this section by considering an application of the above results to the construction of an operational calculus in normed rings.

Let A_F be a ring of functions $\varphi(\zeta)$ analytic on a closed bounded set F of the plane of the complex variable ζ. Suppose, further, that the spectrum of the element x of the normed ring R is contained in F. In § 6 it was shown that A_F can be mapped homomorphically into R in such a way that the function $\varphi(\zeta) \equiv 1$ goes over into the unit element of R and the function $\varphi(\zeta) \equiv \zeta$ into x and that every sequence of analytic functions $\varphi_n(\zeta)$ uniformly convergent on any domain $G \supset F$ goes over into a sequence of elements of R convergent in norm (Theorem 2 of § 6). This makes it possible to give a reasonable definition of an analytic function of an element x in R and one for which it is natural to speak of an *operational calculus on elements x of R.*

This construction can be generalized to the case of several elements of the ring in the following way. *Let $A(F)$ be a ring of functions $f(\zeta)=f(\zeta_1, \ldots, \zeta_n)$ analytic on a closed bounded set F in the complex n-dimensional space $\mathbf{C}^n$. Suppose, further, that the joint spectrum S of the elements $x_1, \ldots, x_n$ of R is contained in F. Then there exists a homomorphism of $A(F)$ into R such that the function $f(\zeta)\equiv 1$ corresponds to the unit element e of R and the function $f_j(\zeta)\equiv\zeta_j$ to the element x_j $(j=1,\ldots,n)$ and such that the uniform convergence of the functions $f_\nu(\zeta)$ $(f_\nu\in A(F))$ on any domain $G\supset F$ implies the convergence of the corresponding elements $\tilde{f}_\nu(x)$ in the norm of R.*

***15.** *Proof*: We assume, to begin with, that the elements $x_1, \ldots, x_n$ are generators of R, so that their joint spectrum S coincides with the space $\mathfrak{M}(R)$ of maximal ideals of R. As was shown under A in the proof of Theorem 1, every analytic function $f(\zeta)$ on S can be represented in the form of a Weil integral

$$f(\tau_1, \ldots, \tau_n)=\frac{1}{(2\pi i)^n}\sum_{1\leq i_1<\ldots<i_n\leq N}\int_{\sigma_{i_1\ldots i_n}}\frac{D_{i_1\ldots i_n}f(\zeta)\,d\zeta}{\prod_{\nu=1}^{n}[P_{i_\nu}(\zeta)-P_{i_\nu}(\tau)]}, \tag{8}$$

and we can associate with the function $f(\zeta)$ the element $\tilde{f}(x)=\tilde{f}(x_1, \ldots, x_n)$ of R defined by the formula

$$\tilde{f}(x)=\frac{1}{(2\pi i)^n}\sum_{1\leq i_1<\ldots<i_n\leq N}\int_{\sigma_{i_1\ldots i_n}}\prod_{\nu=1}^{n}[P_{i_\nu}(\zeta)e-P_{i_\nu}(x)]^{-1}\times$$
$$\times D_{i_1\ldots i_n}(\zeta)f(\zeta)\,d\zeta. \tag{9}$$

Let us verify that this formula realizes the required homomorphism. Obviously, it defines a linear mapping of $A(F)$ onto a certain set of elements of R and the uniform convergence of the functions $f_\nu(\zeta)$ in the domain $G\supset F$ implies the convergence of the corresponding elements $\tilde{f}_\nu(x)$ in the norm of R. We have to show that this mapping carries the functions $1, \zeta_1, \ldots, \zeta_n$ into $e, x_1, \ldots, x_n$, respectively, and the product of the functions $f(\zeta)$ and $g(\zeta)$ into the product of the elements $\tilde{f}(x)$ and $\tilde{g}(x)$.

We observe, first of all, that the integral (8) is equal to zero if we have $|P_\mu(\tau)|>1$ for at least one μ $(1\leq\mu\leq N)$. The proof[8] consists in carrying the integral (8), by means of simple algebraic transformations of the func-

[8] See, for example, B. A. Fuks, *Theory of Analytic Functions of Several Complex Variables,* Providence, 1963.

tions under the integral sign, into a sum of integrals of analytic functions of ζ, which vary over the domain $\{\zeta \in \mathbf{C}^n : |P_\mu(\zeta)| < 1\}$, the integration being over the boundary of this domain; by Cauchy's Theorem, each of these integrals is equal to zero. If the element $P_\mu(x) - P_\mu(\zeta)e$ has an inverse in R for every ζ such that $|P_\mu(\zeta)| \leqq 1$, then our argument can be translated into the language of elements of R and it shows that in this case the integral (9) is also equal to zero. For these algebraic transformations of the expressions under the integral sign can also be carried out on the elements of R; on the other hand, the integral of every abstract analytic function (with values in R) taken over the boundary of its domain of analyticity is equal to zero, because the result of applying an arbitrary linear functional to an abstract analytic function yields an ordinary analytic function.

It is easy to derive from this that the value of the integral (9) does not depend on the special choice of the Weil domain used in constructing it, as long as it is contained in the domain of analyticity of $f(\zeta)$. For on the difference of two such domains one of the polynomials $P_\mu(\zeta)$ is everywhere greater than 1 in absolute value; this means that $P_\mu(\zeta) - P_\mu(\tau)$ does not vanish for $\tau \in S$, so that the element $P_\mu(x) - P_\mu(\zeta)e$ has an inverse in R.

Suppose, in particular, that the function $f(\zeta)$ is of the form $f_1(\zeta_1) \ldots f_n(\zeta_n)$, where $f_j(\zeta_j)$ is analytic for $|\zeta_j| \leqq c_j$. Here we can take for the Weil domain $\{\zeta \in \mathbf{C}^n : |\zeta_j| \leqq c_j \ (j = 1, \ldots, n)\}$; the corresponding Weil integral then becomes a product of n ordinary Cauchy integrals:

$$\tilde{f}(x) = \prod_{j=1}^{n} \left\{ \frac{1}{2\pi i} \int_{|\zeta_j| = c_j} f_j(\zeta_j)(\zeta_j e - x_j)^{-1} d\zeta_j \right\}. \tag{10}$$

But, as was shown in § 6, for ordinary Cauchy integrals the relation

$$\frac{1}{2\pi i} \int_{|\zeta| = c} (\zeta e - x)^{-1} d\zeta = e, \qquad \frac{1}{2\pi i} \int_{|\zeta| = c} \zeta (\zeta e - x)^{-1} d\zeta = x$$

holds. Hence it follows immediately that the Weil integral (9) carries the functions $1, \zeta_1, \ldots, \zeta_n$ into the elements $e, x_1, \ldots, x_n$, respectively, of R.

It remains to show that the product $f(\zeta)g(\zeta)$ goes over into $\tilde{f}(x)\tilde{g}(x)$. In the case where f and g each depend on one coordinate only, $f(\zeta) = f(\zeta_j)$ and $g(\zeta) = g(\zeta_k)$, this follows for $k = j$ from what was proved in § 6 and for $k \neq j$, from the formula (10). Hence the required inclusion is easily derived in the case where $f(\zeta)$ and $g(\zeta)$ are products of functions each depending on one coordinate only. Going over to sums and taking the linear properties of an integral into account, we see that the property in question is true for a wide class of functions, including in particular all polynomials. Furthermore, every function $f(\zeta)$ that is analytic on F and hence in some domain G including F is the sum of a series of polynomials uniformly convergent within G that is obtained from by expanding the expressions

$$1/[P_\mu(\zeta)-P_\mu(\tau)]$$

under the integral sign in (8) into power series in the ratios $P_\mu(\tau)/P_\mu(\zeta)$. Hence the required property $f(\zeta)g(\zeta)\rightarrow\tilde{f}(x)\tilde{g}(x)$ is easily obtained from the corresponding relation for polynomials by passing to the limit. And so our theorem is proved, under the provisional assumption that the elements $x_1, \ldots, x_n$ are generators of R.

We emphasize that in this case there can only be one homomorphism with the properties in question, because the mapping $\zeta_j\rightarrow x_j$ extends in a unique way to all polynomials and, further, by means of a passage to the limit, to all analytic functions on F.

***16.** We now proceed to the discussion of the general case. Let $x_1, \ldots, x_n$ be arbitrary elements of R, and S their joint spectrum. Further, let $G_1\supset G_2\supset\ldots\supset G_\nu\supset\ldots$ be a sequence of domains in $\mathbf{C}^n$ contracting to S. We must assign to every function $f(\zeta)$ analytic in at least one of the domains G_ν an element $\tilde{f}(x)=\tilde{f}(x_1, \ldots, x_n)$ of R for which the homomorphism conditions are satisfied and the uniform convergence $f_m(\zeta)\rightarrow f(\zeta)$ on G_ν implies that $\tilde{f}_m(x)\rightarrow\tilde{f}(x)$ in the norm of R.

As was shown under C) in the proof of Theorem 1, for every $\nu=1, 2, \ldots$ we can define elements $y_{11}, \ldots, y_{1n_1}, y_{21}, \ldots, y_{2n_2}, \ldots, y_{\nu 1}, \ldots, y_{\nu n_\nu}$ of R such that the projection of the set $\mathfrak{M}_\nu$ of maximal ideals of R_ν generated by the elements $x_1, \ldots, x_n, y_{11}, \ldots, y_{\nu n_\nu}$ from the space $\mathbf{C}^{n+n_1+\cdots+n_\nu}$ of the variables $\zeta_1, \ldots, \zeta_n, \zeta_{11}, \ldots, \zeta_{\nu n_\nu}$, where it is naturally located, into the space $\mathbf{C}^n$ lies inside the domain G_ν.

By what has been proved above, there exists a unique homomorphism of the ring $A(\mathfrak{M}_\nu)$ of all functions $f(\zeta)=f(\zeta_1, \ldots, \zeta_n, \zeta_{11}, \ldots, \zeta_{\nu n_\nu})$ analytic on $\mathfrak{M}_\nu$ into R such that the functions $1, \zeta_1, \ldots, \zeta_n, \zeta_{11}, \ldots, \zeta_{\nu n_\nu}$ go over into the elements $e, x_1, \ldots, x_n, y_{11}, \ldots, y_{\nu n_\nu}$, respectively, and the sequence $f_m(\zeta)$, uniformly convergent in the domain $G\supset\mathfrak{M}_\nu$, corresponds to the sequence $\tilde{f}_m(x)$ convergent in norm. In particular, this homomorphism assigns to every function $f(\zeta)=f(\zeta_1, \ldots, \zeta_n)$ analytic in the domain $G_\nu\subset\mathbf{C}^n$ an element $\tilde{f}(x)$ of R, because such a function can be regarded as given in a neighborhood of $\mathfrak{M}_\nu$. The transition from ν to $\nu+1$ extends the domain of definition of this homomorphism from the class of functions analytic in G_ν to that of functions analytic in $G_{\nu+1}$, and by virtue of the uniqueness proved above, the image of the function $f(\zeta)$ analytic in G_ν is the same element of R in both mappings, $A(G_\nu) \to R$ and $A(G_{\nu+1}) \to R$. The homomorphism theorem is thus completely proved. //

***17.** *Remark*: The operational calculus on certain elements of a ring, i.e., the homomorphic mapping of $A(S)$ into R described above, was first constructed in 1954 by L. Waelbroeck [72] following a somewhat different path. Within the domain G containing the joint spectrum S of the elements $x_1, \ldots, x_n$ of R, Waelbroeck constructed a domain defined by the inclusions $\zeta_j \in \Delta_j$ $(j=1, \ldots, n)$, $P_j(\zeta) \in \Delta_j$ $(j=n+1, \ldots, N)$, where $P_j(\zeta)$ are polynomials and Δ_j $(j=1, \ldots, N)$ are domains in the complex plane, possibly not simply connected. To every function $f(\zeta_1, \ldots, \zeta_n)$ analytic in G one can assign the function $F(\zeta)=F(\zeta_1, \ldots, \zeta_N)$ in $\mathbf{C}^N$, equal to $f(\zeta_1, \ldots, \zeta_n)$ at the points where $\zeta_j = P_j(\zeta_1, \ldots, \zeta_n)$ $(j=n+1, \ldots, N)$. By certain theorems of H. Cartan and Oka, $F(\zeta)$ can be extended uniquely (as an analytic function) to the whole set

$$E=\{\zeta_1 \in \Delta_1, \ldots, \zeta_n \in \Delta_n, \zeta_{n+1} \in \Delta_{n+1}, \ldots, \zeta_N \in \Delta_N\},$$

whereupon it can be represented in the form of a multiple Cauchy integral

$$F(\zeta)=\prod_{j=1}^{N}\left\{\frac{1}{2\pi i}\int_{\Gamma_j}\frac{F(\tau_j)\,d\tau_j}{\tau_j-\zeta_j}\right\}, \tag{11}$$

where the Γ_j are contours bounding the domains Δ_j $(j=1, \ldots, N)$. The formula (11) defines a homomorphism of the ring of functions $A(E)$ into R. It can be proved by means of another theorem of Cartan that the kernel of this homomorphism is the ideal J generated by the functions $\zeta_j - P_j(\zeta_1, \ldots, \zeta_n)$ $(j=n+1, \ldots, N)$. The factor ring $A(E)/J$ is then isomorphic to the ring $A(S)$ of analytic functions on S, which in this way turns out to be mapped one-to-one onto a set of elements of R.

§ 14. Decomposition of a Normed Ring into a Direct Sum of Ideals

1. DEFINITION 1. A normed ring R is a *direct sum* of the two ideals I_1 and I_2 if:

a) the intersection $I_1 \cap I_2$ consists of the element 0 only;

b) every element $x \in R$ can be represented in the form of a sum $x = x_1 + x_2$, where $x_1 \in I_1$, $x_2 \in I_2$.

From a) it follows easily that the representation b) is unique. For if $x = x_1 + x_2 = y_1 + y_2$, then $0 = (x_1 - y_1) + (x_2 - y_2)$. Since 0 and $x_1 - y_1$ belong to I_1, we see that $x_2 - y_2 \in I_1$; but on the other hand, $x_2 - y_2 \in I_2$; therefore $x_2 - y_2 = 0$, $x_2 = y_2$, and hence $x_1 = y_1$ as well.

Furthermore, since I_1 and I_2 are ideals, we have, for arbitrary elements $x_1 \in I_1$ and $x_2 \in I_2$, $x_1 x_2 \in I_1 \cap I_2$ and hence, by a), $x_1 x_2 = 0$.

Let R be the direct sum of its ideals I_1 and I_2. We consider the components of the unit element: $e = e_1 + e_2$ $(e_1 \in I_1, e_2 \in I_2)$. Since $e_1 e_2 = 0$, we have $e = e^2 = e_1^2 + e_2^2$, and hence, by the uniqueness of the decomposition, $e_1^2 = e_1$, $e_2^2 = e_2$. The element e_1 in the ideal I_1—and likewise, e_2 in I_2—plays the role of the unit element: if $x_1 \in I_1$, then $x_1 e_2 = 0$ and therefore we have $x_1 e_1 = x_1(e - e_2) = x_1$. The ideals I_1 and I_2 are closed: if $x_n \in I_1$ and $x_n \to x$, then $x_n = x_n e_1 \to x e_1$ and consequently $x = x e_1 \in I_1$. *Thus, I_1 and I_2 are normed rings with the unit elements e_1 and e_2, respectively.*

Suppose, conversely, that there exists an element $e_1 \in R$, different from the zero and the unit element, that has the property $e_1^2 = e_1$. We consider the element $e_2 = e - e_1$; obviously,

$$e_1 e_2 = e_1(e - e_1) = e_1 - e_1^2 = 0 \quad \text{and} \quad e_2^2 = (e - e_1)e_2 = e_2.$$

Let I_1 and I_2 be the ideals generated by the elements e_1 and e_2, respectively. Let $x \in I_1 \cap I_2$; this means that $x = x' e_1 = x'' e_2$ and hence it follows that $x = x(e_2 + e_1) = x' e_1 e_2 + x'' e_2 e_1 = 0$. Thus, $I_1 \cap I_2$ consists of 0 only. Furthermore, every element $x \in R$ can be represented in the form of a sum $x = x e_1 + x e_2$, where $x e_1 \in I_1$ and $x e_2 \in I_2$. We see that the conditions a) and b) of Definition 1 are satisfied, so that R is the direct sum of the ideals I_1 and I_2.

2. THEOREM 1: *If a commutative normed ring R is split into the direct sum of two ideals I_1 and I_2, then the space $\mathfrak{M}(R)$ of maximal ideals of R splits into the sum of disjoint closed sets F_1 and F_2 that serve as the spaces of maximal ideals for the rings I_1 and I_2, respectively.*

Proof: Let e_1 and e_2 be the unit elements of I_1 and I_2. It follows from the equations $e_1^2 = e_1$, $e_2^2 = e_2$ that the functions $e_1(M)$ and $e_2(M)$ assume the values 0 and 1 only. Let F_1 be the set of all those maximal ideals for which $e_1(M) = 1$, and F_2, that of all those maximal ideals for which $e_2(M) = 1$. These sets are closed. Since $e_1 + e_2 = e$, we have

$$e_1(M) + e_2(M) = 1;$$

therefore F_1 and F_2 are disjoint and together constitute $\mathfrak{M}(R)$. Thus, to the decomposition of R into the direct sum of the ideals I_1 and I_2 there corresponds a decomposition of the space $\mathfrak{M}(R)$ of maximal ideals of R into the sum of two closed disjoint sets, $\mathfrak{M}(R) = F_1 + F_2$, where

$$F_1 = \{M \in \mathfrak{M}(R) : e_1(M) = 1\}, \quad F_2 = \{M \in \mathfrak{M}(R) : e_2(M) = 1\}.$$

Let us show that $F_1=\mathfrak{M}(I_1)$ and $F_2=\mathfrak{M}(I_2)$. For let $M_1\in F_1$. By assigning to every element $x_1\in I_1$ the number $x_1(M_1)$ we obtain a homomorphic mapping of I_1 into the field of complex numbers which is not trivial, because e_1 goes over into 1. Thus, every maximal ideal $M_1\in F_1$ of R defines a certain maximal ideal of I_1. Moreover, two distinct maximal ideals $M_1, M_1'\in F_1$ correspond to distinct maximal ideals of I_1, because for every element $x\in R$ we have that $M_1(x)=M_1(xe_1)$ and $M_1'(x)=M_1'(xe_1)$, so that M_1 and M_1' must differ at least for the element $xe_1\in I_1$. Conversely, let M' be a maximal ideal of I_1. By assigning to every element $x\in R$ the number $x_1(M')$, where $x_1=xe_1$, we obtain a non-trivial homomorphic mapping of R into the field of complex numbers. The maximal ideal defined by this homomorphic mapping must be contained in F_1, because for every maximal ideal $M_2\in F_2$ we have $x_1(M_2)=0$. Thus, every maximal ideal M' of I_1 defines a certain maximal ideal M_1 of R, and obviously distinct maximal ideals $M'\subset I_1$ generate distinct maximal ideals $M_1\subset R$; and if M' generates M_1, then M_1 generates, in the way indicated above, that same maximal ideal M'. Thus we have established a one-to-one correspondence between F_1 and $\mathfrak{M}(I_1)$. From the equation $x(M')=x_1(M_1)$ it follows that the topological spaces F_1 and $\mathfrak{M}(I_1)$ are homeomorphic. In the same way we can satisfy ourselves that F_2 and $\mathfrak{M}(I_2)$ coincide. //

THEOREM 2: *Let R be a normed ring and let $\mathfrak{M}(R)$ be representable in the form of a sum of two disjoint closed sets F_1 and F_2. Then R splits into the direct sum of two ideals I_1 and I_2 that are rings with unit elements having F_1 and F_2, respectively, as their spaces of maximal ideals. Furthermore, the ideals I_1 and I_2 are uniquely defined by the sets F_1 and F_2.*

Proof: For the proof of the first statement it is sufficient, by Theorem 1, to establish the existence of an element e_1 with the properties:

$$\text{a) } e_1^2=e_1; \qquad \text{b) } \{M\in\mathfrak{M}(R):\ e_1(M)=1\}=F_1.$$

The function $f(M)$ equal to 1 on F_1 and to 0 on F_2 satisfies the conditions of Theorem 1 of § 13; by virtue of this theorem there exists a $z\in R$ such that $z(M)=1$ on F_1 and $z(M)=0$ on F_2. If $z^2=z$, we put $e_1=z$ and $e_2=e-z$. But if $z^2\neq z$ (z^2 and z may differ, in general, by an element of the radical), we proceed as follows. All the values of $z(M)$ on F_1 lie within the circle $|1-\lambda|<1/3$, and all the values of $z(M)$ on F_2 within the circle $|\lambda|<1/3$. Let D denote the set of these circles. The spectrum of the element z is contained in D, and the function $e_1(\lambda)$ equal to 1 within the circle $|1-\lambda|<1/2$ and to 0 within the circle $|\lambda|<1/2$ is analytic in D. We put

$$e_1 = e_1(z) = \frac{1}{2\pi i} \int_{|1-\lambda|=\frac{1}{2}} (\lambda e - z)^{-1} d\lambda.$$

Since $(e_1(\lambda))^2 = e_1(\lambda)$, we have, by equation (2) of § 6, $e_1{}^2 = e_1$. Furthermore, by Theorem 1 of § 6, $e_1(M)$ is equal to 1 on F_1 and to 0 on F_2, and hence e_1 is different from 0 and e. Thus, the first statement of the theorem is proved.

For the proof of the second statement it is sufficient to satisfy ourselves of the following: If e_1 and e_1' are such that $e_1{}^2 = e_1$, $e_1'^2 = e_1'$ and $e_1(M) = 1$ if and only if $e_1'(M) = 1$, then $e_1' = e_1$.

Since the function $e_1(M) - e_1'(M)$ is equal to 0 everywhere on $\mathfrak{M}(R)$, the powers of $e_1 - e_1'$ must tend to zero. But, as is easy to verify,

$$(e_1 - e_1')^{2n+1} = e_1 - e_1'$$

for every integer $n \geqq 0$. Hence it follows that $e_1 = e_1'$. //

3. The concept of the direct sum of ideals can be generalized in an obvious way to the case of an arbitrary finite number of components $I_1, \ldots, I_n$: every element $x \in R$ must be representable, in a unique way, in the form of a sum $x = x_1 + \ldots + x_n$, where $x_i \in I_i$ $(i = 1, \ldots, n)$. From Theorem 2 it follows, in particular, that *a normed ring R with a finite number of maximal ideals splits into the direct sum of a finite number of ideals that are rings with one maximal ideal.*

§ 15. The Normed Space Adjoint to a Normed Ring

1. Like every normed space, a normed ring R has an adjoint space R' consisting of all linear functionals on R.

We have already studied the 'multiplicative' functionals $M(x) = x(M)$ in a normed ring; since every multiplicative functional has the norm 1, the multiplicative functionals are located on the surface of the unit sphere in the conjugate space.

To every element $x \in R$ there corresponds the linear operator x^* in R' that is adjoint to the operator of multiplication by x in R; the operator x^* acts on the functional f by the formula

$$\langle y, x^*f \rangle = \langle xy, f \rangle \qquad (f \in R', y \in R).$$

We shall say that $P \subset R'$ is an *invariant subspace* of R' if P is invariant under every operator x^*. If $P \subset R'$ is a closed subspace, then for it to be invariant it is necessary and sufficient that it is invariant under the operators

x^* corresponding to the generators of R; for from $x_n \rightarrow x$ it follows that $x_n^* f \rightarrow x^* f$.

The set $A^{\perp}$ of all linear functionals that are equal to zero on every element x of a given set $A \subset R$, as the reader knows, is called the *orthogonal complement* of A. The orthogonal complement of a set A is always a weakly closed subspace of R'; and conversely, every weakly closed subspace $P \subset R'$ is the orthogonal complement of a certain set $A \subset R$, namely the (weakly closed) set of all elements $x \in R$ on which every functional $f \in P$ vanishes; we shall call the latter the orthogonal complement to P in R and shall denote it by $P^{\perp}$.

The orthogonal complement $J^{\perp}$ of an ideal $J \subset R$ is an invariant subspace of R'. For if $f \in J^{\perp}$, then for every $y \in J$ we have $\langle y, x^* f \rangle = \langle xy, f \rangle = 0$, no matter what the element $x \in R$; therefore $x^* f$ occurs in $J^{\perp}$ whenever f does. Conversely, if $P \subset R'$ is an invariant subspace of R', then its orthogonal complement $P^{\perp} \subset R$ is an ideal in R, because it follows from $f \in P$, $x \in R$, $y \in P^{\perp}$ that $\langle xy, f \rangle = \langle y, x^* f \rangle = 0$.

Thus, the weakly closed ideals in R and the weakly closed invariant subspaces of R' are orthogonal complements of each other.

The invariant subspace $M^{\perp} \subset R'$ orthogonal to a maximal ideal $M \subset R$ is one-dimensional and consists of the multiples of the functional $M(x)$; conversely, every one-dimensional invariant subspace $P \subset R'$ has as its orthogonal complement a certain maximal ideal $M \subset R$.

The one-dimensional invariant subspaces $P \subset R'$ are, naturally, minimal invariant subspaces of R'.

Every non-zero weakly closed invariant subspace $P \subset R'$ contains a one-dimensional invariant subspace. For the proof, we consider the ideal $P^{\perp} \subset R$ orthogonal to P. It is contained in a certain maximal ideal M. Therefore, P as the orthogonal complement of the ideal $P^{\perp}$ contains the one-dimensional invariant subspace orthogonal to the maximal ideal M.

***2.** By way of illustration, we consider the space W' adjoint to the ring W of absolutely convergent Fourier series (Example 3 of § 1). W is isomorphic to the space l_1 of absolutely convergent numerical series $\sum_{n=-\infty}^{\infty} c_n$ and, as is well known, the space l_1' adjoint to l_1 can be identified with the space of all bounded sequences $\{\ldots, f_{-1}, f_0, f_1, \ldots\}$, so that if

$$x = \{\ldots, c_{-1}, c_0, c_1, \ldots\} \in l_1,$$

then

$$\langle x, f \rangle = \sum_{n=-\infty}^{\infty} f_n c_n. \tag{1}$$

On the basis of this we shall identify W' with the latter space.

A closed invariant subspace of W' can be defined as a closed subspace that is invariant under the operators adjoint to the operators of multiplication by $\exp(it)$ and by $\exp(-it)$ in W (because $\exp(it)$ and $\exp(-it)$ are generators of this ring). The operator of multiplication by $\exp(it)$ shifts the sequence of coefficients of the series $x(t) = \sum_{n=-\infty}^{\infty} c_n \exp(int)$ one place to the right; the operator of multiplication by $\exp(-it)$ shifts the same sequence one place to the left. From the formula (1) it now follows easily that the operators $[\exp(it)]^*$ and $[\exp(-it)]^*$ shift an arbitrary sequence $f = \{\ldots, f_{-1}, f_0, f_1, \ldots\} \in W'$ one place to the left and to the right, respectively. Thus, a closed invariant subspace of W' is a closed subspace that is invariant under right and left shifts.

As we know, every maximal ideal M_0 of W consists of the functions $x(t) = \sum_{n=-\infty}^{\infty} c_n \exp(int)$ that vanish at a fixed point $t_0 \in [0, 2\pi)$, so that $x(t_0) = \sum_{n=-\infty}^{\infty} c_n \exp(int_0) = 0$. Hence it follows that the one-dimensional invariant subspace $M_0^{\perp} \subset W'$ orthogonal to M_0 consists of the multiples of the sequence $\{\exp(int_0)\}$.

Since, as we have shown, every weakly closed invariant subspace $P \subset W'$ contains a one-dimensional invariant subspace, we obtain the following result: *Every weakly closed subspace of the space of bounded two-sided sequences that is invariant under shifts contains a subsequence of the form* $\{\exp(int_0)\}$. This fact forms the content of a well-known theorem of Beurling [5, 6].

Beurling has also proved that *every weakly closed invariant subspace* $P \subset W'$ *containing only one sequence* $\{\exp(int_0)\}$ *is one-dimensional and consists of the multiples of this sequence.* This result can also be obtained from ring arguments. Namely, the orthogonal complement in W to a subspace $P \subset W'$ satisfying Beurling's conditions is a weakly closed ideal in W containing only one maximal ideal M_{t_0}. But, as will be proved in § 39, not only every weakly closed, but also every strongly closed ideal of W, if it is contained in only one maximal ideal, coincides with this maximal ideal. Hence it follows that P is one-dimensional and therefore consists of the multiples of the sequence $\{\exp(int_0)\}$.

PART TWO

CHAPTER III

THE RING OF ABSOLUTELY INTEGRABLE FUNCTIONS AND THEIR DISCRETE ANALOGUES

§ 16. The Ring V of Absolutely Integrable Functions on the Line

1. We denote by $L^1(-\infty, \infty)$, or simply by L^1, the space of all measurable complex functions of a real variable that are absolutely integrable on the line; and by introducing the norm

$$\| x \| = \int_{-\infty}^{\infty} | x(t) | \, dt, \tag{1}$$

we turn it into a Banach space.

THEOREM 1: *If $x(t)$ and $y(t) \in L^1$, then the integral*

$$(x * y)(t) = \int_{-\infty}^{\infty} x(t-\tau) y(\tau) d\tau \tag{2}$$

exists for almost all t and also belongs to L^1; moreover

$$\| x * y \| \leqq \| x \| \, \| y \|. \tag{3}$$

The operation of convolution $*$ defined by the formula (2) *is bilinear, associative, and commutative, so that L^1 is a normed ring with respect to the multiplication defined, by formula* (2), *as convolution.*

Proof: Since $x(t)$ is Lebesgue measurable on the line, the function $x(t-\tau)$ can be proved to be Lebesgue measurable on the plane (t, τ). Consequently, $x(t-\tau)y(\tau)$ is also a measurable function of (t, τ). But then, by Fubini's Theorem,[1] the existence of the first of the integrals

[1] See S. Saks, *Theory of the Integral,* Warsaw, 1937, Chap. III, § 9, or P. R. Halmos, *Measure Theory,* New York, 1950, § 36.

$$\int_{-\infty}^{\infty}\int_{-\infty}^{\infty} x(t-\tau)\, y(\tau)\, dt\, d\tau,$$

$$\int_{-\infty}^{\infty}\left(\int_{-\infty}^{\infty} x(t-\tau)\, y(\tau)\, dt\right) d\tau, \quad \int_{-\infty}^{\infty}\left(\int_{-\infty}^{\infty} x(t-\tau)\, y(\tau)\, d\tau\right) dt,$$

and for $x(t) \geqq 0$, $y(\tau) \geqq 0$ the existence of any one of them, implies the existence of the other two and the equality of all three integrals. But the second integral exists:

$$\int_{-\infty}^{\infty}\left(\int_{-\infty}^{\infty} x(t-\tau)\, y(\tau)\, dt\right) d\tau = \int_{-\infty}^{\infty}\left(\int_{-\infty}^{\infty} x(t-\tau)\, dt\right) y(\tau)\, d\tau =$$

$$= \int_{-\infty}^{\infty}\left(\int_{-\infty}^{\infty} x(t)\, dt\right) y(\tau)\, d\tau = \int_{-\infty}^{\infty} x(t)\, dt \int_{-\infty}^{\infty} y(\tau)\, d\tau.$$

When we replace $x(t)$ and $y(\tau)$ in this equation by $|x(t)|$ and $|y(\tau)|$, we conclude that the first, and hence also the third, integral exists, i.e., the integral (2) exists for almost all t, is measurable as a function of t and is absolutely integrable, and

$$\int_{-\infty}^{\infty}\left(\int_{-\infty}^{\infty} x(t-\tau)\, y(\tau)\, d\tau\right) dt = \int_{-\infty}^{\infty} x(t)\, dt \int_{-\infty}^{\infty} y(\tau)\, d\tau.$$

Replacing $x(t)$ and $y(\tau)$ here by $|x(t)|$ and $|y(\tau)|$, we obtain

$$\|x * y\| = \int_{-\infty}^{\infty}\left|\int_{-\infty}^{\infty} x(t-\tau)\, y(\tau)\, d\tau\right| dt \leqq$$

$$\leqq \int_{-\infty}^{\infty}\left(\int_{-\infty}^{\infty} |x(t-\tau)|\,|y(\tau)|\, d\tau\right) dt = \|x\|\,\|y\|,$$

which is the inequality (3). Applying Fubini's Theorem again and making the substitution $\tau \to \tau + \sigma$, we obtain

$$x * (y * z) = \int_{-\infty}^{\infty} x(t-\tau)\left(\int_{-\infty}^{\infty} y(\tau-\sigma)\, z(\sigma)\, d\sigma\right) d\tau =$$

$$= \int_{-\infty}^{\infty}\left(\int_{-\infty}^{\infty} x(t-\tau)\, y(\tau-\sigma)\, d\tau\right) z(\sigma)\, d\sigma =$$

$$= \int_{-\infty}^{\infty}\left(\int_{-\infty}^{\infty} x(t-\sigma-\tau)\, y(\tau)\, d\tau\right) z(\sigma)\, d\sigma = (x * y) * z,$$

so that the convolution is associative. Further, the substitution $\tau \to t - \tau$ gives

$$x * y = \int_{-\infty}^{\infty} x(t-\tau)\, y(\tau)\, d\tau = \int_{-\infty}^{\infty} x(\tau)\, y(t-\tau)\, d\tau = y * x,$$

so that the convolution is commutative. Finally, the bilinearity of the convolution is obvious. //

***2.** Fubini's Theorem, which has been used in the proof of Theorem 1, can by no means be attributed to the elementary part of the theory of the Lebesgue integral. Moreover, the applicability of this theorem is based on the fact that $x(t-\tau)$ is measurable as a function of two variables, and the verification of this, although elementary, is rather tedious (which is why we have omitted it). Let us give a proof of Theorem 1 that only uses, instead of Fubini's Theorem, its completely elementary analogue for continuous functions in a finite rectangular domain and some elementary properties of the Lebesgue integral.

Proof: Let L be the vector subspace of L^1 formed by all continuous 'finite' functions, i.e., those that are equal to zero for all values of the variable of sufficiently large absolute value. The 'convolution' (2) of the functions $x(t)$ and $y(t) \in L$ (where the integration is actually between finite limits) is easily seen also to belong to L. The operation of convolution (2) thus defined in L is obviously bilinear; and the substitution $\tau \to t - \tau$ in (2), as before, shows that the operation is commutative. Further, just as above but relying on the theorem of the interchange of the order of integration for continuous functions in a finite rectangular domain, we can satisfy ourselves of the fact that convolution is also associative. Finally, on the basis of the same theorem but otherwise exactly as before, we can establish the inequality (3), which shows that the convolution in L is continuous in the norm (1) jointly in the two 'factors.'

Since L is dense in L^1, it now follows that there exists in L^1 a uniquely defined multiplication $\cdot$ that coincides on L with the convolution and preserves all the properties of the latter. Thus, L^1 is a normed ring with respect to this multiplication and it only remains to show that the multiplication so defined can be expressed by the same formula (2).

Suppose, to begin with, that $y(t)$ is bounded, $|y(t)| \leq C$, so that the lemma to be proved below is applicable. This lemma implies, in particular, that the function $(x * y)(t)$ is integrable in any finite interval. We now take $x_n, y_n \in L$, so that

$$\| x - x_n \| < \frac{1}{n^2}$$

and

$$\| y - y_n \| < \frac{1}{n^2 \max | x_n(t) |}.$$

Then

$$\int_{-n}^{n} | (x * y)(t) - (x_n * y_n)(t) | \, dt \leq$$

$$\leq \int_{-n}^{n} \left| \int_{-\infty}^{\infty} | x(t-\tau) - x_n(t-\tau) | \, | y(\tau) | \, d\tau \right| dt +$$

$$+ \int_{-n}^{n} \left| \int_{-\infty}^{\infty} | x_n(t-\tau) | \, | y(\tau) - y_n(\tau) | \, d\tau \right| dt \leq$$

$$\leq 2Cn \| x - x_n \| + 2n \max | x_n(t) | \| y - y_n \| < \frac{2(C+1)}{n}. \tag{4}$$

Hence it follows, first of all, that

$$\int_{-n}^{n} | (x * y)(t) | \, dt \leq \int_{-n}^{n} | (x_n * y_n)(t) | \, dt + \frac{2(C+1)}{n} \leq$$

$$\leq \| x_n * y_n \| + \frac{2(C+1)}{n}. \tag{5}$$

But since $x_n \to x$ and $y_n \to y$, we see by the inequality (3) that $\| x_n * y_n \|$ is bounded. Therefore the inequality (5) shows that $x * y \in L^1$, and by now passing to the limit $n \to \infty$ in (4), we obtain that $\| x * y - x \cdot y \| = 0$, i.e., $x \cdot y = x * y$.

Since every function $z \in L^1$ can be represented in the form of a linear combination of non-negative real functions of L^1,

$$z = \frac{|\Re z| + \Re z}{2} - \frac{|\Re z| - \Re z}{2} + i \frac{|\Im z| + \Im z}{2} - i \frac{|\Im z| - \Im z}{2},$$

and the operations $*$ and $\cdot$ are bilinear, in order to complete the proof that the two operations coincide it is sufficient to show that $x * y$ exists and coincides with $x \cdot y$ for any two non-negative real functions $x, y \in L^1$. For this purpose, we put

$$y_N(t) = \begin{cases} y(t), & \text{where } y(t) \leq N, \\ N, & \text{where } y(t) > N. \end{cases}$$

For $N \to \infty$, we have $\| y - y_N \| \to 0$, and therefore $\| x \cdot y_N - x \cdot y \| \to 0$. On the other hand, by what has been shown above, $x \cdot y_N = x * y_N$ (because y_N is bounded), and since $y_N(\tau) \nearrow y(\tau)$, i.e., $y_N(\tau)$ tends to $y(\tau)$ monotonically at every point, we have $x(t-\tau) y_N(\tau) \nearrow x(t-\tau) y(\tau)$, and consequently

$$(x * y_N)(t) = \int_{-\infty}^{\infty} x(t-\tau)\, y_N(\tau)\, d\tau \nearrow \int_{-\infty}^{\infty} x(t-\tau)\, y(\tau)\, d\tau = (x * y)(t).$$

In conjunction with the preceding, this shows that $(x * y)(t)$ is finite for almost all t, belongs to L^1, and coincides in L^1 with $x \cdot y$. //

3. THEOREM 2: *The ring L^1 does not contain a unit element.*

The proof of this theorem will be based on the following lemma, which will also be required elsewhere.[2]

LEMMA: *The convolution $(x * y)(t)$ of the function $x \in L^1$ with an arbitrary bounded measurable function $y(t)$* (which obviously exists for all values of t) *is a continuous function of t.*

Proof of the Lemma: Let $|y(t)| \leqq C$. Then

$$|(x * y)(t+h) - (x * y)(t)| \leqq$$

$$\leqq C \int_{-\infty}^{\infty} |x(t+h-\tau) - x(t-\tau)|\, d\tau = C \int_{-\infty}^{\infty} |x(\tau+h) - x(\tau)|\, d\tau.$$

But the latter integral tends to zero for $h \to 0$. In fact, if $x(t)$ is a continuous 'finite' function, i.e., a continuous function that is equal to zero outside a finite interval, then this follows from its uniform continuity and from the boundedness of the set on which it is different from zero. This property of 'absolute continuity' of the Lebesgue integral carries over to the remaining functions of L^1 by virtue of the fact that the set of continuous finite functions is dense in L^1 and that L^1 is invariant with respect to the 'translations' $x(\tau) \to x(\tau + h)$. This proves the lemma.

Proof of Theorem 2: If L^1 were to contain a unit element $e(t)$ for the convolution, then every bounded function $x \in L^1$ would have to coincide almost everywhere with the function $(e * x)(t)$ which, by the lemma, is continuous. But in $L^1(-\infty, \infty)$ there are bounded functions that differ from every continuous function on a set of positive measure: we can take, as a simple example, the characteristic function of a finite interval. //

The normed ring obtained by formal adjunction of a unit element to L^1 will be denoted by V.

Thus, V is formed by the elements $\mathfrak{z} = \lambda e + x(t)$, where e is the unit element adjoined, λ is an arbitrary complex number, and $x(t)$ is an arbitrary function of L^1, and where $\|\mathfrak{z}\| = |\lambda| + \|x\|$.

---

[2] It has already been used just above, in § 16.2.

Now let $L^1(0, \infty)$ be the space of all absolutely integrable measurable complex functions on the half-line $0 \leqq t < \infty$ with the norm

$$\| x \| = \int_0^\infty | x(t) | \, dt.$$

It is a normed ring when convolution is taken as the ring multiplication, where 'convolution' is defined by the formula

$$(x * y)(t) = \int_0^t x(t-\tau) y(\tau) d\tau. \tag{2'}$$

In fact,

$$x(t) \to \hat{x}(t) = \begin{cases} x(t) & \text{if } t \geqq 0, \\ 0 & \text{if } t < 0 \end{cases} \tag{6}$$

is an isometric imbedding of $L^1(0, \infty)$ into $L^1(-\infty, \infty)$, and

$$\int_0^t x(t-\tau) y(\tau) d\tau = \int_{-\infty}^\infty \hat{x}(t-\tau) \hat{y}(\tau) d\tau.$$

Thus, $L^1(0, \infty)$ identified with its image $L^1{}_+$ in $L^1(-\infty, \infty)$ under the mapping (6), is a subring of L^1. The subring of V obtained by adjoining a unit element to $L^1{}_+$ will be denoted by V_+.

***4.** The multiplication in the ring I of Example 5 of § 1 was given by the same formula (2′). But there t only ranged over the interval $[0, 1]$. Thus, if we extend every function $x(t) \in L^1(0, 1)$ to the half-line $(0, \infty)$ or the whole line $(-\infty, \infty)$ by putting $x(t)$ equal to zero everywhere outside the interval $[0, 1]$, then the product in $L^1(0, 1)$ can be regarded as the restriction to the interval $[0, 1]$ of the product of the corresponding functions in $L^1(0, \infty)$ or $L^1(-\infty, \infty)$.[3]

5. Similarly, the space $L^1(0, T)$ of all absolutely integrable measurable complex functions on an arbitrary interval $[0, T]$, with the norm

$$\| x \| = \int_0^T | x(t) | \, dt,$$

is a normed ring when the ring multiplication is convolution, as defined by the formula

$$(x * y)(t) = \int_0^t x(t-\tau) y(\tau) d\tau \qquad (0 \leqq t \leqq T).$$

[3] Hence we can derive once more all the properties of the convolution in $L^1(0, 1)$ that were established in § 1.

The ring obtained by adjoining a unit element to $L^1(0, T)$ will be denoted by $I(0, T)$.

§ 17. Maximal Ideals of the Rings V and V_+

1. Obviously, L^1 is a maximal ideal in V. We denote it by M_∞; we shall see below that it plays the role of the point at infinity in the space of maximal ideals of the ring V.

Let us find the rest of the maximal ideals of V. Let s be an arbitrary real number. For every element $\mathfrak{z} = \lambda e + x(t) \in V$ we put

$$\mathfrak{z}^\sim(s) = \lambda + \int_{-\infty}^{\infty} x(t)\, e^{ist}\, dt.$$

It is easy to verify that $\mathfrak{z} \to \mathfrak{z}^\sim(s)$ is a homomorphic mapping of V into the field of complex numbers. For the only thing that needs to be verified is that the product of elements of the ring goes over into the product of the corresponding numbers. To do this it is sufficient to show that the convolution of the functions $x, y \in L^1$ corresponds to the multiplication of their *Fourier transforms*

$$x^\sim(s) = \int_{-\infty}^{\infty} x(t)\, e^{ist}\, dt, \qquad y^\sim(s) = \int_{-\infty}^{\infty} y(t)\, e^{ist}\, dt.$$

Since $|x^\sim(s)| \leqq \|x\|$ and the set L of finite continuous functions is dense in L^1, we can restrict ourselves to the case where $x, y \in L$. But since the interchange of the order of integration is valid for continuous functions in a finite rectangular domain, we then have

$$\int_{-\infty}^{\infty} (x * y)(t)\, e^{ist}\, dt = \int_{-\infty}^{\infty} \left(\int_{-\infty}^{\infty} x(t-\tau)\, y(\tau)\, d\tau \right) e^{ist}\, dt =$$

$$= \int_{-\infty}^{\infty} \left(\int_{-\infty}^{\infty} x(t-\tau)\, e^{is(t-\tau)}\, y(\tau)\, e^{is\tau}\, d\tau \right) dt =$$

$$= \int_{-\infty}^{\infty} \left(\int_{-\infty}^{\infty} x(t-\tau)\, e^{is(t-\tau)}\, dt \right) y(\tau)\, e^{is\tau}\, d\tau =$$

$$= \int_{-\infty}^{\infty} x(t)\, e^{ist}\, dt \int_{-\infty}^{\infty} y(\tau)\, e^{is\tau}\, d\tau.$$

The maximal ideal generated by the mapping $\mathfrak{z} \to \mathfrak{z}^\sim(s_0)$, i.e., the totality of all elements $\mathfrak{z} = \lambda e + x(t) \in V$ for which

$$\mathfrak{z}^{\sim}(s_0) = \lambda + x^{\sim}(s_0) = \lambda + \int_{-\infty}^{\infty} x(t)\, e^{is_0 t}\, dt = 0,$$

will be denoted by M_{s_0}. Thus, $\mathfrak{z}(M_{s_0}) = \mathfrak{z}^{\sim}(s_0)$. If $s_1 \neq s_2$, then the maximal ideals M_{s_1} and M_{s_2} are distinct; for then there exists a $t = t_0$ such that $e^{is_1 t_0} \neq e^{is_2 t_0}$, and by taking for $x(t)$ the characteristic function of a sufficiently small neighborhood of the point t_0, we have $x^{\sim}(s_1) \neq x^{\sim}(s_2)$. Moreover, all these maximal ideals are different from M_∞, since $x(M_\infty) = 0$ for every function $x \in L^1$, whereas for every s_0 there exists a function $x \in L^1$ such that $x(M_{s_0}) = x^{\sim}(s_0) \neq 0$ (for example, the characteristic function of a sufficiently small neighborhood of the point $t = 0$).

2. We shall now show that *the maximal ideals M_s and M_∞ exhaust all the maximal ideals of the ring V.*

We denote by $x_{a,b}(t)$ the characteristic function of the interval (a, b). The functions $x_{a,b}(t)$ are generators of L^1, and hence also of V.

LEMMA: *A closed ideal of the ring V containing the function $z(t) \in L^1$ also contains all its 'shifts' $z(t - \lambda)$, namely*

$$z(t-\lambda) = \lim_{h \to 0} \left\{ z(t) * \frac{x_{c,\lambda+h}(t) - x_{c,\lambda}(t)}{h} \right\} \qquad (c < \lambda), \qquad (1)$$

where the limit is to be understood in the sense of convergence in norm.

Proof: The functions $\frac{x_{c,\lambda+h}(t) - x_{c,\lambda}(t)}{h}$ (which obviously do not depend on c) are bounded in norm, for they all have the norm 1. Hence it follows that it is sufficient to prove the limit relation (1) for the functions $z(t) = x_{a,b}(t)$, which are, as we have said above, generators of V. But for $z(t) = x_{a,b}(t)$ the product under the limit sign is easy to calculate: its graph is given in the figure. The difference between this product and $x_{a,b}(t-\lambda)$

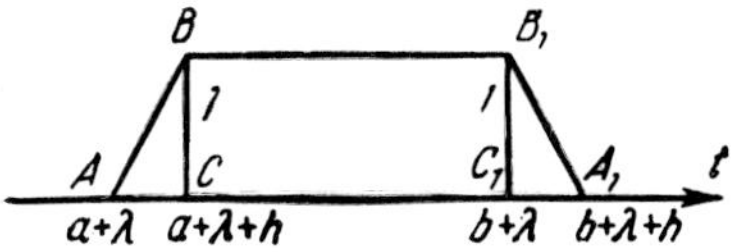

has the norm h, equal to the sum of the areas of the triangles ABC and $A_1B_1C_1$. Thus, the limit of the product for $h \to 0$ is indeed the function $x_{a,b}(t-\lambda)$. //

THEOREM 1: *If a maximal ideal M of V is different from M_∞, then there exists a real number s such that for every $\mathfrak{z} = \lambda e + x(t) \in V$ we have the equality*

$$\mathfrak{z}(M) = \lambda + \int_{-\infty}^{\infty} x(t)\, e^{ist}\, dt = \mathfrak{z}^{\sim}(s); \tag{2}$$

in other words, M coincides with a maximal ideal M_s.

Proof: Let M be a maximal ideal of V different from M_∞. Then there exists a function $z(t) \in L^1$ such that $z(M) \neq 0$. We put $z_\lambda(t) = z(t-\lambda)$. The application of the functional $M(x) = x(M)$ to both sides of equation (1) shows that

$$\lim_{h \to 0} M\left\{\frac{x_{c,\lambda+h}(t) - x_{c,\lambda}(t)}{h}\right\} = \frac{d}{d\lambda} M(x_{c,\lambda}) = \chi(\lambda) \tag{3}$$

exists for all λ, where

$$\chi(\lambda) = M(z_\lambda)/M(z), \tag{4}$$

so that, in particular,

$$\chi(0) = 1. \tag{5}$$

Since $\left\| \frac{x_{c,\lambda+h} - x_{c,\lambda}}{h} \right\| = 1$ and $|M(x)| \leqq \|x\|$, it follows from the formulas (3) that

$$\chi(\lambda) \leqq 1. \tag{6}$$

Further, since

$$z_{\lambda+\mu} * z = z_\lambda * z_\mu,$$

we have

$$M(z_{\lambda+\mu})\, M(z) = M(z_\lambda)\, M(z_\mu).$$

Dividing by $[M(z)]^2$ and bearing formula (4) in mind, we obtain

$$\chi(\lambda+\mu) = \chi(\lambda)\, \chi(\mu). \tag{7}$$

In particular, for $\mu = -\lambda$, taking (5) into account, we have

$$1 = \chi(\lambda)\, \chi(-\lambda);$$

hence, by (6), it follows that

$$|\chi(\lambda)| = 1. \tag{8}$$

It follows from formula (4) that the function $\chi(\lambda)$ is continuous, because z_λ is (in norm) a continuous function of λ (see the proof of the lemma of § 16). But every continuous function that satisfies the functional equation (7) and the condition (8) is of the form $e^{is\lambda}$, where s is a real number. Therefore we have

$$\chi(\lambda) = e^{is\lambda}.$$

Upon integrating the second of the equations (3) over the interval $a \leqq \lambda \leqq b$ and observing that $M(x_{c,b}) - M(x_{c,a}) = M(x_{a,b})$, we obtain

$$x_{a,b}(M) = M(x_{a,b}) = \int_a^b e^{is\lambda}\,d\lambda = \int_{-\infty}^{\infty} x_{a,b}(t)\,e^{ist}\,dt = x_{a,b}^{\sim}(s).$$

Thus, the equation (2) is established for the functions $\mathfrak{z} = x_{a,b}(t)$. Since these functions are generators of V, (2) also holds for all the elements $\mathfrak{z} \in V$ and Theorem 1 is proved. //

Using Theorems 3′ and 1 of §§ 4 and 6, we derive the following corollaries of Theorem 1.

COROLLARY 1: *The element $\mathfrak{z} = \lambda e + x(t) \in V$ has an inverse in V if and only if the expression*

$$\lambda\left(\lambda + \int_{-\infty}^{\infty} x(t)\,e^{ist}\,dt\right)$$

is distinct from zero for every s.

COROLLARY 2: *If $X(s)$ is the Fourier transform of the function $x \in L^1$ and $F(z)$ is an analytic function, regular on the closure of the set of values of the function $X(s)$ and vanishing at the point $z = 0$, then $Y(s) = F(X(s))$ is also the Fourier transform of a function $y \in L^1$.*

If instead of Theorem 1 of § 6 we use the stronger Theorem 1 of § 13 and if, identifying the maximal ideals of the ring V with the corresponding points, we consider $\hat{V}$ as a ring of functions on the line $-\infty < s < \infty$, complemented by the point $s = \infty$, then we obtain the following result.

COROLLARY 2′: *If $Y(s)$ ($|s| \leqq \infty$) is a function that can be expanded in the neighborhood of every point s_0 (including $s_0 = \infty$) in a power series of the form*

$$Y(s) = \sum_{n=0}^{\infty} a_n\,[X_{s_0}(s) - X_{s_0}(s_0)]^n,$$

where $X_{s_0}(s) \in \hat{V}$, then $Y(s) \in \hat{V}$.

***3.** In particular, if $X(s) \in \hat{V}$ does not vanish for any s ($|s| \leqq \infty$) and the total increment of $\arg X(s)$ in running through the line $-\infty < s < \infty$ is equal to zero, then the function $\log X(s)$ exists in $\hat{V}$, i.e., a function $Y(s)$ exists which satisfies the equation $\exp(Y(s)) = X(s)$. This fact is used in the theory of integral equations (see also the end of § 35). It cannot be obtained immediately from the preceding Corollary 2, because the spectrum of the element $X(s)$ may not fit into any sheet of the Riemann surface of the function $\log \zeta$.

4. Let us determine the structure of the space $\mathfrak{M}(V)$ of maximal ideals of V. To be specific, let us show that $\mathfrak{M}(V)$ *is a projective line*, i.e., is homeomorphic to a circle. The projective line is obtained by adjoining the point at infinity to the real line $-\infty < s < \infty$, where the neighborhoods of the ordinary points remain as before and a neighborhood of the point at infinity is defined as the set of points (including the point at infinity itself) that satisfy the inequality $|s| > A$ for all possible $A > 0$. We shall show that *if we identify the maximal ideals M_s with the corresponding points s of the real line $-\infty < s < \infty$ and the maximal ideal M_∞ with the point at infinity, then the topology of $\mathfrak{M}(V)$ as the space of maximal ideals coincides with the topology of the projective line.*

On account of the compactness of the projective line, it is sufficient for this purpose, by Theorem 1′ of § 5, to show that $\mathfrak{z}(M) = \lambda + x(M)$, considered as functions on the projective line, are continuous for all $\mathfrak{z} = \lambda e + x(t) \in V$. Since $x(M_s) = x^\sim(s)$ and $x(M_\infty) = 0$, we only have to show that the Fourier transform $x^\sim(s)$ of the absolutely integrable function $x(t)$ is continuous and tends to zero for $|s| \to \infty$. But the Fourier transform $x_{a,b}{}^\sim(s)$ of the function $x_{a,b}(t)$ has these properties, as is clear from the equations

$$x_{a,b}{}^\sim(s) = \int_{-\infty}^{\infty} x_{a,b}(t)\, e^{ist}\, dt = \int_a^b e^{ist}\, dt = \frac{e^{isb} - e^{isa}}{is},$$

and since the $x_{a,b}(t)$ are generators of V, we easily conclude, bearing in mind the obvious inequality $|x^\sim(s)| \leqq \| x \|$, that all the functions $x^\sim(s)$ have the properties stated.

On the basis of this we can identify the maximal ideals of V with the corresponding points of the projective line, extending the Fourier transform of the function $x(t) \in L^1$ to the point at infinity by defining $x^\sim(\infty)$ to be zero.

5. In the theory of the Fourier integral great importance attaches to the uniqueness theorem, according to which *an absolutely integrable function is uniquely determined by its Fourier transform.* By Theorem 1 this is nothing but the statement that in the ring V there are no generalized nilpotent elements $x(t) \in L^1$ other than zero. Since an element $\mathfrak{z} = \lambda e + x(t)$ for which $\lambda \neq 0$ is obviously not generalized nilpotent (because $\mathfrak{z}(M_\infty) = \lambda \neq 0$), we see that *the uniqueness theorem for the Fourier transform of an absolutely integrable function is equivalent to the theorem on the absence of a radical in the ring V.* In Chapter IV this theorem will be proved for a wide class of rings including V.

***6.** Let us apply the results just obtained to an analysis of the problem of the solvability of an integral equation with a kernel that depends on the difference of its arguments.

We consider the integral equation

$$\varphi(t) + \lambda \int_{-\infty}^{\infty} k(t-\tau)\, \varphi(\tau)\, d\tau = f(t), \tag{9}$$

where $f(t),\ k(t) \in L^1(-\infty, \infty)$. This equation can be written in the form

$$\varphi(t) + \lambda (k * \varphi)(t) = f(t).$$

Assuming that $\varphi(t)$ also belongs to L^1, we go over from the equation (9) to the corresponding equation on the set of maximal ideals; in other words, we take the Fourier transform (extended to the point at infinity) of all the terms of the equation (9). We arrive at the equation

$$\varphi^{\sim}(s) + \lambda k^{\sim}(s)\, \varphi^{\sim}(s) = f^{\sim}(s),$$

or

$$\varphi^{\sim}(s)\,[1 + \lambda k^{\sim}(s)] = f^{\sim}(s).$$

For this equation to be solvable for the Fourier transforms $f^{\sim}(s)$ of arbitrary functions $f(t) \in L^1$ it is necessary and sufficient that the function $1 + \lambda k^{\sim}(s)$ should not vanish for any real value s. Indeed, the necessity of this condition is obvious, since if $1 + \lambda k^{\sim}(s_0) = 0$, then equation (9) is unsolvable for those $f(t)$ for which $f^{\sim}(s_0) = 0$. But if the condition is satisfied, then the function $g(s) = 1/[1 + \lambda k^{\sim}(s)]$ is also an element of the ring V. Taking the unit element out as a separate term, we obtain

$$\frac{1}{1 + \lambda k^{\sim}(s)} = 1 + Q(s, \lambda),$$

where $Q(s, \lambda)$ is the Fourier transform of a function $q(t, \lambda) \in L^1$. As a result, we have

$$\varphi^{\sim}(s) = \frac{1}{1 + \lambda k^{\sim}(s)} f^{\sim}(s) = [1 + Q(s, \lambda)]\, f^{\sim}(s) =$$
$$= f^{\sim}(s) + Q(s, \lambda)\, f^{\sim}(s),$$

since (by the uniqueness theorem for the Fourier transforms of absolutely integrable functions)

$$\varphi(t) = f(t) + \int_{-\infty}^{\infty} q(t-\tau, \lambda)\, f(\tau)\, d\tau. \tag{10}$$

Thus, *if the function $k^{\sim}(s) = \int_{-\infty}^{\infty} k(t) e^{ist} dt$ does not assume the value $-1/\lambda$,* *then the equation* (9) *is solvable and the formula for its solution has the form* (10), *where* $q(t, \lambda)$ *is again a function that is absolutely integrable with respect to* t. Thus we have obtained a fundamental result of the theory of integral equations on the line with a kernel that depends on a difference.

7. We shall now discuss the subring V_+ of V obtained by adjoining a unit element to the ring $L^1{}_+$ of those functions of $L^1(-\infty, \infty)$ that are equal to zero for all $t < 0$. This ring plays a role in problems connected with the Laplace transform. Let us find the maximal ideals of V_+.

$L^1{}_+$ forms a maximal ideal in V_+; as before, we shall denote it by M_∞. Let M be a maximal ideal of V_+ other than M_∞. Since the operation $x(t) \to x(t-\lambda)$ is defined on the entire ring $L^1{}_+$ only for the values $\lambda \geqq 0$, the arguments given in the proof of Theorem 1 show that the function

$$\chi(\lambda) = \frac{d}{d\lambda} M(x_0, \lambda)$$

exists and is continuous for all $\lambda \geqq 0$ and satisfies (for these λ) the conditions

$$|\chi(\lambda)| \leqq 1$$

and

$$\chi(\lambda + \mu) = \chi(\lambda)\chi(\mu).$$

Hence it follows that

$$\chi(\lambda) = e^{is\lambda},$$

where s can now be an arbitrary complex number with a non-negative imaginary part; and, in the same way as at the end of the proof of Theorem 1, we obtain

$$(\lambda e + x)(M) = \lambda + \int_0^\infty x(t)\, e^{ist}\, dt$$

for all $\lambda e + x \in V_+$. It is easy to see that, conversely, the mapping

$$\lambda e + x \to \lambda + \int_0^\infty x(t)\, e^{ist}\, dt$$

for arbitrary s with $\Im s \geqq 0$ is also a homomorphism of V_+ into the field of complex numbers and therefore determines a maximal ideal M_s of V_+; moreover, if $s_1 \neq s_2$, then $M_{s_1} \neq M_{s_2}$, and all $M_s \neq M_\infty$. Thus, the set of maximal ideals of V_+ can be identified with the half-plane $\Im s \geqq 0$ supplemented by the 'improper' point (corresponding to the maximal ideal M_∞).

***8.** We leave it to the reader to formulate the analogues of the Corollaries 1, 2, and 2′ for the ring V_+ and to prove the homeomorphism of the space $\mathfrak{M}(V_+)$ with the half-plane $\mathfrak{I}s \geqq 0$, extended to a compactum by adjoining the point at infinity.

§ 18. The Ring of Absolutely Integrable Functions With a Weight

***1.** With hardly any change, the arguments given in the two preceding sections enable us to generalize considerably the results obtained in those sections.

Let $\alpha(t)$ be a positive function defined and continuous for all real values of t, and satisfying the condition

$$\alpha(t_1 + t_2) \leqq \alpha(t_1)\,\alpha(t_2), \tag{1}$$

whatever values t_1 and t_2 may have.

By repeating the arguments given in § 16 with the appropriate modifications we can satisfy ourselves of the fact that the set[4] $L[\alpha]$ of all measurable complex functions $x(t)$ $(-\infty < t < \infty)$ for which

$$\| x \| = \int_{-\infty}^{\infty} | x(t) | \, \alpha(t)\, dt < \infty,$$

forms a normed ring with the usual linear operations and with the convolution

$$(x * y)(t) = \int_{-\infty}^{\infty} x(t-\tau)\, y(\tau)\, d\tau$$

as multiplication, and that this ring has no unit element. The normed ring obtained from $L[\alpha]$ by adjoining a unit element will be denoted by $V[\alpha]$.

We denote the maximal ideal of $V[\alpha]$ formed by all the functions of $L[\alpha]$ by M_∞. We preface the search for the remaining maximal ideals by the following remarks concerning the 'weight function' $\alpha(t)$.

If $\alpha(t)$ satisfies the condition (1), then so does $\alpha(-t)$. By arguments similar to those in footnote 4, § 3.1, applied to the functions $\log \alpha(t)$ and $\log \alpha(-t)$, we can show that

$$\tau_1 = \sup_{t>0}\, [1/(-t)] \cdot \log \alpha(t) = \lim_{t \to +\infty}\, [1/(-t)] \cdot \log \alpha(t) \tag{2}$$

and

$$\tau_2 = \inf_{t>0}\, [1/t] \cdot \log \alpha(-t) = \lim_{t \to +\infty}\, [1/t] \cdot \log \alpha(-t). \tag{3}$$

---

[4] Another notation (used in the original Russian-language edition of the present book) is $L^{\langle\alpha\rangle}$

Moreover, since it follows from (1) that

$$[1/(-t)] \log \alpha(t) \leqq (1/t) \log \alpha(-t) - (1/t) \log \alpha(0)$$

for all $t > 0$, we have

$$-\infty < \tau_1 \leqq \tau_2 < \infty. \tag{4}$$

Now let M be a maximal ideal of the ring $V[\alpha]$ other than M_∞. It is easy to see that the operation $x(t) \rightarrow x_\lambda(t) = x(t-\lambda)$ is defined in $L[\alpha]$ for all real λ, that the lemma of § 17 remains valid, and that x_λ is (in norm) a continuous function of λ. Repeating the arguments given in the proof of Theorem 1 of § 17, we arrive at the conclusion that the function

$$\chi(\lambda) = \frac{d}{d\lambda} M(x_{c,\lambda}) = \lim_{h \to 0} M\left\{\frac{x_{c,\lambda+h} - x_{c,\lambda}}{h}\right\} \qquad (c < \lambda)$$

exists for all real λ, is continuous, and satisfies the conditions

$$\chi(\lambda + \mu) = \chi(\lambda)\chi(\mu) \tag{5}$$

and

$$|\chi(\lambda)| \leqq \alpha(\lambda), \tag{6}$$

of which the latter follows from the fact that

$$\lim_{h \to 0} \left\| \frac{x_{c,\lambda+h} - x_{c,\lambda}}{h} \right\| = \alpha(\lambda).$$

From (5) and the continuity of the function $\chi(\lambda)$ it follows that

$$\chi(t) = \exp(izt),$$

where $z = \sigma + i\tau$ is a fixed complex number. The inequality (6) shows that

$$\exp(-\tau t) \leqq \alpha(t) \tag{7}$$

for all real values of t or, what is the same,

$$[1/(-t)] \cdot \log \alpha(t) \leqq \tau \leqq [1/t] \cdot \log \alpha(-t)$$

for all $t > 0$. Bearing in mind the formulas (2) and (3), we conclude that

$$\tau_1 \leqq \tau \leqq \tau_2. \tag{8}$$

Thus, z lies in a finite strip (in virtue of (4)) of a width determined by the inequality (8) and, just as at the end of the proof of Theorem 1 of § 17, we obtain

$$(\lambda e + x)(M) = \lambda + \int_{-\infty}^{\infty} x(t) \exp(izt) dt \tag{9}$$

for all $\lambda e + x \in V[\alpha]$, where the integral in (9) converges absolutely in virtue of the inequality (7).

Conversely, by assigning to every element $\lambda e + x \in V[\alpha]$ the number that expresses the right-hand side of (9), where $z = \sigma + i\tau$ is an arbitrary fixed point of the strip $\tau_1 \leqq \tau \leqq \tau_2$, we obtain a homomorphic mapping of the ring $V[\alpha]$ into the field of complex numbers, which therefore determines a maximal ideal of this ring.

Thus, the maximal ideals $M \neq M_\infty$ fill out the strip $\tau_1 \leqq \Im z \leqq \tau_2$; the maximal ideal M_∞ closes this, making it a compact space. Obviously, propositions similar to the Corollaries 1, 2, and 2′ to Theorem 1 of § 17 hold.

***2.** Let us now consider the set $L_+[\alpha]$ of all measurable complex functions $x(t)$ $(0 \leqq t < \infty)$ satisfying the condition

$$\| x \| = \int_0^\infty | x(t) | \alpha(t) dt < \infty,$$

where $\alpha(t)$ is a positive continuous function defined only for $t \geqq 0$ and satisfying the condition (1) for these values of t. This set forms a normed ring with the usual linear operations and the multiplication defined by the formula

$$(x * y)(t) = \int_0^t x(t - \tau) y(\tau) d\tau.$$

The normed ring obtained by adjoining a unit element to $L_+[\alpha]$ will be denoted by $V_+[\alpha]$.

***3.** The same arguments by which we have found the maximal ideals of the ring $V[\alpha]$ now lead to the following result: *The set of maximal ideals of the ring $V_+[\alpha]$ is the half-plane $\Im z \geqq \tau_1 = \lim_{t \to +\infty} (1/-t) \cdot \log \alpha(t)$, which can be closed in a natural way to a compact space by the maximal ideal $M_\infty = L_+[\alpha]$.* Note that the possibility $\tau_1 = +\infty$ is not excluded (it is realized, for example, for $\alpha(t) = t^{-t}$); the half-plane $\Im z \geqq \tau_1$ then degenerates and M_∞ remains as the only maximal ideal of the ring $V_+[\alpha]$; in this case, precisely those elements $\lambda e + x \in V_+[\alpha]$ have inverses for which $\lambda \neq 0$.

We apply this result to derive the following theorem of Tauberian type:

Let $\alpha(t)$ be a positive continuous function defined for all $t \geqq 0$ and satisfying the condition (1) and let $F(t)$ be a function that is measurable and bounded on every finite interval of the positive axis. Let $x_0(t) \in L_+[\alpha]$ be such that

$$\int_0^\infty x_0(t) \exp(izt) dt \neq -1 \tag{10}$$

for all $z=\sigma+i\tau$ *with* $\tau \geqq \lim_{t\to+\infty} [1/(-t)] \log \alpha(t)$ *and*

$$\left(F(t)+\int_0^t x_0(t-s)F(s)ds\right)\alpha(t)\to 0 \qquad \textit{for} \qquad t\to\infty. \tag{11}$$

Then

$$F(t)\alpha(t)\to 0 \quad \textit{for} \quad t\to\infty$$

as well.

Proof: To every element $\mathfrak{z}=\lambda e+x(t)\in V_+[\alpha]$ there corresponds the operation $\mathfrak{z}*$ that carries an arbitrary function $\Phi(t)$, measurable and bounded on every finite interval $0\leqq t\leqq\infty$, into the function

$$\mathfrak{z}*\Phi(t)=\lambda\Phi(t)+\int_0^t x(t-s)\Phi(s)\,ds,$$

which is also measurable and bounded on every finite interval. Obviously,

$$\mathfrak{z}_1*(\mathfrak{z}_2*\Phi(t))=(\mathfrak{z}_1*\mathfrak{z}_2)*\Phi(t).$$

Within the ring $V_+[\alpha]$, the operation $\mathfrak{z}*$ is that of multiplication by $\mathfrak{z}$. It is easy to show that it follows from $\Phi(t)\alpha(t)\to 0$ (for $t\to\infty$) that

$$(\mathfrak{z}*\Phi(t))\alpha(t)\to 0$$

(for $t\to\infty$) for every $\mathfrak{z}\in V_+[\alpha]$. By (10), the element $e+x_0$ does not belong to any maximal ideal of $V_+[\alpha]$ and hence has an inverse in $V_+[\alpha]$. Furthermore, the condition (11) can be written in the form

$$[(e+x_0)*F(t)]\,\alpha(t)\to 0.$$

But then, by what has been proved above, we also have

$$(e+x_0)^{-1}*[(e+x_0)*F(t)]\,\alpha(t)=$$
$$=[(e+x_0)^{-1}*(e+x_0)]*F(t)\,\alpha(t)=F(t)\,\alpha(t)\to 0,$$

which is what we wished to prove. //

This theorem was first established in the special case $\alpha(t)\equiv 1$ by Paley and Wiener ([43], pp. 59-60).

§ 19. Discrete Analogues to the Rings of Absolutely Integrable Functions

1. In § 16 we discussed the rings of absolutely integrable functions V, V_+, and $I(0,T)$ in which multiplication was given as convolution by the formulas

$$\text{(a)}\ (x * y)(t) = \int_{-\infty}^{\infty} x(t-\tau)\, y(\tau)\, d\tau$$

in the ring V;

$$\text{(b)}\ (x * y)(t) = \int_{0}^{\infty} x(t-\tau)\, y(\tau)\, d\tau = \int_{0}^{t} x(t-\tau)\, y(\tau)\, d\tau$$

in the ring V_+;

$$\text{(c)}\ (x * y)(t) = \int_{0}^{T} x(t-\tau)\, y(\tau)\, d\tau = \int_{0}^{t} x(t-\tau)\, y(\tau)\, d\tau$$

$(0 \leqq t \leqq T)$ in the ring $I(0, T)$.

In this section we shall discuss discrete analogues of the rings V, V_+ and $I(0, T)$ (and also of the rings $V[\alpha]$ and $V_+[\alpha]$ of § 18).

2. Instead of functions $x(t)$ of a continuous argument t we shall consider functions a_n of an integral argument n, which ranges in the first case from $-\infty$ to $+\infty$, in the second case from 0 to $+\infty$, and in the third from 0 to m. As norms we naturally take the quantities

$$\| x \| = \sum_n | a_n |,$$

where the sum extends over the corresponding sets of values n. We shall give multiplications by formulas analogous to the formulas (a)-(c):

$$\text{(a')} \qquad a_n * b_n = \sum_{k=-\infty}^{\infty} a_{n-k} b_k ;$$

$$\text{(b')} \qquad a_n * b_n = \sum_{k=0}^{\infty} a_{n-k} b_k = \sum_{k=0}^{n} a_{n-k} b_k ;$$

$$\text{(c')} \qquad a_n * b_n = \sum_{k=0}^{m} a_{n-k} b_k = \sum_{k=0}^{n} a_{n-k} b_k \qquad (0 \leqq n \leqq m).$$

The ring of sequences $\{a_n\}$ $(-\infty < n < \infty)$ with the multiplication given by formula (a′) is obviously isomorphic to the ring W of absolutely convergent Fourier series; the isomorphism can be established by means of the correspondence

$$\{a_n\} \to \sum_{n=-\infty}^{\infty} a_n \exp(int).$$

The ring of finite sequences $\{a_n\}$ $(0 \leqq n \leqq m)$ with the multiplication given by formula (c′) is isomorphic to the ring $I^{(m)}$ of Example 4 of § 1.

The ring of sequences $\{a_n\}$ $(0 \leqq n < \infty)$ with the multiplication given by formula (b′) will be denoted by W_+. We have not come across this ring before.

Occasionally it is convenient to represent the rings W, W_+, $I^{(m)}$ as rings of formal power series: $\sum_{n=-\infty}^{\infty} a_n X^n$ (for W), $\sum_{n=0}^{\infty} a_n X^n$ (for W_+), $\sum_{n=0}^{m} a_n X^n$ (for $I^{(m)}$). With this notation the operation of multiplication turns into the ordinary operation of multiplication of power series (in the last case, subject to the condition that $X^p = 0$ for $p > m$).

3. Let us find the maximal ideals of W_+.

Suppose that an element X of W_+ goes over under the canonical homomorphism $W_+ \to W_+/M$ into the complex number ζ. By the condition g) of § 4 we have $|\zeta| \leqq \| X \| = 1$, and $x = \sum_{n=0}^{\infty} a_n X^n$ goes over into $\sum_{n=0}^{\infty} a_n \zeta^n$. Conversely, every mapping $\sum_{n=0}^{\infty} a_n X^n \to \sum_{n=0}^{\infty} a_n \zeta^n$ with $|\zeta| \leqq 1$ is obviously a homomorphism of W_+ into the field of complex numbers. Thus, the set of maximal ideals of W_+ is the disc $|\zeta| \leqq 1$ of the complex plane; here $x(M) = x(\zeta) = \sum_{n=0}^{\infty} a_n \zeta^n$. And since it follows from this that $x(M) = 0$ only for $x = 0$, we conclude that W_+ is isomorphic to the ring of analytic functions having a Taylor expansion that is absolutely convergent in the disc $|\zeta| \leqq 1$.

W_+ can also be regarded as a subring of W, formed by those elements of the latter for which all a_n with $n < 0$ are equal to zero. We recall that the set of maximal ideals of W is the circle $\zeta = \exp(it)$ $(0 \leqq t < 2\pi)$. Thus, on transition from W to W_+ the set of maximal ideals is extended by the adjunction of all the interior points of the disc $|\zeta| \leqq 1$.

We have observed a similar process on transition from the ring V to its subring V_+: to the line $-\infty < s < \infty$ (complemented by the point at infinity) we had to adjoin all the points of the upper half-plane.

4. Now let us proceed to the analogues of the rings $V[\alpha]$ and $V_+[\alpha]$ of § 18.

Let α_n be a positive function of the integral argument n $(-\infty < n < \infty)$ satisfying the condition

$$\alpha_{k+l} \leqq \alpha_k \alpha_l. \tag{1}$$

The same condition is then also satisfied by $\alpha' = \alpha_{-n}$. By applying the results obtained in footnote 4 of § 3.2 to α_n and α_{-n}, we obtain

$$\rho_1 = \sup_{n>0} \frac{1}{\sqrt[n]{\alpha_{-n}}} = \lim_{n\to+\infty} \frac{1}{\sqrt[n]{\alpha_{-n}}},$$

$$\rho_2 = \inf_{n>0} \sqrt[n]{\alpha_n} = \lim_{n\to+\infty} \sqrt[n]{\alpha_n};$$

moreover, since it follows from (1) that

$$\frac{1}{\sqrt[n]{\alpha_{-n}}} \leqq \frac{1}{\sqrt[n]{\alpha_0}} \sqrt[n]{\alpha_n}$$

for all $n > 0$, we have

$$(0 <) \ \varrho_1 \leqq \varrho_2 \ (< +\infty).$$

We denote by $W[\alpha]$ the set of all formal series $x = \sum_{n=-\infty}^{\infty} a_n X^n$ for which

$$\|x\| = \sum_{n=-\infty}^{\infty} |a_n| \alpha_n < \infty. \tag{2}$$

From (1) it follows that if $W[\alpha]$ contains any two series $x = \sum_{k=-\infty}^{\infty} a_k X^k$ and $y = \sum_{l=-\infty}^{\infty} b_l X^l$, it also contains their formal product

$$\sum_{m=-\infty}^{\infty} c_m X^m = \sum_{m=-\infty}^{\infty} \left(\sum_{l=-\infty}^{\infty} a_{m-l} b_l \right) X^m,$$

and that

$$\begin{aligned} \|xy\| &= \sum_{m=-\infty}^{\infty} \left| \sum_{l=-\infty}^{\infty} a_{m-l} b_l \right| \alpha_m \leqq \\ &\leqq \sum_{m=-\infty}^{\infty} \sum_{l=-\infty}^{\infty} |a_{m-l}| \alpha_{m-l} |b_l| \alpha_l = \\ &= \sum_{l=-\infty}^{\infty} \left(\sum_{m=-\infty}^{\infty} |a_{m-l}| \alpha_{m-l} \right) |b_l| \alpha_l = \|x\| \, \|y\|. \end{aligned}$$

Thus, $W[\alpha]$ is a normed ring under the usual operations on power series.

Let M be a maximal ideal of $W[\alpha]$ and $X(M) = \zeta = \varrho \exp(i\varphi)$. Then $X^n(M) = \varrho^n \exp(in\varphi)$ $(n = 0, \pm 1, \pm 2, \ldots)$; and since $\| X^n \| = \alpha_n$, we have $\varrho^n \leqq \alpha_n$ $(n = 0, \pm 1, \pm 2, \ldots)$; hence

$$\varrho_1 \leqq \varrho \leqq \varrho_2. \tag{3}$$

Thus, for every $x = \sum_{n=-\infty}^{\infty} a_n X^n \in W[\alpha]$, we have

$$x(M) = \sum_{n=-\infty}^{\infty} a_n \zeta^n, \tag{4}$$

where $\zeta = \varrho \exp(i\varphi)$ is a point of the annulus defined by the inequalities (3).

Conversely, by assigning to every element $x = \sum_{n=-\infty}^{\infty} a_n X^n \in W[\alpha]$ the sum of the series (4), where $\zeta = \varrho \exp(i\varphi)$ is an arbitrary fixed point of the annulus (3), we obtain a homomorphic mapping of $W[\alpha]$ into the field of complex numbers, thereby determining a maximal ideal of this ring.

Thus, $W[\alpha]$ has the domain (3) as its set of maximal ideals and is isomorphic to the ring of functions analytic in this domain for which the Laurent series (4) satisfies the condition (2).

For $\varrho_1 = \varrho_2$ the annulus (3) degenerates into a circle, and the necessary and sufficient condition for this is, obviously, the equality

$$\inf_{n>0} \sqrt[n]{\alpha_n} \cdot \inf_{n>0} \sqrt[n]{\alpha_{-n}} = 1.$$

If both factors on the left-hand side are equal to 1, then $\varrho_1 = \varrho_2 = 1$ and $W[\alpha]$ assumes the form of the ring of absolutely convergent Fourier series $\sum_{n=-\infty}^{\infty} a_n \exp(in\varphi)$ with the weights α_n.

Now let us consider the set $W_+[\alpha]$ of all formal power series $x = \sum_{n=0}^{\infty} a_n X^n$ satisfying the condition

$$\| x \| = \sum_{n=0}^{\infty} | a_n | \alpha_n < \infty \tag{5}$$

where α_n is a positive function of n that is defined only for the integers $n \geqq 0$ and satisfies the condition (1) only for such values of k and l. This set forms a normed ring under the usual operations on power series and, exactly as before, we can state that this ring $W_+[\alpha]$ is isomorphic to the ring of all functions of a complex variable, analytic in the interior of the disc $|\zeta| \leqq \varrho_2 = \lim_{n\to\infty} \sqrt[n]{\alpha_n}$, whose Taylor series $\sum_{n=0}^{\infty} a_n \zeta^n$ satisfies the condition (5). However, in contrast to the case of the ring $W[\alpha]$, the possibility $\varrho_2 = 0$ is not excluded here; in this (and only this) case $W[\alpha]$ is a ring with a single maximal ideal.

We leave it to the reader to formulate for the rings W_+, $W[\alpha]$, and $W_+[\alpha]$ the analogues of the Corollaries 1, 2, and 2′ to Theorem 1 of § 17.

***5.** B. S. Mitjagin [40] has given a complete description of the spaces of maximal ideals of the n-dimensional analogues of the rings $W_+[\alpha]$ (as closed n-fold multicircular domains of a complex n-dimensional space having closed domains of regularity).

CHAPTER IV

HARMONIC ANALYSIS ON COMMUTATIVE LOCALLY COMPACT GROUPS

1. In § 16 we discussed the ring V obtained by formal adjunction of a unit element to the ring L^1 of all absolutely integrable measurable complex functions on the real line, with the norm $\| x \| = \int_{-\infty}^{\infty} | x(t) | dt$ and the multiplication ('convolution')

$$(x * y)(t) = \int_{-\infty}^{\infty} x(t - \tau)y(\tau)d\tau.$$

We found that the maximal ideals of V, other than L^1, can be put into one-to-one correspondence with the points of the line $-\infty < s < \infty$ so that for the maximal ideal M_s corresponding to the point s we have

$$x(M_s) = \int_{-\infty}^{\infty} x(t) \exp(ist)dt = x^{\sim}(s). \tag{1}$$

At the same time, we observed that the Fourier transform $x^{\sim}(s)$ of an absolutely integrable function $x(t)$ can be treated as the canonical representation of $x(t)$ in the form of a function $x(M)$ on the maximal ideals of V.

In the definition of V we have made use only of the fact that on the real line there exists a measure that is invariant under translation (Lebesgue measure). As Haar [24] (see also [4]) has shown, a measure having these properties exists on every locally compact group satisfying the second axiom of countability; and this result was subsequently extended by Weil [73] to all locally compact groups. For every commutative group G of this kind a natural analogue of V is the ring $V(G)$ of all complex functions $x(g)$ $(g \in G)$, measurable and absolutely integrable with respect to the Haar measure, with the convolution[1]

$$(x * y)(g) = \int x(g - h)y(h)dh$$

[1] We use the additive notation for the group operation in G. Whenever the domain of integration of an integral is not indicated, it is assumed that this domain is the entire group G.

as multiplication and with a unit element adjoined if necessary. The problem now arises: Can the canonical representation of $x(g)$ as a function on the maximal ideals of $V(G)$ be written in the form of an integral of type (1) and what should be the analogue here to the functions $\exp(ist)$?

The proof of the fact that the convolution of functions $x(t)$ corresponds to the multiplication of their Fourier transforms is based only on the following property of the functions $\exp(ist)$: $\exp(is(u+v)) = \exp(isu) \cdot \exp(isv)$. Moreover, the existence of the integral (1) for all $x \in L^1$ is based on the boundedness of the functions $\exp(ist)$—namely, $|\exp(ist)| = 1$. The properties referred to, in conjunction with the continuity of the functions $\exp(ist)$, express the fact that these functions are characters of the additive topological group of real numbers.

In general, following Pontrjagin,[2] we define a *character* of a commutative topological group G as a homomorphic mapping of the group into the group K of all rotations of a circle, topologized in the natural fashion. The group K can be represented analytically as the additive group of real numbers reduced modulo 2π and as the multiplicative group of complex numbers of absolute value 1. In the additive treatment of the group K, a character χ of G is given in the form of a continuous function $\chi(g)$ of the argument $g \in G$, assuming real values reduced modulo 2π and satisfying the condition

$$\chi(g+h) = \chi(g) + \chi(h) \qquad (\text{mod } 2\pi)$$

('additive character'). In the multiplicative treatment of K a character χ is given in the form of a continuous complex function $\exp(i\chi(g))$, obviously satisfying the conditions

$$\exp(i\chi(g+h)) = \exp(i\chi(g)) \cdot \exp(i\chi(h)), \qquad |\exp(i\chi(g))| = 1.$$

Thus, we may expect that between the maximal ideals of $V(G)$ (other than the set $L^1(G)$ of all absolutely integrable measurable functions, in case it does not contain a unit element with respect to convolution) and the characters of G there exists a one-to-one correspondence such that, for the maximal ideal M_χ corresponding to the character χ, we have

$$x(M_\chi) = \int x(g)\, e^{i\chi(g)}\, dg = x^\sim(\chi).$$

This would enable us to treat the Fourier transform $x^\sim(\chi)$ of an absolutely integrable function $x(g)$ as its canonical representation in the form of a

[2] L. S. Pontrjagin, *Topologische Gruppen* [translated from the 2nd Russian edition], Teil 2, Leipzig, 1958, p. 7, Definition 36. (See also Pontrjagin, *Topological Groups*, Princeton, 1939, p. 127, Definition 34, which is slightly less general.)

function on the maximal ideals of $V(G)$. We shall not assume here a previous knowledge of the existence of non-trivial characters on an arbitrary commutative locally compact group G. We shall construct these characters by means of the maximal ideals of $V(G)$ and shall show that there are sufficiently many of them.

On this basis we shall also obtain a fundamental proposition of harmonic analysis—the theorem of Plancherel, which then enables us to give a purely analytical proof of the Pontrjagin duality law for commutative locally compact groups. In conclusion, we shall discuss positive-definite functions, which also play an important role in harmonic analysis.

§ 20. The Group Ring of a Commutative Locally Compact Group

1. In all that follows G will denote an arbitrary locally compact commutative group; (B), the field of all Borel sets on G, i.e., the smallest collection of sets in G containing all closed sets and invariant with respect to the operations of taking complements and countable unions (and consequently, countable intersections).

On the field (B) there exists a *Haar measure,* i.e., a completely additive function of sets $m(E) \geqq 0$ that is finite on all compact sets, different from zero for all open sets, and *invariant under translation*:

$$m(E+g)=m(E) \quad \text{for all} \quad E \in (B) \text{ and } g \in G.$$

This measure can be extended in the usual way (with preservation of invariance) to all sets that are *measurable* with respect to it, the measure of every such set (and in particular, of every set of (B)) coinciding with the lower bound of the measures of the open sets containing it and the upper bound of the measures of the compact sets contained in it (*regularity* of the Haar measure). Finally, the Haar measure on G is also *invariant with respect to reflection*: for all measurable sets $E \subset G$, we have

$$m(-E)=m(E).$$

Thus, the Haar measure on G has all the fundamental properties of the Lebesgue measure on the line (and coincides with it when G is taken to be the additive topological group of real numbers).

2. Henceforth, we shall always understand measurability or integrability of a function on G as meaning measurability or integrability with respect to the Haar measure. $L^p(G)$ $(p \geqq 1)$ denotes the space of all measurable complex functions $x(g)$ on G that satisfy the condition

$$\| x \|_p = \left(\int | x(g) |^p dg \right)^{\frac{1}{p}} < \infty, \tag{1}$$

where the integration is taken with respect to Haar measure and the domain of integration is the whole group G. $L^p(G)$, furnished with the norm (1), is a Banach space. For $p = 1$, we shall write simply $\| x \|$ instead of $\| x \|_1$. $L^{1,2}(G)$ shall mean $L^1(G) \cap L^2(G)$.

From the invariance of the measure under translations there follows also the *invariance of the integral*:

$$\int x(g+h)dg = \int x(g)dg \qquad \text{for all} \qquad x \in L^1(G) \quad \text{and} \quad h \in G.$$

In precisely the same way, the invariance of the measure under reflection implies the *invariance of the integral under reflection*:

$$\int x(-g)dg = \int x(g)dg \qquad \text{for all} \qquad x \in L^1(G).$$

The *support* of a function on a topological space is defined as the closure of the set of points at which it is different from zero; *finite* functions are functions with a compact support. The continuous finite functions on G form a vector space; we shall denote this space by $L(G)$. It is contained as an everywhere-dense subset in every space $L^p(G)$ $(p \geqq 1)$.

Every element $g_0 \in G$ generates a *translation operator* T_{g_0}:

$$T_{g_0} x(g) = x(g - g_0).$$

From the invariance of the integral it follows that if $x \in L^p(G)$, then $T_h x \in L^p(G)$ for every $h \in G$ and $\| T_h x \| = \| x \|$. Moreover, for every $x \in L^p(G)$,

$$\| T_h x - x \|_p \to 0 \qquad \text{for} \qquad h \to 0; \tag{2}$$

in other words, *for every* $\varepsilon > 0$ *there exists a neighborhood* U *of the zero element of* G *such that*

$$\int | x(g-h) - x(g) |^p dg < \varepsilon \qquad \textit{for all} \qquad h \in U \tag{2'}$$

(*absolute continuity* of the integral with respect to Haar measure).[3]

---

[3] See the proof of the lemma in § 16. A proof of the existence of the integral and the Haar measure and a substantiation of all their properties enumerated here can be found in [73], [35], or [42], and an axiomatic treatment of Haar measure in [49].

3. THEOREM 1: *If x and $y \in L^1(G)$, then the integral*

$$(x * y)(g) = \int x(g-h)y(h)dh \tag{3}$$

exists for almost all g and also belongs to $L^1(G)$; moreover,

$$\| x * y \| \leqq \| x \| \| y \|,$$

and the operation of convolution $*$ defined by the formula (3) *is bilinear, associative, and commutative.*

Proof: With the help of the Haar measure m given on G we can construct a completely additive measure μ (also a Haar measure) on $G \times G$, which is uniquely determined by the following properties: for every 'rectangle' $X \times Y$, where $X, Y \in (B)$, the equation $\mu(X \times Y) = m(X)m(Y)$ holds. It can be shown that $x(g-h)$, and hence $x(g-h)y(h)$ as well, is measurable as a function of the point $(g, h) \in G \times G$ with respect to the measure μ. Hence Fubini's Theorem is applicable to $x(g-h)y(h)$ and we can therefore repeat word for word the proof of the theorem in question given in § 16.2 for the case where G is the additive topological group of real numbers. //[4]

Theorem 1 shows that $L^1(G)$ *is a commutative normed ring with convolution as multiplication.*

4. If the group G is discrete, then obviously all the one-point sets have the same finite measure $\varrho > 0$. Without loss of generality, we may assume that $\varrho = 1$ (for this can always be achieved by a suitable 'normalization' of the Haar measure on G, namely, by dividing the measure of every set by ϱ). Then the integral of the function $x(g) \in L^1(G)$ simply turns out to be the sum of all the values of $x(g)$, so that $L^1(G)$ is nothing other than the space $l^1(G)$ of all 'absolutely summable' functions $x(g)$, i.e., the complex functions on G that differ from zero at not more than a countable set of points and such that

$$\| x \| = \sum_{g \in G} | x(g) | < \infty.$$

The convolution of the functions $x, y \in l^1(G)$ proceeds according to the formula

$$(x * y)(g) = \sum_{h \in G} x(g-h)y(h).$$

[4] With minor modifications, the proof given in § 16 that does not depend on Fubini's Theorem turns out to be suitable in the general case under discussion as well.

$l^1(G)$ contains the 'δ-functions,' i.e., non-negative real functions of norm 1 the whole integral of which is concentrated at a single point; these are simply the functions

$$e_{g_0}(g) = \begin{cases} 1 & \text{for} \quad g = g_0, \\ 0 & \text{for} \quad g \neq g_0. \end{cases}$$

Multiplication by $e_{g_0}(g)$ in the ring $l^1(G)$ is equivalent to translation by g_0:

$$T_{g_0}x = x * e_{g_0} \qquad \text{for all} \qquad x \in l^1(G). \tag{4}$$

In particular, $e_0(g)$ is the unit element of $l^1(G)$. Thus, *if G is a discrete group, then $L^1(G)$ is a ring with a unit element.*

5. In the case where G is a non-discrete group, all its one-point sets have measure 0, because otherwise its infinite compact sets (which necessarily exist in a non-discrete locally compact group) could not have finite measure. Therefore $L^1(G)$ cannot contain δ-functions. For the integral of a δ-function taken over an arbitrary neighborhood of the point g_0 at which this δ-function is concentrated would have to be equal to 1; but since the set formed by these points has measure 0, the Haar measure being regular, there exists a sequence of neighborhoods of g_0 whose measures tend to zero, and then the integrals over these neighborhoods of an arbitrary function of $L^1(G)$ would also have to tend to zero.

In the (general) case where the group G is non-discrete, the role of the δ-functions is taken by the 'δ-sequences' that contract to the elements of the group. Let us interpret $e_A(g)$, where A is a measurable set of positive measure, as an arbitrary function of $L^1(G)$ satisfying the conditions:

1. $e_A(g) \geqq 0$ for all $g \in G$;
2. $e_A(g) = 0$ for all $g \notin A$;
3. $\| e_A \| = 1$.

An example of such a function is $f_A(g)/m(A)$, where $f_A(g)$ is the characteristic function of the set A of finite positive measure $m(A)$.

6. DEFINITION 1: Let g_0 be a fixed element of G and $\{U\}$, a fundamental system of neighborhoods of zero and let every $U \in \{U\}$ be associated with the function $e_{g_0+U}(g)$. We shall then say that these functions form a *δ-sequence contracting to the element* g_0.

As an explanation of this term we remark that if G has a countable fundamental system of neighborhoods of zero $\{U_n\}$, then the functions $e_{g_0+U_n}(g)$ actually form a sequence in the usual sense of the word. In the general case the set $\{U\}$, partially ordered by the inclusion relation $\supset$, is a *directed* set (i.e., for arbitrary U_1, $U_2 \in \{U\}$ there exists a $U_3 \in \{U\}$ such that $U_1 \supset U_3$ and $U_2 \supset U_3$) and the functions $e_{g_0+U}(g)$ form a *generalized sequence.*

As is well known, the concept of limit can be carried over immediately to generalized sequences (limit in the sense of Shatunovskii [54] and Moore-Smith [41]). In particular, a number a is called the limit of a generalized sequence of numbers f_U whose argument U ranges over a fundamental system of neighborhoods of zero $\{U\}$ of G if, for every $\varepsilon > 0$, there exists a $U_\varepsilon \in \{U\}$ such that $|f_U - a| < \varepsilon$ for all $U \in \{U\}$ contained in U_ε. We shall use the notation $\lim_{U \to 0} f_U = a$ to designate this.

7. A fundamental role in the sequel will be played by the following lemma, which establishes a relation for non-discrete groups G that takes the place of formula (4).

LEMMA: *Let $\{e_{g_0+U}\}$ be an arbitrary δ-sequence contracting to the element $g_0 \in G$. Then for every $x \in L^1(G)$*

$$\lim_{U \to 0} \| T_{g_0} x - x * e_{g_0+U} \| = 0, \tag{4'}$$

where the limit relation (4′) is satisfied for every given $x \in L^1(G)$ uniformly for all δ-sequences $\{e_{g_0+U}\}$ and all $g_0 \in G$.

Proof: By (2′), for every $\varepsilon > 0$ there exists a neighborhood of zero U_ε such that $\int |x(g) - x(g-h)| \, dg < \varepsilon$ for all $h \in U_\varepsilon$. Upon making the substitutions $g \to g + g_0$ and $h \to h + g_0$, applying Fubini's Theorem, and observing that $e_{g_0+U}(h + g_0)$ is $e_U(h)$, we obtain

$$\begin{aligned} &\| T_{g_0} x - x * e_{g_0+U} \| = \\ &= \int \left| \int [x(g-g_0) - x(g-h)] \, e_{g_0+U}(h) \, dh \right| dg = \\ &= \int \left| \int [x(g) - x(g-h)] \, e_{g_0+U}(h+g_0) \, dh \right| dg \leqq \\ &\leqq \int \left(\int | x(g) - x(g-h) | \, dg \right) e_U(h) \, dh \leqq \\ &\qquad \leqq \sup_{h \in U} \int | x(g) - x(g-h) | \, dg \leqq \varepsilon \end{aligned}$$

for all $U \subset U_\varepsilon$, no matter what the element $g_0 \in G$ and the δ-sequence $\{e_{g_0+U}\}$. //

As a first application of this lemma, we prove the following theorem:

THEOREM 2: *If G is non-discrete, then the ring $L^1(G)$ does not have a unit element.*

Proof: Let $\{e_U\}$ be a δ-sequence contracting to the zero element of G and suppose, in contradiction to the statement of the theorem, that there exists in $L^1(G)$ a unit element $e(g)$ with respect to convolution. By the lemma, $\lim_{U \to 0} \| e - e * e_U \| = 0$. But since $e(g)$ is the unit element, we have $(e * e_U)(g) = e_U(g)$ for almost all g. Thus,

$$\lim_{U \to 0} \int | e(g) - e_U(g) | \, dg = 0.$$

Since all the $e_U(g)$, where $U \subset V$, are zero outside V, this means that $\int_{CV} | e(g) | \, dg = 0$ and therefore $\int_V | e(g) | \, dg = 1$, no matter what neighborhood of zero V is. But, by what has been shown above, in the case of a non-discrete group G this is impossible. //

8. DEFINITION 2: We define the *group ring* $V(G)$ of the locally compact group G as follows: in the case in which G is discrete, it is the ring $L^1(G)$; in the case in which G is non-discrete, it is $L^1(G)$ with a formally adjoined unit element. Thus, in the latter case, $V(G)$ consists of the elements $\mathfrak{z} = \lambda e + x$, where $x \in L^1(G)$, λ assumes arbitrary complex values, e is the unit element adjoined, and $\| \mathfrak{z} \| = | \lambda | + \| x \|$.

By way of explanation of the term 'group ring' we make the following observations.

The group ring of a finite group G is defined as the ring of formal polynomials $\sum_{k=1}^{n} c_k g_k$, where g_k ranges over all elements of the group and c_k are arbitrary complex coefficients. This definition immediately generalizes to the case of an arbitrary discrete group G. The group ring R_G is constructed in this case as follows. Its elements are all the formal sums $x = \sum x(g) g$, where $x(g)$ is an absolutely summable function on G and where the algebraic operations are carried out by the ordinary formal rules for the operations on polynomials followed by the 'collection of like terms'; thus, in particular (in the case of an additive discrete group G),

$$xy = \sum_h x(h) h \sum_k y(k) k = \sum_h \sum_k x(h) y(k)(h + k) = \sum_g z(g) g,$$

where

$$z(g) = \sum_{h+k=g} x(h) y(k) = \sum_h x(h) y(g - h) = \\ = \sum_h x(g - h) y(h). \tag{5}$$

It is easy to verify that R_G, furnished with the norm

$$\| x \| = \sum | x(g) |, \tag{6}$$

satisfies all the axioms of a normed ring. It contains the group G algebraically in the form of the group of monomials $1g = \sum e_g(h)h$, where

$$e_g(h) = \begin{cases} 1 \text{ for } h = g, \\ 0 \text{ for } h \neq g. \end{cases} \tag{7}$$

However, for a non-discrete group G this definition of the group ring is unsuitable, since in its very essence it treats G as a discrete group and the topology existing in G is in no way taken into account. But to specify an element $x = \sum x(g)g$ of R_G is equivalent to specifying the function $x(g)$. Therefore R_G is isomorphic to the ring $l^1(G)$ of absolutely summable functions $x(g)$ in which the norm is defined by formula (6) and in which addition and multiplication by complex numbers are as usual and multiplication of elements is defined in accordance with the formula (5) as the 'convolution'

$$(x * y)(g) = \sum_h x(g-h)y(h). \tag{8}$$

As we have seen, the definition of the group ring in that case generalizes to an arbitrary locally compact group, provided $x(g)$ is taken to be a measurable complex function, absolutely integrable with respect to Haar measure, and the sums (6) and (8) are replaced by the corresponding integrals. Thus, the ring $V(G)$ is a natural generalization of the group ring treated in the theory of finite groups.

§ 21. Maximal Ideals of the Group Ring and the Characters of a Group

1. A connection between the maximal ideals of $V(G)$ and the characters of G appears immediately when G is a *discrete* commutative group.

In this case let M be a maximal ideal of $V(G)$ or, what is the same, the ring $l^1(G)$ of absolutely summable functions on G (see the end of § 20). Then the function of g

$$M(g) = e_g(M), \tag{1}$$

where

$$e_g(h) = \begin{cases} 1 \text{ for } h = g, \\ 0 \text{ for } h \neq g, \end{cases}$$

is a multiplicative character of G.

In fact, since $e_{g+h} = e_g * e_h$, we have

$$M(g+h) = e_{g+h}(M) = e_g(M)\, e_h(M) = M(g)M(h). \tag{2}$$

Further,[5]

[5] Because e_0 is the unit element of the ring $V(G)$.

$$|M(g)| \leqq \|e_g\| = 1 \quad \text{and} \quad M(0) = e_0(M) = 1;$$

hence by (2) (for $h = -g$), it follows that $|M(g)| = 1$.

Since every element x of $l^1(G)$ can be represented in the form $x = \sum x(g) e_g$, where the series converges in norm, it follows from (1) that

$$x(M) = \sum x(g) M(g) = \sum x(g) e^{i\chi_M(g)}, \tag{3}$$

where $\chi_M(g)$ is the additive character of G corresponding to the multiplicative character $M(g)$ generated by the maximal ideal M (i.e., $M(g) = e^{i\chi_M(g)}$).

Conversely, let χ be an (additive) character of G. By assigning to every element $x \in l^1(G)$ the number $\sum x(g) e^{i\chi(g)}$ we obtain a homomorphic mapping of the ring $l^1(G)$ into the field of complex numbers. Indeed, to verify this we only have to go over from the product of the ring elements to the product of the numbers corresponding to them. But

$$\sum_g \left(\sum_h x(g-h)\, y(h)\right) e^{i\chi(g)} = \sum_h \left(\sum_g x(g-h)\, e^{i\chi(g)}\right) y(h) =$$
$$= \sum_h \left(\sum_g x(g-h)\, e^{i\chi(g-h)}\right) e^{i\chi(h)}\, y(h) =$$
$$= \sum_g x(g)\, e^{i\chi(g)} \sum_h y(h)\, e^{i\chi(h)}.$$

Since e_0 goes over into 1, this mapping is not the null mapping. We denote its kernel by M_χ. M_χ is a maximal ideal of $l^1(G)$. Here

$$x(M_\chi) = \sum x(g) e^{i\chi(g)}$$

and, in particular,

$$e_g(M_\chi) = e^{i\chi(g)}. \tag{4}$$

A comparison of the formulas (1) and (4) shows that the maximal ideal M_χ generated by the character χ in turn generates this character:

$$\chi = \chi_{M_\chi}.$$

From (3) it follows that distinct maximal ideals generate distinct characters. For if $\chi_{M_1} = \chi_{M_2}$, then, by (3), $x(M_1) = x(M_2)$ for all $x \in l^1(G)$, i.e., M_1 and M_2 coincide.

From (4) it follows that distinct characters generate distinct maximal ideals. For if $M_{\chi_1} = M_{\chi_2}$ then, by (4), $e^{i\chi_1(g)} = e^{i\chi_2(g)}$ for all $g \in G$, i.e., $\chi_1 = \chi_2$.

We have thus set up a one-to-one correspondence between the characters of the discrete commutative group G and the maximal ideals of its group ring

$l^1(G)$, this one-to-one correspondence being such that by identification of the maximal ideals with the corresponding characters of the functions on the set of maximal ideals of $l^1(G)$ we obtain as the canonical representative of the element $x(g) \in l^1(G)$ its Fourier transform

$$x^{\sim}(\chi) = \sum_{g \in G} x(g) e^{i\chi(g)}.$$

2. This is the situation in the case that is simplest from the topological point of view, where G is a discrete group. From now on, let us take G to be a non-discrete group. In that case, $L^1(G)$ forms a maximal ideal in the group ring $V(G)$; we denote it by M_∞. Thus,

$$x(M_\infty) = 0 \qquad \text{for all} \qquad x \in L^1(G).$$

We shall show that, in basic outline, there exists a correspondence between the characters of G and the maximal ideals of the group ring $V(G)$, other than M_∞, as in the discrete case.

The character of the discrete commutative group G corresponding to the maximal ideal M of its group ring $l^1(G)$ was determined by the values assumed on M by the δ-functions $e_g(h)$. In the group ring $V(G)$ of the non-discrete group G the role of these missing δ-functions will be played by the δ-sequences contracting to the element g of G. Although these δ-sequences do not tend to a limit, the values they assume on every maximal ideal M of $V(G)$, as we shall show, have a limit, which, for every fixed M and variable g, is a character of G, and all the characters of G are obtained in this way.

3. THEOREM 1: *Let M be a maximal ideal of the ring $V(G)$, other than M_∞ and let $\{e_{g+U}\}$ be a δ-sequence contracting to the element $g \in G$. Then*

$$\lim_{U \to 0} e_{g+U}(M) = M(g) \tag{5}$$

exists and does not depend on the choice of the δ-sequence, and the limit relation (5) is satisfied uniformly for all $g \in G$ and all δ-sequences $\{e_{g+U}\}$. $M(g)$ is a multiplicative character of G.

Proof: Since $M \neq M_\infty$, there exists a function $z \in L^1(G)$ such that $z(M) \neq 0$. We put $T_g z = z_g$. By the lemma of the preceding section,

$$|z_g(M) - z(M) e_{g+U}(M)| \leqq \leqq \|z_g - z * e_{g+U}\| \to 0 \quad \text{for } U \to 0, \tag{6}$$

i.e.,

$$e_{g+U}(M) \to \frac{z_g(M)}{z(M)}$$

uniformly for all $g \in G$ and all δ-sequences $\{e_{g+U}\}$. Fixing z, we see that for all δ-sequences $\{e_{g+U}\}$ contracting to g there exists one and the same limit

$$M(g) = \frac{z_g(M)}{z(M)}. \tag{7}$$

Fixing an arbitrary δ-sequence $\{e_{g+U}\}$, we see that $M(g)$ does not depend on the choice of z.

Let us show that $M(g)$ is a multiplicative character of G.

First of all, $M(g)$ is a continuous function of g. For by (7) and the relation (2) of § 20,

$$|M(g') - M(g)| = \left| \frac{(z_{g'} - z_g)(M)}{z(M)} \right| \leqq$$
$$\leqq \frac{\|z_{g'} - z_g\|}{|z(M)|} = \frac{\|z_{g'-g} - z\|}{|z(M)|} \to 0 \quad \text{for } g' \to g.$$

Further, since $|e_{g+U}(M)| \leqq \| e_{g+U} \| = 1$, we also have

$$|M(g)| \leqq 1. \tag{8}$$

Furthermore, as follows from (7),

$$M(0) = 1. \tag{9}$$

It remains to show that

$$M(g+h) = M(g)\, M(h), \tag{10}$$

because, by putting $h = -g$ in (10) and taking (8) and (9) into account, we then obtain $|M(g)| = 1$. But if $\{e_{g+U}\}$ is a δ-sequence contracting to g and $\{e_{h+V}\}$ is a δ-sequence contracting to h, then

$$\{e_{g+U} * e_{h+V}\}$$

is a δ-sequence contracting to $g+h$. Therefore

$$M(g+h) = \lim_{U,V \to 0} (e_{g+U} * e_{h+V})(M) =$$
$$= \lim_{U \to 0} e_{g+U}(M) \lim_{V \to 0} e_{h+V}(M) = M(g)\, M(h).$$

This proves the theorem. //

4. Let $\chi_M(g)$ be the additive character of G corresponding to the multiplicative character $M(g)$, so that $M(g) = e^{i\chi_M(g)}$. We shall say that *the maximal ideal M generates the character χ_M*.

Note: From equation (5) and the relation (6), which is true for all $z \in L^1(G)$, it follows that $z_g(M) = e^{i\chi_M(g)} z(M)$ for an arbitrary function $z \in L^1(G)$.

THEOREM 2: *If $M \neq M_\infty$, then for all $x \in L^1(G)$ we have the equation*

$$x(M) = \int x(g)\, e^{i\chi_M(g)}\, dg, \tag{11}$$

and hence

$$\mathfrak{z}(M) = (\lambda e + x)(M) = \lambda + \int x(g)\, e^{i\chi_M(g)}\, dg. \tag{12}$$

Proof: Obviously, it is sufficient to prove (11) for a non-zero non-negative real function $x \in L(G)$, as the linear combinations of such functions fill out $L(G)$ and hence are dense in $L^1(G)$. By Theorem 1, for every $\varepsilon > 0$ there exists a neighborhood of zero U such that

$$\left| e_{g+U}(M) - e^{i\chi_M(g)} \right| < \varepsilon \tag{13}$$

for all $g \in G$ and all functions e_{g+U}. Since the support D of the function $x(g)$ is a compact set with interior points, it can be decomposed into a finite number of pairwise disjoint measurable subsets D_n with interior points 'small compared with U,' i.e., such that $g_n - g_n' \in U$ for all g_n, $g_n' \in D_n$. Let $x_n(g) = x(g) f_n(g)$, where $f_n(g)$ is the characteristic function of the set D_n, so that $x(g) = \sum x_n(g)$. We put $y_n(g) = x_n(g) / \int x_n(g)\, dg$. Since $y_n(g)$ is $e_{h+U}(g)$ for every $h \in D_n$, we obtain, taking (13) into account,

$$\begin{aligned}
&\left| x_n(M) - \int x_n(g)\, e^{i\chi_M(g)}\, dg \right| = \\
&\quad = \left| y_n(M) - \int y_n(g)\, e^{i\chi_M(g)}\, dg \right| \int x_n(g)\, dg = \\
&\quad = \left| \int \left[y_n(M) - e^{i\chi_M(g)} \right] y_n(g)\, dg \right| \int x_n(g)\, dg \leqq \\
&\quad \leqq \sup_{g \in D_n} \left| y_n(M) - e^{i\chi_M(g)} \right| \cdot \int x_n(g)\, dg \leqq \varepsilon \int x_n(g)\, dg,
\end{aligned}$$

and consequently

$$\begin{aligned}
\left| x(M) - \int x(g)\, e^{i\chi_M(g)} dg \right| &= \left| \sum \left(x_n(M) - \int x_n(g)\, e^{i\chi_M(g)} dg \right) \right| \leqq \\
&\leqq \varepsilon \sum \int x_n(g)\, dg = \varepsilon \int x(g)\, dg.
\end{aligned}$$

Since ε is arbitrary, this implies equation (11). //

COROLLARY: *Distinct maximal ideals generate distinct characters.*

For if $\chi_{M_1} = \chi_{M_2}$, then by (12), $\mathfrak{z}(M_1) = \mathfrak{z}(M_2)$ for all $\mathfrak{z} \in V(G)$, and hence M_1 and M_2 coincide. //

THEOREM 3: *Let χ be a character of the group G. Then*

$$\lambda e + x \to \lambda + \int x(g) e^{i\chi(g)} dg \tag{14}$$

is a homomorphic mapping of the group ring $V(G)$ into the field of complex numbers. The maximal ideal M_χ that is the kernel of this mapping is different from M_∞.

Proof: The linearity of the mapping (14) is obvious and we only have to verify that it is multiplicative, i.e., that the product of elements of the group ring corresponds to the product of the numbers associated with them. We introduce the notation

$$(\lambda e + x)^\sim (\chi) = \lambda + \int x(g) e^{i\chi(g)} dg.$$

Applying Fubini's Theorem, we obtain

$$\begin{aligned}(x * y)^\sim (\chi) &= \int \left(\int x(g-h) y(h) dh \right) e^{i\chi(g)} dg = \\ &= \int \left(\int x(g-h) e^{i\chi(g)} dg \right) y(h) dh = \\ &= \int \left(\int x(g-h) e^{i\chi(g-h)} dg \right) y(h) e^{i\chi(h)} dh = \\ &= \int x(g) e^{i\chi(g)} dg \int y(h) e^{i\chi(h)} dh = x^\sim(\chi) y^\sim(\chi).\end{aligned}$$

Hence it follows immediately that

$$(\mathfrak{z}_1 * \mathfrak{z}_2)^\sim (\chi) = \mathfrak{z}_1^\sim(\chi) \mathfrak{z}_2^\sim(\chi)$$

for arbitrary $\mathfrak{z}_1, \mathfrak{z}_2 \in V(G)$. Thus, the first statement of the theorem is proved. For the proof of the second statement, we take an arbitrary function $x \in L^1(G)$ for which $x(g)\, dg \neq 0$. Then $x(g)e^{-i\chi(g)}$ is not contained in M, because

$$\int x(g) e^{-i\chi(g)} \cdot e^{i\chi(g)} dg = \int x(g) dg \neq 0.$$

Since, on the other hand $x(g)e^{-i\chi(g)} \in M_\infty$, this implies that $M_\chi \neq M_\infty$ //

Obviously, $x^\sim(\chi) = x(M_\chi)$ for every function $x \in L^1(G)$. //

We shall say that *the character χ generates the maximal ideal M_χ*. Theorem 2 shows that *a maximal ideal generates the character generating it*: $M = M_{\chi_M}$.

THEOREM 4: *A character is generated by the maximal ideal that it generates*: $\chi = \chi_{M_\chi}$.

Proof: By the preceding theorem, there exists in $L^1(G)$ a function not contained in M_χ. Let $z(g)$ be such a function, i.e., $z^\sim(\chi) \neq 0$. We have

$$z_g^\sim(\chi) = \int z(h-g)\, e^{i\chi(h)}\, dh = e^{i\chi(g)} \int z(h-g)\, e^{i\chi(h-g)}\, dh =$$

$$= e^{i\chi(g)} \int z(h)\, e^{i\chi(h)}\, dh = e^{i\chi(g)} z^\sim(\chi),$$

and hence

$$e^{i\chi(g)} = \frac{z_g^\sim(\chi)}{z^\sim(\chi)} = \frac{z_g(M_\chi)}{z(M_\chi)} = e^{i\chi_{M_\chi}(g)}. \; //$$

Note: In the proof of Theorem 3 we have only used the measurability of the character χ. Therefore Theorem 4 shows that *the measurability of a character χ implies its continuity.*

From Theorem 4 we obtain the following corollary.

COROLLARY: *Distinct characters generate distinct maximal ideals.*

For if M_{χ_1} coincides with M_{χ_2}, then $\chi_1 = \chi_{M_{\chi_1}}$ coincides with $\chi_2 = \chi_{M_{\chi_2}}$. //

5. Theorems 1-4 establish *the canonical one-to-one correspondence between the characters of a group G and the maximal ideals of its group ring $V(G)$, other than M_∞.* At the same time, these theorems show that by identifying a character χ with the corresponding maximal ideal M_χ *we can regard the Fourier transform*

$$\int x(g)\, e^{i\chi(g)}\, dg$$

of the absolutely integrable function $x(g)$ as its canonical representation in the form of a function on the set of maximal ideals of the group ring:

$$x^\sim(\chi) = \int x(g)\, e^{i\chi(g)}\, dg = x(M_\chi).$$

This new treatment of the Fourier transform will lie at the basis of the development of harmonic analysis on locally compact commutative groups.

§ 22. The Uniqueness Theorem for the Fourier Transform and the Abundance of the Set of Characters

1. In view of the connection, established above, between the characters of a locally compact commutative group and the maximal ideals of its group ring, it is natural to expect that there is also a connection between the question as to the existence of a sufficiently large set of characters in such a group and the question as to the existence of a sufficiently large set of maximal ideals in its group ring.

Every communtative group has at least one character: $\chi(g) \equiv 0$. We shall say that a commutative group G has *an abundant set of characters* if for each of its elements g_0, different from zero, there exists a character χ_0 such that $\chi_0(g_0) \neq 0$. Similarly, we could say that a commutative normed ring R has an abundant set of maximal ideals if for every element $x_0 \in R$, different from zero, there exists a maximal ideal M_0 such that $x_0(M_0) \neq 0$. But this is none other than the statement that R is a ring without radical. In the present section we shall prove this statement for the group ring $V(G)$. By the results obtained in the preceding section this will also prove the uniqueness theorem for the Fourier transform of an absolutely integrable function. Hence we shall derive as a corollary that the groups in question have an abundant set of characters.

The operation

$$\mathfrak{z} = \lambda e + x(g) \to \overline{\lambda} e + \overline{x(-g)} = \mathfrak{z}^* \tag{1}$$

satisfies the conditions a)-c) of Definition 1 of § 8 and hence is an involution in the ring $V(G)$. In order to show that $V(G)$ is a ring without radical it is therefore sufficient, by Theorem 4 of § 8, to show that this involution is essential, i.e., that for every non-zero $\mathfrak{z} \in V(G)$ there exists a positive linear functional f on $V(G)$ such that $f(\mathfrak{z} * \mathfrak{z}^*) \geqq 0$. But first we give three lemmas.

2. LEMMA 1: *The convolution $z * x$ of the function $x \in L^1(G)$ with an arbitrary bounded measurable function z exists everywhere and is bounded and uniformly continuous.*

Proof: Let $|z(g)| \leqq C$. The existence of $(z * x)(g)$ for all $g \in G$ is obvious. Further,

$$|(z * x)(g)| = \left| \int z(g-h)\, x(h)\, dh \right| \leqq C \|x\|,$$

i.e., $z * x$ is bounded. Finally, when we make the substitution $k \to k+h$, we obtain

$$(z * x)(g+h) = \int z(g+h-k)\, x(k)\, dk =$$
$$= \int z(g-k)\, x(k+h)\, dk,$$

and hence

$$|(z * x)(g+h) - (z * x)(g)| =$$
$$= \left| \int z(g-k)\, [x(k+h) - x(k)]\, dk \right| \leqq$$
$$\leqq C \int |x(k+h) - x(k)|\, dk,$$

so that $(z * x)(g)$ is uniformly continuous, because the right-hand side does not depend on g and tends to zero for $h \to 0$ (see § 20.2, (2) and (2′)). //

LEMMA 2: *If $u(h)$ and $v(h) \in L^2(G)$, then the convolution $v^* * u$ exists everywhere and is bounded and uniformly continuous.*

Proof: By the invariance of the integral, for every fixed $g \in G$ we also have $T_g v(h) = v(-g+h) \in L^2(G)$. Therefore

$$(v^* * u)(g) = \int \overline{v(-g+h)}\, u(h)\, dh$$

exists for all $g \in G$ and is equal to $\langle u, T_g v\rangle$, where $\langle f, g\rangle$ denotes the scalar product of the functions $f, g \in L^2(G)$. Hence

$$|(v^* * u)(g)| \leqq \|u\|_2 \|T_g v\|_2 = \|u\|_2 \|v\|_2, \tag{2}$$

so that $(v^* * u)(g)$ is bounded. And further,

$$\begin{aligned} |(v^* * u)(g+h) - (v^* * u)(g)| &= |\langle u, (T_{g+h} - T_g) v\rangle| \leqq \\ &\leqq \|u\|_2 \|(T_{g+h} - T_g) v\|_2 = \|u\|_2 \|T_h v - v\|_2, \end{aligned}$$

so that $(v^* * u)(g)$ is uniformly continuous, because $\| T_h v - v \|_2$ does not depend on g and tends to zero for $h \to 0$ (see § 20.2, (2) and (2′)). //

LEMMA 3: *The convolution $(u * x)(h)$ of any two functions $u \in L^2(G)$ and $x \in L^1(G)$ exists for almost all $h \in G$ and belongs to $L^2(G)$.*

Proof: Lemma 2 implies the existence of the integral

$$\int (v^* * u)(-g) x(g) dg;$$

and by means of the substitution $h \to -g + h$ we obtain

$$\begin{aligned} \int (v^* * u)(-g)\, x(g)\, dg &= \int \left(\int \overline{v(g+h)}\, u(h)\, dh \right) x(g)\, dg = \\ &= \int \left(\int u(-g+h)\, \overline{v(h)}\, dh \right) x(g)\, dg. \end{aligned}$$

Applying Fubini's Theorem, we conclude that then the integral

$$\int u(-g+h)\, x(g)\, dg = (u * x)(h)$$

exists for almost all $h \in G$ and

$$\int (v^* * u)(-g)\, x(g)\, dg = \int (u * x)(h)\, \overline{v(h)}\, dh.$$

Since

$$\left|\int (v^* * u)(-g)\, x(g)\, dg\right| \leqq \sup_{g\in G} |(v^* * u)(-g)| \cdot \|x\|,$$

we obtain, by applying the inequality (2),

$$\left|\int (u * x)(h)\, \overline{v(h)}\, dh\right| \leqq \|x\|\, \|u\|_2 \|v\|_2. \tag{3}$$

Since this is true for every $v \in L^2(G)$, we conclude on the basis of a well-known property of Hilbert spaces that $u * x \in L^2(G)$. //

Note: The inequality (3) also shows that

$$\|u * x\|_2 \leqq \|x\|\, \|u\|_2.$$

3. We shall now indicate a simple method of constructing positive linear functionals on $V(G)$.

Theorem 1: *Let $u(h)$ be an arbitrary non-zero element of $L^2(G)$ and*

$$\varphi(g) = \int u(g+h)\, \overline{u(h)}\, dh. \tag{4}$$

Then the expression

$$J_\varphi(\mathfrak{z}) = J_\varphi(\lambda e + x) = \lambda\varphi(0) + \int \varphi(-g)\, x(g)\, dg \tag{5}$$

is a non-zero positive linear functional on $V(G)$.

Proof: Upon making the substitution $h \to -g+h$ in (4), we see that

$$\varphi(g) = (u^* * u)(g).$$

Thus, by Lemma 2, $\varphi(g)$ exists for all $g \in G$ and is continuous; furthermore, by (4) and (2),

$$|\varphi(g)| \leqq \|u\|_2^2 = \varphi(0). \tag{2'}$$

This shows that

$$|J_\varphi(\mathfrak{z})| \leqq \varphi(0)\, \|\mathfrak{z}\|$$

and hence that $J_\varphi(\mathfrak{z})$ is a linear functional on $V(G)$. Since

$$\varphi(0) = \|u\|_2^2 > 0, \tag{6}$$

this functional is non-zero (not only on $V(G)$, but even on $L^1(G)$). Replacing g by $-g$ in (4) and making the substitution $h \to g+h$, we now obtain

$$\varphi(-g) = \int u(-g+h)\, \overline{u(h)}\, dh = \int u(h)\, \overline{u(g+h)}\, dh,$$

i.e.,

$$\varphi(-g) = \overline{\varphi(g)}. \tag{7}$$

From the invariance of the integral under reflection and the equation (7) it follows that

$$J_\varphi(\mathfrak{z}^*) = \bar{\lambda}\varphi(0) + \int \varphi(-g)\overline{x(-g)}\,dg =$$
$$= \bar{\lambda}\varphi(0) + \int \varphi(g)\overline{x(g)}\,dg,$$

i.e.,

$$J_\varphi(\mathfrak{z}^*) = \overline{J_\varphi(\mathfrak{z})}. \tag{8}$$

Now let x be an arbitrary function of $L^1(G)$. By the properties of associativity and commutativity of the convolution[6] and the property of the involution $x^* * u^* = (u * x)^*$, we have

$$(u^* * u) * (x * x^*) = (u * x)^* * (u * x),$$

i.e.,

$$((u^* * u) * (x * x^*))(g) = ((u * x)^* * (u * x))(g) \tag{9}$$

for almost all $g \in G$. But the function $(u^* * u) * (x * x^*)$ is continuous by the Lemmas 2 and 1, and the function $(u * x)^* * (u * x)$ is continuous by the Lemmas 3 and 2. Consequently equation (9) holds for all $g \in G$ without exception, and in particular for $g = 0$. Since in this case

$$((u^* * u) * (x * x^*))(0) = J_\varphi(x * x^*)$$

and

$$((u * x)^* * (u * x))(0) = \int |(u * x)(h)|^2\,dh,$$

we conclude that

$$J_\varphi(x * x^*) = \int |(u * x)(h)|^2\,dh, \tag{10}$$

and hence

$$J_\varphi(x * x^*) \geqq 0 \qquad \text{for all} \qquad x \in L^1(G).$$

Thus, $J_\varphi(x * y^*) \geqq 0$ for $y = x$. Here $J_\varphi(x * y^*)$ is linear in x and 'semi-linear' in y.[7] Hence $J_\varphi(x * y^*)$ has the properties of a scalar product, from which the inequality

$$|J_\varphi(x * y^*)|^2 \leqq J_\varphi(x * x^*)\,J_\varphi(y * y^*)$$

---

[6] In order to establish these properties in the proof of Theorem 1 of § 16, we have made use only of the existence of the corresponding integrals and none whatever of the fact that all the functions in question belong to the space L^1.

[7] I.e., $J_\varphi(x * (\lambda y_1 + \mu y_2)^*) = \bar{\lambda} J_\varphi(x * y_1^*) + \bar{\mu} J_\varphi(x * y_2^*)$.

can be derived in the usual way. On fixing x, putting $y = e_U^*$, where $\{e_U\}$ is a δ-sequence contracting to the zero element of G, and passing to the limit $U \to 0$, we obtain the inequality

$$|J_\varphi(x)|^2 \leqq \varphi(0)\, J_\varphi(x * x^*). \tag{11}$$

This inequality can be extended to arbitrary elements of $V(G)$, i.e.,

$$|J_\varphi(\mathfrak{z})|^2 \leqq \varphi(0)\, J_\varphi(\mathfrak{z} * \mathfrak{z}^*) \tag{12}$$

for all $\mathfrak{z} = \lambda e + x \in V(G)$. For, by (8) and (11),

$$\begin{aligned} |J_\varphi(\mathfrak{z})|^2 &= |\lambda\varphi(0) + J_\varphi(x)|^2 = \\ &= \lambda\bar{\lambda}\varphi^2(0) + \lambda\varphi(0)\overline{J_\varphi(x)} + \bar{\lambda}\varphi(0)\, J_\varphi(x) + |J_\varphi(x)|^2 \leqq \\ &\leqq \lambda\bar{\lambda}\varphi^2(0) + \lambda\varphi(0)\overline{J_\varphi(x)} + \bar{\lambda}\varphi(0)\, J_\varphi(x) + \varphi(0)\, J_\varphi(x * x^*) = \\ &= \varphi(0)\, J_\varphi((\lambda e + x) * (\lambda e + x)^*) = \varphi(0)\, J_\varphi(\mathfrak{z} * \mathfrak{z}^*). \end{aligned}$$

Taking (6) into account, we conclude from (12) that $J_\varphi(\mathfrak{z} * \mathfrak{z}^*) \geqq 0$. //

3. THEOREM 2: *The involution* (1) *in* $V(G)$ *is essential.*

Proof: Let $\mathfrak{z} = \lambda e + x$. Since $M_\infty(\mathfrak{z} * \mathfrak{z}^*) = |\lambda|^2 \geqq 0$, the linear functional $M_\infty(\mathfrak{z}) = \lambda$ is positive; moreover $M_\infty(\mathfrak{z} * \mathfrak{z}^*) > 0$, whenever $\lambda > 0$. Thus, it remains to discuss the case where $\mathfrak{z} = x(g)$, with $\|x\| > 0$. But if $f(x * x^*) = 0$ for every positive linear functional f and, in particular, every functional (5), then it follows from (10) that $u * x = 0$ for all $u \in L^2(G)$. Taking $u = e_U$, where $\{e_U\}$ $(\in L^{1,2}(G))$ is a δ-sequence contracting to the zero element of G, we conclude from the lemma of § 20 that

$$x = \lim_{U \to 0} e_U * x = 0,$$

contrary to assumption. //

From Theorem 4 of § 8 we derive immediately Theorem 3:

THEOREM 3: *$V(G)$ is a ring without radical.*

As we have already indicated, Theorem 3 is nothing other than the Uniqueness Theorem:

UNIQUENESS THEOREM: *If $x, y \in L^1(G)$ and*

$$\int x(g)\, e^{i\chi(g)}\, dg = \int y(g)\, e^{i\chi(g)}\, dg$$

for all characters χ of G, then $x(g)$ and $y(g)$ coincide for almost all $g \in G$.

For, to say that in a commutative normed ring there are no generalized nilpotent elements other than zero is the same as saying that every element of the ring is completely determined by the values that it assumes on the maximal ideals.

THEOREM 4: *A commutative locally compact group G has an abundant set of characters.*

Proof: Let g_0 be an arbitrary element of G, other than zero. We have to show that there exists a character χ_0 such that $e^{i\chi_0(g_0)} \neq 1$. Since $g_0 \neq 0$, there exists a neighborhood of zero U with a compact closure so small that U and $U + g_0$ do not intersect. Let $f(g)$ be the characteristic function of the set U. Then $T_{g_0}f(g) = f(g - g_0)$ is the characteristic function of the set $U + g_0$ and $\| T_{g_0}f - f \| = 2\| f \| > 0$. Thus, by Theorem 3, $T_{g_0}f - f$ is not a generalized nilpotent element, i.e., there exists a maximal ideal M_0 such that

$$(T_{g_0}f - f)(M_0) = (T_{g_0}f)(M_0) - f(M_0) \neq 0,$$

or

$$(T_{g_0}f)(M_0) \neq f(M_0). \tag{13}$$

Obviously $M_0 \neq M_\infty$, because $(T_{g_0}f)(M_\infty) = f(M_\infty) = 0$. By the remark following Theorem 1 of § 21, the inequality (13) can be written in the form

$$M_0(g_0) f(M_0) \neq f(M_0).$$

Hence it follows, first of all, that $f(M_0) \neq 0$; cancelling out $f(M_0)$ and letting χ_0 denote the character generated, in accordance with Theorem 1 of § 21, by the maximal ideal M_0, we obtain

$$e^{i\chi_0(g_0)} = M_0(g_0) \neq 1,$$

and the theorem is proved. //

§ 23. The Group of Characters

1. If χ_1 and χ_2 are two (additive) characters of the commutative topological group G, then their difference

$$(\chi_1 - \chi_2)(g) = \chi_1(g) - \chi_2(g) \qquad (\text{mod } 2\pi) \tag{1}$$

is also a character of the group. Hence the set X of all characters of G is in turn a group. Following Pontrjagin,[8] we can topologize this group by taking the collection of sets of the form

$$\{\chi \in X : | e^{i\chi(g)} - e^{i\chi_0(g)} | < \delta \qquad \text{for all} \qquad g \in F\}, \tag{2}$$

[8] L. S. Pontrjagin, *Topologische Gruppen* [translated from the 2nd Russian edition], Teil 2, Leipzig, 1958, p. 7, Definition 36. (See also Pontrjagin, *Topological Groups*, Princeton, 1939, p. 127, Definition 34, which is slightly less general.)

where F are all possible compact sets of G and δ are arbitrary positive numbers, as a fundamental system of neighborhoods of the character χ_0. It is easy to verify that the group operation (1) is continuous in the topology defined by the neighborhoods (2), so that X furnished with this topology is a topological group. We call it the *group of characters* of G.

Now let G be a locally compact commutative group. In § 21 we established a canonical one-to-one correspondence between the characters of G and the maximal ideals of its group ring $V(G)$, other than M_∞. By identifying the characters with the corresponding maximal ideals we can introduce into the set X of characters of G a topology induced from the space $\mathfrak{M}$ of maximal ideals of $V(G)$. In this topology a fundamental system of neighborhoods of the character χ_0 is formed by the sets of the form

$$\{\chi \in \mathrm{X}: \left|\int x_k(g)\,[e^{i\chi(g)} - e^{i\chi_0(g)}]\,dg\right| < \varepsilon \quad (k = 1, \ldots, n)\}, \tag{3}$$

where $x_1, \ldots, x_n$ are all the possible finite sets of functions of $L^1(G)$ and the ε are arbitrary positive numbers. Moreover, X, obtained from $\mathfrak{M}$ by removal of the single point M_∞, is a locally compact space.

2. The question naturally arises: what is the connection between the Pontrjagin topology of the group X defined by the neighborhoods (2) and the 'ring' topology defined by the neighborhoods (3)? We shall show that these two topologies coincide.

THEOREM 1: *The group ring $V(G)$ of a locally compact commutative group G, furnished with the involution*

$$\mathfrak{z} = \lambda e + x(g) \rightarrow \bar{\lambda} e + \overline{x(-g)} = \mathfrak{z}^*,$$

is symmetric.

Proof: $\mathfrak{z}^*(M_\infty) = \bar{\lambda} = \overline{\mathfrak{z}(M_\infty)}$; but if $M \neq M_\infty$, then by Theorem 2 of § 21, by the invariance of the integral under reflection, and by the equation $\chi_M(-g) = -\chi_M(g)$ (mod 2π), we have

$$\mathfrak{z}^*(M) = \bar{\lambda} + \int \overline{x(-g)}\, e^{i\chi_M(g)}\, dg =$$
$$= \bar{\lambda} + \int \overline{x(g)}\, \overline{e^{i\chi_M(g)}}\, dg = \overline{\mathfrak{z}(M)}. //$$

LEMMA 1: *For every fixed $g \in G$ $e^{i\chi(g)}$ is a continuous function of χ in the ring topology defined by the neighborhoods* (3).

Proof: Let M be the maximal ideal corresponding to the character χ. Then by formula (7) of § 21,

$$e^{i\chi(g)} = \frac{z_g(M)}{z(M)}, \tag{4}$$

where $z \in L^1(G)$, $z(M) \neq 0$, and $z_g = T_g z$. But the expression on the right-hand side is a continuous function of M on $\mathfrak{M}$. Consequently $e^{i\chi(g)}$ in the topology in X induced by $\mathfrak{M}$ is a continuous function of χ. //

LEMMA 2: *Every set* K *of characters having a compact closure* $\overline{\mathrm{K}}$ *in the ring topology defined by the neighborhoods* (3) *is uniformly equicontinuous, i.e., for every* $\varepsilon > 0$ *there exists a neighborhood* U *of the zero element of* G *such that*

$$| e^{i\chi(h+g)} - e^{i\chi(h)} | < \varepsilon$$

for all $\chi \in \mathrm{K}$, $g \in U$, $h \in G$.

Proof: By the formula (4),

$$| e^{i\chi(h+g)} - e^{i\chi(h)} | = | e^{i\chi(g)} - 1 | =$$

$$= \left| \frac{z_g(M) - z(M)}{z(M)} \right| \leqq \frac{\| T_g z - z \|}{| z(M) |}. \tag{5}$$

Since $M_\infty \notin \overline{\mathrm{K}}$ and since, by Theorem 1, the ring $V(G)$ is symmetric, we know by part 4 of the proof of Theorem 1′ of § 7 that there exists a function $\varphi(M)$ that is continuous on $\mathfrak{M}$ and equal to 1 on $\overline{\mathrm{K}}$ and to 0 in M_∞ and also an element $\mathfrak{z} = \lambda e + x \in V(G)$ such that

$$| \varphi(M) - \mathfrak{z}(M) | < 1/4 \qquad \text{for all} \qquad M \in \mathfrak{M}.$$

Taking $M = M_\infty$ here, we obtain $| \lambda | < 1/4$, and hence

$$| \varphi(M) - x(M) | < 1/2 \qquad \text{for all} \qquad M \in \mathfrak{M}.$$

This shows that $| x(M) | > 1/2$ for all $M \in \mathrm{K}$. Upon taking this function x for our z in (5), we obtain

$$| e^{i\chi(h+g)} - e^{i\chi(h)} | \leqq 2 \| T_g x - x \| \qquad \text{for all} \quad \chi \in \mathrm{K} \quad \text{and} \quad h \in G,$$

and the statement of the lemma follows from the formulas (2) and (2′) of § 20. //

3. THEOREM 2: *The ring topology in* X *defined by the neighborhoods* (3) *coincides with the Pontrjagin topology defined by the neighborhoods* (2).

Proof: Since the translation $\chi \to \chi - \chi_0$ carries every neighborhood of the form (2) or (3) of the character χ_0 into a neighborhood of the same form of the null character, it is sufficient to confine ourselves to the case where $\chi_0 = 0$.

1. *Every given neighborhood of the form* (2) *contains a neighborhood of the form* (3).

For since, by Lemma 1, $e^{i\chi(h)}$ is a continuous function of χ in the ring topology for every h and $e^{i\chi(h)} = 1$ for $\chi = 0$, there exists, for every $h \in G$, a neighborhood V_h of the zero element of X of the form (3) such that $|e^{i\chi(h)}-1|<\delta/2$ for all $\chi\in V_h$. Here we may assume, since the space X is locally compact in the ring topology, that V_h has a compact closure in X. But then, by Lemma 2, there exists a neighborhood U_h of the zero element of G such that $|e^{i\chi(g)}-e^{i\chi(h)}|<\delta/2$ and consequently $|e^{i\chi(g)}-1|<\delta$ for all $g\in U_h+h$ and $\chi\in V_h$. Now, since F is compact, there exist elements $h_1,\ldots,h_n\in G$ such that $F\subset\bigcup_{k=1}^{n}(U_{h_k}+h_k)$. But then $V=\bigcap_{k=1}^{n}V_{h_k}$ is a neighborhood (3) contained in the neighborhood (2) in question.

2. *Every given neighborhood of the form* (3) *contains a neighborhood of the form* (2).

For from the regularity of the Haar measure there follows the existence of a compact set $F\subset G$ such that

$$\int_{CF}|x_k(g)|<\varepsilon/4 \qquad (k=1,\ldots,n).$$

The neighborhood (2) with these F and $\delta=\varepsilon/(2\max_{1\leqq k\leqq n}\|x_k\|)$ is contained in the neighborhood (3) in question. Indeed, for every χ of this neighborhood (3) we have

$$\left|\int x_k(g)\,e^{i\chi(g)}\,dg-\int x_k(g)\,dg\right|\leqq$$
$$\leqq\int|e^{i\chi(g)}-1||x_k(g)|\,dg\leqq 2\int_{CF}|x_k(g)|\,dg+$$
$$+\|x_k\|\max_{g\in F}|e^{i\chi(g)}-1|<\frac{\varepsilon}{2}+\|x_k\|\,\delta\leqq\varepsilon \qquad (k=1,\ldots,n).$$

This completes the proof of the theorem. //

Corollary: *The set* X *of characters of a locally compact commutative group G, furnished with the ring topology or, what is the same, the Pontrjagin topology, is a locally compact group.*

§ 24. The Invariant Integral on the Group of Characters

1. Since the group X is locally compact, there is a Haar measure on it in turn. Therefore, in the same way as we have constructed the group of characters X of G, we can construct the group of characters G^* of X. Now, there is a natural embedding of G in G^*. In order to see this, we observe that every element g of the original group G generates a character of X, namely $g(\chi)=\chi(g)$, where g is fixed and χ variable. First of all, $g(\chi)$ is real and uniquely determined modulo 2π; further, by the definition of the group operation in X,

$$g(\chi_1-\chi_2)=(\chi_1-\chi_2)(g)=$$
$$=\chi_1(g)-\chi_2(g)=g(\chi_1)-g(\chi_2) \qquad (\mathrm{mod}\ 2\pi);$$

finally, by Lemma 1 of § 23, $g(\chi)$ is a continuous function of χ. We note that distinct elements of G generate distinct characters of X, since for every pair of distinct elements g, h of G there exists a character χ_0 such that $e^{i\chi_0(g)}\neq e^{i\chi_0(h)}$, i.e., $g(\chi_0)\neq h(\chi_0)$ (mod 2π) (abundance of the set of characters: Theorem 4 of § 22). Thus, the elements of G can be identified with the characters of X that they generate, and then G is contained algebraically in G^*. One of the most important propositions of Pontrjagin's theory of characters is the 'duality law,' which states that under the indicated identification G coincides with G^* both as regards its set of elements and its topology. The following investigations have an important connection with the proof of this Pontrjagin duality law within the context of the theory to be expounded (the proof itself will be given in § 26).

This proof will be based on the close connection between the Haar measures or, what is the same, the invariant integrals on the groups X and G. Of course, the existence of an invariant integral on X follows from the fact that the group is locally compact. But the construction of Haar-Weil that establishes this existence cannot yield anything of value towards our goal. We shall establish the connection between the invariant integrals on X and on G by the actual construction of an invariant integral on X from the given invariant integral on G. This will be done in the present section.

***2.** In establishing this construction we shall start out from the idea that *a connection between the invariant integrals on* X *and on* G *is included in the inversion formulas for Fourier transforms.* That is, in analogy to what holds when G is the additive topological group of real numbers, we may expect that the Fourier transform

$$\varphi^{\sim}(\chi)=\int_G \varphi(g)\exp(i\chi(g))dg$$

of every function φ of the form $u^* * u$, where $u\in L^{1,2}(G)$, which is equal to

$$\left|\int_G u(g)\exp(i\chi(g))dg\right|^2,$$

is an absolutely integrable function on X and that for a suitable normalization of the Haar measure on X we have the inversion formula

$$\varphi(g)=\int_X \varphi^{\sim}(\chi)\exp(-ig(\chi))d\chi.$$

Then, in particular,

$$\varphi(0) = \int_{X} \varphi^{\sim}(\chi) d\chi. \tag{1}$$

But on the left-hand side of the latter (hypothetical) equation there is a value of the function given on the original group G. This enables us to take the formula (1) as the *definition* of the integral of the functions $\varphi^{\sim}(\chi)$ in question and as the starting point for the extension of the integration to continuous finite functions, after which the construction of the whole class of integrable functions on X runs along the same lines as the standard theory.

This was precisely the way in which Krein [28] first achieved the construction of the invariant integral on X from the given invariant integral on G. In the paper by Raikov [43], the formula (1) served as the starting point for a direct construction of a Haar measure on X from the invariant integral on G. However, this direct method of constructing the invariant integral of a Haar measure on X proved to be technically rather complicated. This is mainly due to the fact that the functional

$$I(x^{\sim}) = x(0),$$

whose value for $x = \varphi$ appears on the left-hand side of (1) is not defined for all the elements $x \in L^1(G)$.

Here we shall present another, technically much simpler, method which consists in replacing the direct discussion of the functional I by the discussion of the 'regularizing' functionals

$$I_{\varphi}(\mathfrak{z}^{\sim}) = J_{\varphi}(\mathfrak{z}) = (\varphi * \mathfrak{z})(0) = I(\varphi^{\sim}\mathfrak{z}^{\sim}),$$

that are defined for all possible functions φ of the form $u^* * u$, where $u \in L^{1,2}(G)$. Every such functional is also defined for all $\mathfrak{z} \in V(G)$, and since, by Theorem 1 of § 22, $J_{\varphi}(\mathfrak{z})$ is positive, $I_{\varphi}(\mathfrak{z}^{\sim})$ extends uniquely to all of $C(\mathfrak{M})$ with preservation of positivity. This makes it possible, for every continuous function $h(\chi)$ on X that tends to a finite limit for $\chi \to M_{\infty}$, to take

$$I(\varphi^{\sim} h) = I_{\varphi}(h).$$

Now if $f(\chi)$ is an arbitrary continuous finite function on X and Δ its (compact) support, then by choosing φ so that $\varphi^{\sim}(\chi)$ is different from zero on all of Δ (which, as we have seen, is always possible), and by taking

$$h(\chi) = \begin{cases} \dfrac{f(\chi)}{\varphi^{\sim}(\chi)}, & \text{when } \varphi^{\sim}(\chi) > 0, \\ 0 & \text{when } \varphi^{\sim}(\chi) = 0, \end{cases}$$

we obtain

$$I(f) = I_{\varphi}(f/\varphi^{\sim}).$$

This formula will be taken below as the starting point for the construction of the invariant integration on X.

3. In carrying out the proofs, we shall usually identify the characters of G with the corresponding maximal ideals of its group ring $V(G)$ and so regard X as part of the space $\mathfrak{M}$ of maximal ideals of that ring. Since X is obtained from $\mathfrak{M}$ by the removal of the single point M_∞, every continuous function $h(M)$ on $\mathfrak{M}$ is uniquely defined by its restriction $h(\chi)$ to X, and we shall not distinguish between the functions $h(M)$ and $h(\chi)$. However, in the case of functions $\mathfrak{z}(M)$, where $\mathfrak{z} \in V(G)$, we shall write $\mathfrak{z}^\sim(\chi)$ or $\mathfrak{z}^\sim$, as before, instead of $\mathfrak{z}(\chi)$.

We denote by P the set of all functions φ of the form $u^* \ast u$, where $u \in L^{1,2}(G)$, so that if $\varphi \in P$, then

$$\varphi^\sim(\chi) = |u^\sim(\chi)|^2 \geqq 0 \qquad \text{for all} \qquad \chi \in \mathrm{X}.$$

LEMMA: *For every compact set* $\Delta \subset \mathrm{X}$ *there exists a function* $\varphi \in P$ *such that* $\varphi^\sim(\chi) > 0$ *for all points* $\chi \in \Delta$.

Proof: By Lemma 2 of § 23, there exists a neighborhood V of the zero element of G such that

$$|e^{i\chi(g)} - 1| < 1/2 \qquad \text{for all} \qquad \chi \in \Delta \quad \text{and} \quad g \in V. \tag{2}$$

We choose a neighborhood U of the zero element of G satisfying the condition $U - U \subset V$ and put $\varphi = e_U^* \ast e_U$, where $e_U \in L^{1,2}(G)$ (and where it has the meaning given in the definition of § 20.5), so that $\varphi \in P$. It is easy to see that $\varphi = e_V$, and therefore

$$\varphi^\sim(\chi) = \int_V \varphi(g) e^{i\chi(g)}\, dg = 1 + \int_V \varphi(g) [e^{i\chi(g)} - 1]\, dg,$$

and hence, by (2), $|\varphi^\sim(\chi) - 1| < 1/2$ for all $\chi \in \Delta$. Consequently, $\varphi^\sim(\chi) = |\varphi^\sim(\chi)| > 1/2$ for all points $\chi \in \Delta$. //

4. For every $\varphi \in P$ and every $\mathfrak{z} = \lambda e + x \in V(G)$ we now put

$$J_\varphi(\mathfrak{z}) = \lambda \varphi(0) + \int \varphi(-h)\, x(h)\, dh.$$

According to Theorem 1 of § 22, J_φ is a positive linear functional on $V(G)$. But $V(G)$ is symmetric, by Theorem 1 of § 23. Therefore, according to Theorem 5 of § 8, J_φ generates a positive linear functional $\hat{J}_\varphi$ on the space $\hat{V}(G)$ of functions $\mathfrak{z}(M)$ defined by the equation

$$\hat{J}_\varphi(\mathfrak{z}(M)) = J_\varphi(\mathfrak{z}),$$

where $\hat{J}_\varphi$ extends uniquely to a positive linear functional on the whole of $C(\mathfrak{M})$. We shall denote this extended functional by I_φ and, in accordance with the convention made above, we shall regard it also as a functional on the space of functions $h(\chi)$ that are the restrictions to X of the functions $h(M) \in C(\mathfrak{M})$. Thus, in particular, for every $\mathfrak{z} = \lambda e + x \in V(G)$ we have

$$I_\varphi(\mathfrak{z}^\sim) = \lambda\varphi(0) + \int \varphi(-h)\, x(h)\, dh = (\varphi * \mathfrak{z})(0). \tag{3}$$

In addition, let $\psi \in P$. By (3), we then have

$$I_\varphi(\mathfrak{z}^\sim\psi^\sim) = (\varphi * \mathfrak{z} * \psi)(0) = I_\psi(\mathfrak{z}^\sim\varphi^\sim). \tag{4}$$

Since $I_\varphi(\mathfrak{z}^\sim\psi^\sim)$ and $I_\psi(\mathfrak{z}^\sim\varphi^\sim)$ are continuous functionals of $\mathfrak{z}^\sim$ or, what is the same, of $\mathfrak{z}(M)$ and since the set of functions $\mathfrak{z}(M)$ is dense in $C(\mathfrak{M})$, we conclude from (4) that

$$I_\varphi(h\psi^\sim) = I_\psi(h\varphi^\sim) \tag{5}$$

for every $h \in C(\mathfrak{M})$.

In accordance with the notation of § 20, we shall understand $L(\mathrm{X})$ to mean the set of all continuous finite functions on X, i.e., the set of continuous functions $f(\chi)$ with a compact support Δ_f.[9] For every function $f \in L(\mathrm{X})$ we denote by P_f the set of those $\varphi \in P$ for which $\varphi^\sim(\chi) > 0$ for all points $\chi \in \Delta_f$. *P_f is not empty.* For if $f(\chi) \equiv 0$, then Δ_f is empty and $P_f = P$; but if $f(\chi) \not\equiv 0$, then Δ_f is a non-empty compact set and, by the lemma proved above, P_f is not empty.

By taking, in (5), $\varphi, \psi \in P_f$ and

$$h(\chi) = \begin{cases} \dfrac{f(\chi)}{\varphi^\sim(\chi)\psi^\sim(\chi)} & \text{for} \quad \chi \in \Delta_f, \\ 0 & \text{for} \quad \chi \notin \Delta_f, \end{cases}$$

we obtain

$$I_\varphi\left(\frac{f}{\varphi^\sim}\right) = I_\psi\left(\frac{f}{\psi^\sim}\right). \tag{6}$$

For every function of $L(\mathrm{X})$ we now put

$$I(f) = I_\varphi\left(\frac{f}{\varphi^\sim}\right) \qquad (\varphi \in P_f). \tag{7}$$

By equation (6), $I(f)$ is a uniquely defined functional on $L(\mathrm{X})$.

We denote by $L_+(\mathrm{X})$ the set of all non-negative real functions in $L(\mathrm{X})$.

---

[9] We recall that the support of a function is the closure of the set of those points at which it has a non-zero value.

In accordance with the notation of § 20, T_ξ for every fixed $\xi \in X$ will mean the operator of translation by ξ defined by the equation $T_\xi f(\chi) = f(\chi - \xi)$.

5. THEOREM 1: *The functional $I(f)$ defined on $L(X)$ by the formula (7) is an invariant integral on X, i.e.,*

1. *$I(f)$ depends linearly on f;*
2. *$I(f(\chi - \xi)) = I(f(\chi))$ for every fixed $\xi \in X$;*
3. *$I(f) \geqq 0$ if $f \in L_+(X)$;*
4. *$I(f) > 0$ if $f \in L_+(X)$ and $f(\chi) \not\equiv 0$.*

Proof: 1. Let $f_1, f_2 \in L(X)$ and let λ_1, λ_2 be arbitrary complex numbers. Since $\Delta_{f_1} \cup \Delta_{f_2}$ is compact, by the lemma there exists a function $\varphi \in P$ such that $\varphi^\sim(\chi) > 0$ for all points $\chi \in \Delta_{f_1} \cup \Delta_{f_2}$. Then

$$\varphi \in P_{\lambda_1 f_1 + \lambda_2 f_2} \cap P_{f_1} \cap P_{f_2},$$

and therefore

$$I(\lambda_1 f_1 + \lambda_2 f_2) = I_\varphi\left(\frac{\lambda_1 f_1 + \lambda_2 f_2}{\varphi^\sim}\right) =$$
$$= \lambda_1 I_\varphi\left(\frac{f_1}{\varphi^\sim}\right) + \lambda_2 I_\varphi\left(\frac{f_2}{\varphi^\sim}\right) = \lambda_1 I(f_1) + \lambda_2 I(f_2).$$

2. Let $\mathfrak{z} = \lambda e + x \in V(G)$ and $\xi \in X$. We put $\mathfrak{z}_\xi = \lambda e + e^{-i\xi(g)} x(g)$. Then

$$\mathfrak{z}_\xi^\sim(\chi) = \lambda + \int x(g) e^{i(\chi - \xi)g}\, dg = \lambda + x^\sim(\chi - \xi) =$$
$$= \mathfrak{z}^\sim(\chi - \xi) = T_\xi \mathfrak{z}^\sim(\chi).$$

But if $\varphi(g) \in P$, then we also have $\varphi_\xi(g) = e^{-i\xi(g)}\varphi(g) \in P$, because, as is easy to verify,

$$(u^* * u)_\xi = u_\xi^* * u_\xi$$

for all $\xi \in X$ and $u \in L^1(G)$. Therefore

$$I_\varphi(\mathfrak{z}^\sim) = \lambda\varphi(0) + \int \varphi(-g)\, x(g)\, dg =$$
$$= \lambda\varphi_\xi(0) + \int \varphi_\xi(-g)\, x_\xi(g)\, dg = I_{\varphi_\xi}(\mathfrak{z}_\xi^\sim) = I_{\varphi_\xi}(T_\xi \mathfrak{z}^\sim).$$

Since $\hat{V}(G)$ is dense in $C(\mathfrak{M})$, we conclude from this that

$$I_\varphi(h) = I_{\varphi_\xi}(T_\xi h) \qquad \text{for all} \qquad h \in C(\mathfrak{M}).$$

Now let $f \in L(X)$. Taking $\varphi \in P_f$ and hence $\varphi_\xi \in P_{T_\xi f}$, we obtain

$$I(T_\xi f) = I_{\varphi_\xi}\left(\frac{T_\xi f}{\varphi_\xi^\sim}\right) = I_{\varphi_\xi}\left(T_\xi \frac{f}{\varphi^\sim}\right) = I_\varphi\left(\frac{f}{\varphi^\sim}\right) = I(f).$$

3. If $f\in L_+(\mathrm{X})$ and $\varphi\in P_f$ then $f/\varphi^\sim\in L_+(\mathrm{X})$, and since I_φ is a positive functional, $I(f)=I_\varphi(f/\varphi^\sim)\geqq 0$.

4. For the proof of the last statement of the theorem it is sufficient to establish the existence of at least one function $f_0\in L_+(\mathrm{X})$ for which $I(f_0)>0$. Indeed, if $f\in L_+(\mathrm{X})$ and $f(\chi)\not\equiv 0$, then there exists an open set $v\subset\mathrm{X}$ such that $f(\chi)>a>0$ for all $\chi\in V$ and points $\chi_1,\ldots,\chi_n\in\mathrm{X}$ such that $\Delta_{f_0}\subset\bigcup_{k=1}^{n}(V+\chi_k)$. Putting $b=\max_{\chi\in\mathrm{X}} f_0(\chi)$, we then obtain

$$\frac{b}{a}\sum_{k=1}^{n} f(\chi-\chi_k)\geqq f_0(\chi)\qquad\text{for all}\qquad \chi\in\mathrm{X},$$

and therefore, by the properties of the functional I already proved,

$$\frac{b}{a}nI(f)=I\left(\frac{b}{a}\sum_{k=1}^{n} f(\chi-\chi_k)\right)\geqq I(f_0),$$

and hence $I(f)>0$.

Suppose now that $u\in L^{1,2}(G)$ and $u(g)\geqq 0$ for all $g\in G$ and $\|u\|=\int u(g)dg\neq 0$. Suppose further that $\varphi=u^*\ast u$, so that $\varphi\in P$ and $\varphi^\sim(\chi)\geqq 0$ and $\varphi^\sim(\chi)\not\equiv 0$, because $\varphi^\sim(0)=\|u\|^2>0$. We put $c=I_\varphi(\varphi^\sim)$. By what we have established in the proof of Theorem 1 of § 22, $\varphi(g)$ is continuous, $\varphi(-g)=\overline{\varphi(g)}$, and $\varphi(0)>0$, and we therefore have

$$c=\int\varphi(-g)\varphi(g)\,dg=\int|\varphi(g)|^2\,dg>0.$$

Let α denote the smaller of the (positive) numbers $\max_{\chi\in\mathrm{X}}\varphi^\sim(\chi)$ and $c/\varphi(0)$ and let A be the set of those $M\in\mathfrak{M}$ for which $\varphi(M)\geqq 2\alpha/3$ and B the set of those $M\in\mathfrak{M}$ for which $\varphi(M)\leqq\alpha/3$. A and B are not empty and are closed, by the continuity of the function $\varphi(M)$, and $A\subset\mathfrak{M}\setminus B$. By what has been shown under 5. in the proof of Theorem 1′ of § 7, there exists a function $h\in C(\mathfrak{M})$ such that $h(M)=1$ on A, $h(M)=0$ on B, and $0\leqq h(M)\leqq 1$ on the entire space $\mathfrak{M}$. We have

$$c=I_\varphi(\varphi^\sim)=I_\varphi(h\varphi^\sim)+I_\varphi((1-h)\varphi^\sim).$$

But since $[1-h(M)]\varphi(M)<2\alpha/3$ for all $M\in\mathfrak{M}$,

$$I_\varphi((1-h)\varphi^\sim)\leqq I_\varphi\left(\frac{2}{3}\alpha\right)=\frac{2}{3}\alpha\varphi(0)<c.$$

Therefore $I_\varphi(h\varphi^\sim)>0$. And now, the function $f_0(\chi)=h(\chi)[\varphi^\sim(\chi)]^2$ has the required properties. For since $\varphi(M_\infty)=0$ and $\overline{\mathfrak{M}\setminus B}$ is the closed set of all points at which $\varphi(M) \geqq \alpha/3$, $\overline{\mathfrak{M} \setminus B}$ is a compact set in X; and since $\Delta_{f_0} \subset \Delta_h \subset \overline{\mathfrak{M} \setminus B}$ and $\varphi^{\sim}(\chi) > 0$ on $\overline{\mathfrak{M} \setminus B}$, we conclude that $f_0 \in L_+(X)$ and $\varphi \in P_{f_0}$. But then

$$I(f_0) = I_\varphi\left(\frac{f_0}{\varphi^{\sim}}\right) = I_\varphi(h\varphi^{\sim}) > 0.$$

This completes the proof of the theorem. //

6. The integral $I(f)$ generates a completely additive measure μ in a standard way on the field of Borel sets of the space X, where for a compact set $\Delta \subset X$ we have $\mu(\Delta) = \inf I(f)$ over all $f \in L_+(X)$ for which $f(\chi) > 1$ on Δ, so that $\mu(\Delta) < \infty$, and for an open set $O \subset X$ we have $\mu(O) = \sup I(f)$ over all $f \in L_+(X)$ for which $f(\chi) \leqq f_O(\chi)$, where f_O is the characteristic function of the set O, from which it follows easily on the basis of 4. of Theorem 1 that $\mu(O) > 0$. The invariance of $I(f)$ implies that this measure is invariant and hence that it coincides with a suitably normalized Haar measure on X.[1] Moreover, for all $f \in L(X)$ we have $I(f) = \int f(\chi)\,d\chi$, where the integral on the right-hand side is defined by this measure and the completion of $L(X)$ in the norm $\| f \|_1 = \int | f(\chi) |\, d\chi$ coincides with the space $L^1(X)$ of all absolutely integrable functions on X with respect to the Haar measure (which are defined to within the values on sets of measure 0).

§ 25. Inversion Formulas for the Fourier Transform

1. The main object of the present section is the proof of the fundamental proposition of harmonic analysis, the Theorem of Plancherel. The Theorem of Plancherel extends the concept of a Fourier transform to all measurable square-integrable functions and establishes that this transform is an isometric isomorphic mapping of the space $L^2(G)$ onto the space $L^2(X)$; the inverse is given here by the 'conjugate' Fourier transform, in which every character $e^{i\chi(g)}$ of G is replaced by the conjugate character $e^{-ig(\chi)}$ of X. The proof of this theorem will be based on the inversion formula for the Fourier transform of the functions of the class P introduced in the preceding section, that is, of the functions on G that can be represented in the form $(u^* * u)(g) = \int u(g+h)\overline{u(h)}\,dh$, where $u \in L^{1,2}(G)$.

As in the preceding section, we denote by Δ_f the support of the function $f(\chi)$, i.e., the closure of the set of those points $\chi \in X$ at which $f(\chi) \neq 0$, and by

[1] It is uniquely determined by the properties just mentioned apart from an arbitrary constant factor (see [73], [42], or [48]).

P_f the set of those functions $\varphi \in P$ for which $\varphi^{\sim}(\chi) = \int \varphi(g) e^{i\chi(g)} dg > 0$ for all points $\chi \in \Delta_f$.

2. THEOREM 1: *If $\varphi \in P$, i.e., $\varphi(g) = \int u(g+h)\overline{u(h)}dh$, where $u \in L^{1,2}(G)$, then*

$$\varphi^{\sim}(\chi) = \int \varphi(g) e^{i\chi(g)} dg \in L^1(X)$$

and

$$\varphi(g) = \int \varphi^{\sim}(\chi) e^{-i\chi(g)} d\chi.$$

Proof: For $\varphi(g) \equiv 0$, the statement is trivial. Suppose that $\varphi(g) \not\equiv 0$. Then by the uniqueness theorem for the Fourier transform (§ 22), we also have $\varphi^{\sim}(\chi) \not\equiv 0$. Since $\varphi^{\sim}(\chi) = |u^{\sim}(\chi)|^2 \geqq 0$, we have

$$c = \max_{\chi \in X} \varphi^{\sim}(\chi) > 0.$$

Let C_n be the set of those $M \in \mathfrak{M}$ for which $\varphi(M) \geqq c/n$ and D_n the set of those $M \in \mathfrak{M}$ for which $\varphi(M) \leqq c/(n+1)$ $(n = 1, 2, \ldots)$. The sets C_n and D_n are not empty and they are disjoint and closed. Therefore, by 5. of the proof of Theorem 1′ of § 7, there exists a function $h_n \in C(\mathfrak{M})$ such that $h_n(M) = 1$ on C_n, $h_n(M) = 0$ on D_n, and $0 \leqq h_n(M) \leqq 1$ on the entire space $\mathfrak{M}$. Since $\varphi^{\sim}(\chi) \geqq c/(n+1)$ on $\overline{CD}_n$, and $\varphi(M_\infty) = 0$, we have $\overline{CD}_n \subset X$. But $\Delta_{h_n} \subset \overline{CD}_n$; consequently $h_n \in L(X)$ and $\varphi^{\sim} \in P_{h_n}$, so that

$$\int h_n(\chi) \varphi^{\sim}(\chi) d\chi = I(h_n\varphi^{\sim}) = I_\varphi(h_n) \leqq I_\varphi(1) = \varphi(0). \tag{1}$$

Since the functions $h_n(\chi)\varphi^{\sim}(\chi)$ form a monotonic increasing sequence tending to $\varphi^{\sim}(\chi)$ on the whole of X, it follows from (1) that $\varphi^{\sim} \in L^1(X)$ and that

$$\int \varphi^{\sim}(\chi) d\chi \leqq \varphi(0). \tag{2}$$

Suppose that also $\psi \in P$, i.e., $\psi = v^* * v$, where $v \in L^{1,2}(G)$. Since $h_n\psi^{\sim} \in L(X)$ and $\varphi \in P_{h_n\psi^{\sim}}$, we have, taking formula (5) of § 24 into account,

$$\int h_n(\chi) \varphi^{\sim}(\chi) \psi^{\sim}(\chi) d\chi = I(h_n\varphi^{\sim}\psi^{\sim}) = I_\varphi(h_n\psi^{\sim}) = I_\psi(h_n\varphi^{\sim}). \tag{3}$$

But $h_n\varphi^{\sim}\psi^{\sim}$ and $h_n\varphi^{\sim}$ tend monotonically to $\varphi^{\sim}\psi^{\sim}$ and $\varphi^{\sim}$, respectively, for $n \to \infty$. Taking into account that the functionals I and I_φ are positive and passing to the limit $n \to \infty$ in (3), we therefore conclude that

$$\int \varphi^{\sim}(\chi) \psi^{\sim}(\chi) d\chi = I_\psi(\varphi^{\sim}) = \int \varphi(g) \overline{\psi(g)} dg. \tag{4}$$

Since $\varphi(g)$ is continuous, for every $\varepsilon > 0$ there exists a neighborhood V of the zero element of G such that $\Re\varphi(g) \geqq \varphi(0) - \varepsilon$ for all $g \in G$. Let U be a neighborhood of the zero element of G satisfying the condition $U - U \subset V$. We choose $v = e_U$, where $e_U \in L^{1,2}(G)$ has the meaning given in § 20.5. Then $\psi = e_U^* * e_U = e_V$; and since, by formula (4), the integral $\int \varphi(g)\overline{\psi(g)}dg$ is real, we have

$$\int \varphi(g)\overline{\psi(g)}\,dg = \int_V \Re\varphi(g) \cdot e_V(g)\,dg \geqq \varphi(0) - \varepsilon;$$

on the other hand, $\psi^\sim(\chi) = |\tilde{e_U}(\chi)|^2 \leqq 1$. From the formulas (2) and (4), we therefore obtain

$$\varphi(0) - \varepsilon \leqq \int \varphi(g)\overline{\psi(g)}\,dg = \int \varphi^\sim(\chi)\psi^\sim(\chi)\,d\chi \leqq$$
$$\leqq \int \varphi^\sim(\chi)\,d\chi \leqq \varphi(0),$$

from which it follows, inasmuch as ε is arbitrary, that

$$\int \varphi^\sim(\chi)\,d\chi = \varphi(0). \tag{2'}$$

Now we put

$$\omega(g) = (1 + |\lambda|^2)\varphi(g) + \lambda\varphi(g+h) + \bar{\lambda}\varphi(g-h) =$$
$$= ((u + \lambda T_{-h}u)^* * (u + \lambda T_{-h}u))(g).$$

Since $u + \lambda T_{-h}u \in L^{1,2}(G)$, we have $\omega \in L^{1,2}(G)$; and by what has just been proved,

$$\omega(0) = (1 + |\lambda|^2)\varphi(0) + \lambda\varphi(h) + \bar{\lambda}\varphi(-h) = \int \omega^\sim(\chi)\,d\chi =$$
$$= (1 + |\lambda|^2)\int \varphi^\sim(\chi)\,d\chi + \lambda\int\left(\int \varphi(g+h)\,e^{i\chi(g)}\,dg\right)d\chi +$$
$$+ \bar{\lambda}\int\left(\int \varphi(g-h)\,e^{i\chi(g)}\,dg\right)d\chi = (1 + |\lambda|^2)\int \varphi^\sim(\chi)\,d\chi +$$
$$+ \lambda\int \varphi^\sim(\chi)\,e^{-i\chi(h)}\,d\chi + \bar{\lambda}\int \varphi^\sim(\chi)\,e^{i\chi(h)}\,d\chi,$$

or, when we take (2′) into account,

$$\lambda\varphi(h) + \bar{\lambda}\varphi(-h) = \lambda\int \varphi^\sim(\chi)\,e^{-i\chi(h)}\,d\chi + \bar{\lambda}\int \varphi^\sim(\chi)\,e^{i\chi(h)}\,d\chi.$$

Dividing by λ and putting, successively, $\lambda = 1$ and $\lambda = i$, we obtain

$$\varphi(h) = \int \varphi^{\sim}(\chi)\, e^{-i\chi(h)}\, d\chi.$$

This completes the proof of the theorem. //

3. THEOREM 2 (Generalized Theorem of Plancherel): *The operator*

$$Tx = \int x(g)\, e^{i\chi(g)}\, dg \qquad (x \in L^2(G)),$$

where the integral on the right-hand side is to be understood as the limit in $L^2(X)$ *of the integrals* Tx_n *for sequences* $\{x_n\} \subset L^{1,2}(G)$ *that converge to* x *in* $L^2(G)$, *is uniquely defined for all* $x \in L^2(G)$ *and maps* $L^2(G)$ *isometrically onto* $L^2(X)$; *and its inverse is the operator*

$$T^{-1}f = \int f(\chi)\, e^{-i\chi(g)}\, d\chi \qquad (f \in L^2(X)) \tag{5}$$

(*which is defined similarly*), *so that we have the equations*

$$x(g) = \int \left(\int x(h)\, e^{i\chi(h)}\, dh \right) e^{-i\chi(g)}\, d\chi \tag{5'}$$

and

$$f(\chi) = \int \left(\int f(\xi)\, e^{-i\xi(g)}\, d\xi \right) e^{i\chi(g)}\, dg.$$

Proof: Let $u \in L^{1,2}(G)$. Then $u^* * u \in P$; and by Theorem 1, $(u^* * u)^{\sim} = |u^{\sim}(\chi)|^2 \in L^1(X)$, so that $u^{\sim} \in L^2(X)$ and, by the same theorem,

$$(u^* * u)(g) = \int u(g+h)\, \overline{u(h)}\, dh =$$
$$= \int \left| \int u(h)\, e^{i\chi(h)}\, dh \right|^2 e^{-i\chi(g)}\, d\chi.$$

Putting $g = 0$ in this equation, we obtain

$$\int |u(h)|^2\, dh = \int \left| \int u(h)\, e^{i\chi(h)}\, dh \right|^2 d\chi.$$

Thus, the operator T maps $L^{1,2}(G)$ isometrically into $L^2(X)$. Since $L^{1,2}(G)$ is everywhere dense in $L^2(G)$, it extends uniquely to the whole of $L^2(G)$ with preservation of the isometry of the mapping (and hence, with preservation of its property of being one to one).

Let us show that the image of $L^2(G)$ is the whole of $L^2(X)$.

For every $\xi \in X$ and every function $z(g)$ on G we put $z_\xi(g) = e^{-i\xi(g)} z(g)$; hence $z_\xi^{\sim}(\chi) = z^{\sim}(\chi - \xi)$ (see § 24.5). If $x \in L^1(G)$ and $\varphi \in P$, then we have that $x_\xi \in L^1(G)$ also and, as we have proved in § 24.5, $\varphi_\xi \in P$. Moreover, since $|\varphi_\xi(g)| \leq \varphi_\xi(0)$ by formula (2′) of § 22, we have

$$|(\varphi_\xi * x_\xi)(g)| \leqq \|x_\xi\| \varphi_\xi(0) = \|x\| \varphi(0),$$

so that $\varphi_\xi * x_\xi$, as a bounded function in $L^1(G)$, belongs to $L^2(G)$ and $(\varphi_\xi * x_\xi)^\sim \in L^2(X)$.

Suppose now that the function $f \in L^2(X)$ is orthogonal to $TL^2(G)$. By what has just been proved, we have, for arbitrary $\xi \in X$, $x \in L^1(G)$, and $\varphi \in P$,

$$\int f(\chi+\xi)\, \varphi^\sim(\chi)\, x^\sim(\chi)\, d\chi = \int f(\chi)\, \varphi^\sim(\chi-\xi)\, x^\sim(\chi-\xi)\, d\chi =$$

$$= \int f(\chi)\, (\varphi_\xi * x_\xi)^\sim(\chi)\, d\chi = 0.$$

Multiplying the first integral by $\overline{f(\xi)}$, integrating with respect to ξ, and applying Fubini's Theorem, we obtain

$$\int (f^* * f)(\chi)\, \varphi^\sim(\chi)\, x^\sim(\chi)\, d\chi = 0. \tag{6}$$

But by Lemma 2 of § 22 (applied to $L^2(X)$ instead of $L^2(G)$), $(f^* * f)(\chi)$ is a bounded continuous function. Therefore $g(\chi) = (f^* * f)(\chi)\varphi^\sim(\chi)$ is a continuous function on X, tending to zero for $\chi \to M_\infty$; and consequently, as the restriction to X of a function $g(M) \in C(\mathfrak{M})$, it is equal to zero at the point M_∞. By the symmetry of the ring $V(G)$ and Theorem 1 of § 7, for every $\varepsilon > 0$ there exists a $\mathfrak{z} = \lambda e + x \in V(G)$ such that

$$|g(M) - \lambda - x(M)| < \varepsilon \qquad \text{for all} \qquad M \in \mathfrak{M};$$

taking $M = M_\infty$ here, we obtain $|\lambda| < \varepsilon$ and consequently

$$|g(\chi) - x^\sim(\chi)| < 2\varepsilon \qquad \text{for all} \qquad \chi \in X.$$

In consequence of Theorem 1, $g \in L^1(X)$ and hence, in virtue of its boundedness, $g \in L^2(X)$; therefore, by taking (6) into account, we obtain

$$\int g^2(\chi)\, d\chi = \int g^2(\chi)\, d\chi - \int g(\chi)\, x^\sim(\chi)\, d\chi =$$

$$= \int g(\chi)\, [g(\chi) - x^\sim(\chi)]\, d\chi \leqq 2\varepsilon \int g(\chi)\, d\chi.$$

Inasmuch as ε is arbitrary, we conclude that $\int g^2(\chi) d\chi = 0$, and hence, being continuous, $g(\chi) = (f^* * f)(\chi)\varphi^\sim(\chi) \equiv 0$. But it follows from the lemma of § 24 that for every $\chi_0 \in X$ there exists a $\varphi \in P$ such that $\varphi^\sim(\chi_0) \neq 0$. Consequently $(f^* * f)(\chi) \equiv 0$. Putting $\chi = 0$ here, we obtain $\int |f(\xi)|^2 d\xi = 0$, i.e., $f = 0$. Since $TL^2(G)$ is closed in $L^2(X)$ because of the completeness of $L^2(G)$ and the isometry of T, we have therefore proved that

$$TL^2(G) = L^2(X).$$

It remains to show that the operator inverse to T is defined by formula (5), i.e., that formula (5′) holds. But for the functions $x \in P$, i.e., $x = u^* * u$, where $u \in L^{1,2}(G)$, the truth of (5′) was established in Theorem 1. Since

$$v^* * u = \frac{1}{4}[(u+v)^* * (u+v) - (u-v)^* * (u-v) +$$
$$+ i(u+iv)^* * (u+iv) - i(u-iv)^* * (u-iv)],$$

(5′) is then also true for all functions $x = v^* * u$, where $u, v \in L^{1,2}(G)$. But the set of these functions is dense in $L^2(G)$. For, on the one hand, the set of all bounded functions of $L^{1,2}(G)$ is dense in $L^2(G)$. On the other hand, if $u \in L^{1,2}(G)$ and $|u(g)| \leqq C$, then taking $v = e_U^*$, where $\{e_U\}$ $(\subset L^{1,2}(G))$ is a δ-sequence contracting to the zero element of G, we have

$$|(v^* * u)(g)| \leqq C,$$

and therefore

$$\|v^* * u - u\|_2 = \left(\int |(e_U * u)(g) - u(g)|^2 \, dg\right)^{\frac{1}{2}} \leqq$$
$$\leqq (2C \|e_U * u - u\|_1)^{\frac{1}{2}},$$

so that, by the lemma of § 20, $\|v^* * u - u\|_2 \to 0$ for $U \to 0$. Thus, for every $x \in L^2(G)$ there exists a sequence $\{x_n\} \subset L^2(G)$ converging to x such that

$$x_n(g) = \int \left(\int x_n(h) e^{i\chi(h)} \, dh\right) e^{-i\chi(g)} \, d\chi,$$

where the integrals are understood in the usual sense. Moreover, by what has already been proved, $x_n^{\sim} \to x^{\sim}$ in $L^2(X)$, and we therefore conclude that the formula (5′) is also true for x, provided the integrals in (5′) are interpreted in the generalized sense defined in the statement of Theorem 2. //

§ 26. The Pontrjagin Duality Law

1. In the construction of the Haar measure on the group of characters we have been guided by the inversion formulas for the Fourier transform. But for functions of the class L^2 these formulas are symmetric with respect to the groups G and X, so that G, as it were, plays the role of the group of characters of X in these formulas. In the present section we shall obtain just this 'duality law,' which states that G is in fact the group of characters of X; we shall start from the inversion formulas for the Fourier transform that were proved in the preceding section.

2. As already mentioned at the beginning of § 24, G can be regarded as an algebraic subgroup of the group G^* of characters of X.

THEOREM 1 (Pontrjagin's Duality Law): *G coincides with G^* both in its set of elements and in its topology.*

Proof: Capital italic letters with an asterisk shall denote sets of G^*; the same letters without an asterisk shall denote the projections of these sets onto G. $f^{\sim}(\chi)$, where $f \in L^1(\mathrm{X})$, shall denote $\int f(\chi) e^{ig^*(\chi)} d\chi$ $(g^* \in G^*)$; f_E, where E is a set of G or G^*, shall denote the characteristic function of this set.

The proof will be divided into several steps.

1. *G is homeomorphic to part of G^*.*

For the sets

$$U^* = \{g^* \in G^* : \quad |f_k^{\sim}(g^*) - f_k^{\sim}(g_0)| < \varepsilon \qquad (k = 1, \ldots, n;\ f_k \in L^1(\mathrm{X}))\} \tag{1}$$

form a basis of neighborhoods of the element $g_0 \in G$ in G^*. But the functions $f^{\sim}(g)$, where $f \in L^1(\mathrm{X})$, are continuous on G. In fact, for every $\varepsilon > 0$ there exists a compact set $\Delta \subset \mathrm{X}$ such that $\int_{C\Delta} |f(\chi)| d\chi < \varepsilon$; by Lemma 2 of § 23, there exists a neighborhood U of the zero element of G such that

$$|e^{ih(\chi)} - 1| < \varepsilon \text{ for all } \chi \in \Delta \text{ and } h \in U;$$

therefore

$$|f^{\sim}(g+h) - f^{\sim}(g)| \leqq 2 \int_{C\Delta} |f(\chi)| d\chi + \int_{\Delta} |e^{ih(\chi)} - 1| |f(\chi)| d\chi < \varepsilon(2 + \|f\|)$$

for all $h \in U$ and $g \in G$. Hence it follows that the projection

$$U = \{g \in G : \quad |f_k^{\sim}(g) - f_k^{\sim}(g_0)| < \varepsilon \quad (k = 1, \ldots, n; f_k \in L^1(\mathrm{X}))\}$$

of every neighborhood (1) is an open set in G. On the other hand, for every neighborhood V of the zero element of G there exists a neighborhood U^* of the zero element of G^* such that $U \subset V$. For there exists a neighborhood W of zero with a compact closure for which $W - W \subset V$. We put

$$\varphi(g) = \int f_W(g+h) f_W(h) dh.$$

Since $\varphi \in P$, we have by Theorem 1 of § 25, $\varphi(g) = f^{\sim}(-g)$, where $f(\chi) = |f_W^{\sim}(\chi)|^2 \in L^1(\mathrm{X})$. Now we choose

$$U^* = \{g^* \in G^*: |f^\sim(-g^*) - f^\sim(0)| < \varphi(0)\}.$$

Then

$$U = \{g \in G: |\varphi(g) - \varphi(0)| < \varphi(0)\} \subset W - W \subset V,$$

since $\varphi(g) = 0$ obviously holds outside $W - W$. Thus, we have proved that the topology induced in G by G^* coincides with the topology given in G.

2. *The ring* $V(X)$ *is regular*, i.e., for every point $g_0^* \in G^*$ and neighborhood V^* of that point, there exists a function $f \in L^1(X)$ such that $f^\sim(g_0^*) \neq 0$ and $f^\sim(g^*) = 0$ everywhere outside V^*.

For let W^* be a neighborhood of the zero element of G^* having a compact closure and such that $g_0^* + W^* - W^* \subset V^*$. We put

$$\varphi(g^*) = \int f_{W^*}(g^* - g_0^* + h^*) f_{W^*}(h^*)\, dh^*$$

(where the integration is taken with respect to the Haar measure on G^*). Obviously, $\varphi(g_0^*) > 0$ and $\varphi(g^*) = 0$ everywhere outside $g_0^* + W^* - W^*$, and hence, a fortiori, outside V^*. But since $\varphi \in L^2(G^*)$, by Theorem 2 of § 25 applied to the groups X and G^*, we have $\varphi(g^*) = f^\sim(g^*)$, where

$$f(\chi) = \int \varphi(g^*) e^{-ig^*(\chi)}\, dg^* = e^{-ig_0^*(\chi)} \left| \int f_{W^*}(h^*) e^{-ih^*(\chi)}\, dh^* \right|^2;$$

moreover $f \in L^1(X)$, because $f_{W^*} \in L^2(G^*)$ and hence, by the same theorem, $\int f_{W^*}(h^*) e^{-ih^*(x)}\, dh^* \in L^2(X)$. Note that $f(\chi)$ is bounded.

3. *If* $f \in L^1(X)$ *is bounded and* $f^\sim(g) = 0$ *for all* $g \in G$, *then* $f = 0$.

For, in virtue of the fact that it is bounded, $f \in L^2(X)$; and therefore, on the basis of Theorem 2 of § 25, for almost all $\chi \in X$ we have the equation

$$f(\chi) = \int e^{i\chi(g)} f^\sim(-g)\, dg = 0.$$

4. *G is everywhere dense in* G^*.

For if there were an open set $V^* \subset G^*$ that does not intersect G, then by 2. there would exist a bounded function $f \in L^1(X)$ such that $f^\sim(g) = 0$ for all $g \in G$ and at the same time $f^\sim(g_0^*) \neq 0$ for some $g_0^* \in V^*$. But from 3. it follows that $f(\chi)$ would have to be equal to zero almost everywhere and hence $f^\sim(g^*)$ would be equal to zero for all $g^* \in G^*$, in contradiction to the definition of f.

5. $G = G^*$.

For by 1., there exists a neighborhood U^* of the zero element of G^* such that its projection U onto G has a compact closure $\overline{U}$ in V. By 4., U and hence $\overline{U}$ as well, is dense in U^*. But since by 1. the embedding of G in G^* is continuous, $\overline{U}$ is compact and hence closed in G^*. Therefore $U^* \subset \overline{U}$.

Now let g_0^* be an arbitrary element of G^*. By 4., its neighborhood $g_0^* - U^*$ contains an element $g_0 \in G$. But then $g_0^* \in g_0 + U^* \subset g_0 + \overline{U} \subset G$, i.e., $g_0^* \in G$.

Hence G coincides with G^* both as far as the elements are concerned (5.) and in the topology (5. and 1.). //

§ 27. Positive-Definite Functions

***1.** The functions by means of which we constructed the positive linear functionals on $V(G)$ in §§ 22 and 24 belong to the class of the so-called *positive-definite functions,* i.e., the class of functions $\varphi(g)$ on G that are characterized by the following structural property: for every finite set of elements $g_1, \ldots, g_n$ of G and complex numbers $\xi_1, \ldots, \xi_n$ we have the inequality

$$\sum_{k=1}^{n} \sum_{l=1}^{n} \varphi(g_k - g_l)\bar{\xi}_k \xi_l \geqq 0. \tag{1}$$

For if

$$\varphi(g) = \int u(g+h)\overline{u(h)}dh, \tag{2}$$

where $u \in L^2(G)$, then, by substituting the form (1) for φ, we obtain

$$\sum_{k=1}^{n} \sum_{l=1}^{n} \varphi(g_k - g_l)\bar{\xi}_k\xi_l = \sum_{k=1}^{n} \sum_{l=1}^{n} \bar{\xi}_k\xi_l \int u(g_k - g_l + h)\overline{u(h)}\, dh =$$

$$= \sum_{k=1}^{n} \sum_{l=1}^{n} \bar{\xi}_k\xi_l \int u(g_k + h)\overline{u(g_l + h)}\, dh =$$

$$= \int \left| \sum_{k=1}^{n} \bar{\xi}_k u(g_k + h) \right|^2 dh \geqq 0.$$

Now it turns out that the functionals J_φ generated by functions of the form (2) are positive, because these functions are positive definite.

THEOREM 1: *The expression*

$$J_\varphi(\mathfrak{z}) = J_\varphi(\lambda e + x) = \lambda\varphi(0) + \int \varphi(-g)x(g)dg \tag{3}$$

is a positive linear functional on $V(G)$ for every continuous positive-definite function $\varphi(g)$.

Proof: That the functional (3) generated by the function (2) is positive was established in the proof of Theorem 1 of § 22, essentially on the basis of the following properties of this function: $\varphi(g)$ is continuous,

$$\varphi(-g) = \overline{\varphi(g)}, \tag{4}$$

$$|\varphi(g)| \leqq \varphi(0), \tag{5}$$

and

$$J_\varphi(x^* * x) = \int \varphi(-g) \left(\int x(g+h)\overline{x(h)}d(h)\right)dg =$$
$$= \int\int \varphi(g-h)\overline{x(g)}x(h)dg\,dh \geqq 0 \tag{6}$$

for all $x, y \in L^1(G)$. Thus, we only have to show that every continuous positive-definite function on G has the properties (4), (5), and (6).

Putting $n=2$, $g_1=g$, $g_2=0$, $\xi_1=1$, $\xi_2=\lambda$ in (1), we obtain the inequality

$$\varphi(0) + \varphi(g)\lambda + \varphi(-g)\bar{\lambda} + \varphi(0)\cdot|\lambda|^2 \geqq 0, \tag{7}$$

which is true for all complex values of λ. For $\lambda=0$ this gives $\varphi(0)\geqq 0$. Therefore $\varphi(g)\lambda + \varphi(-g)\bar{\lambda}$ is real for all λ; and since $\varphi(g)\lambda + \overline{\varphi(g)}\bar{\lambda}$ is also real for all λ, we find by subtraction that $[\varphi(-g) - \overline{\varphi(g)}]\bar{\lambda}$ is real for all λ, and (4) follows for $\lambda = i[\varphi(-g) - \overline{\varphi(g)}]$.

Now if $\varphi(0) > 0$, then putting $\bar{\lambda} = -\varphi(g)/\varphi(0)$ in (7), we obtain $\varphi(0) - |\varphi(g)|^2/\varphi(0) \geqq 0$, and this implies (5). But if $\varphi(0)=0$, then by putting $\bar{\lambda} = -\varphi(g)$ in (7) we obtain $-2|\varphi(g)|^2 \geqq 0$, and hence $\varphi(g)=0$; and so (5) is again true.

Finally, the inequality (6), which is an integral analogue of the inequality (1), follows from the latter by means of the following device which is due to Riesz [51].

Suppose, to begin with, that $x \in L(G)$, that K is the support of x, and that $x \neq 0$, so that $\mu = m(K) > 0$. We put $\xi_k = x(g_k)$ $(k=1, \ldots, n)$ in (1) and, treating $g_1, \ldots, g_n$ as independent variables, we take the n-fold integral of both sides of the inequality

$$\sum_{k=1}^{n}\sum_{l=1}^{n} \varphi(g_k - g_l)\overline{x(g_k)}x(g_l) \geqq 0 \tag{1'}$$

with respect to the set K. We obtain

$$n\varphi(0)\mu^{n-1}\int_K |x(g)|^2 dg +$$
$$+ n(n-1)\mu^{n-2}\int\int_{K\times K} \varphi(g-h)\overline{x(g)}x(h)dg\,dh \geqq 0.$$

Dividing by $n(n-1)\mu^{n-2}$ and passing to the limit $n\to\infty$, we arrive at the inequality

$$\int\int_{K\times K} \varphi(g-h)\overline{x(g)}x(h)dg\,dh \geqq 0;$$

i.e., bearing in mind that $x(g)=0$ outside K, we have the inequality (6), which is then proved for every function $x \in L(G)$ (for $x(g)\equiv 0$ it is trivial). And since $L(G)$ is dense in $L^1(G)$, we see, by the continuity of the functional (3), which follows from (5), that the inequality (6) extends to all the functions of $L^1(G)$. //

Thus, the absence of a radical in the group ring, the uniqueness theorem for the Fourier transform of an absolutely integrable function, and the existence of an abundant set of characters are ultimately consequences of the existence of a sufficiently large set of positive-definite functions in the groups under discussion.

***2.** Theorem 1 turns out to have a converse.

LEMMA 1: *If $\Phi(\mathrm{P})$ is a completely additive regular non-negative real function of the set P on the field of Borel sets of the space X of characters of the locally compact commutative group G, then*

$$\varphi(g)=\int_{\mathrm{X}} e^{-i\chi(g)}d\Phi(\chi) \tag{8}$$

is a continuous positive-definite function on G.

Proof: By Lemma 1 of § 23, the integral (8) exists for every $g\in G$. Further, by substituting the form (1) in (8) and bearing in mind that $\Phi(\mathrm{P})\geqq 0$, we obtain

$$\sum_{k=1}^{n}\sum_{l=1}^{n}\varphi(g_k-g_l)\,\bar{\xi}_k\xi_l=\int_{\mathrm{X}}\left(\sum_{k=1}^{n}\sum_{l=1}^{n}e^{-i\chi(g_k-g_l)}\,\bar{\xi}_k\xi_l\right)d\Phi(\chi)=$$

$$=\int_{\mathrm{X}}\left|\sum_{k=1}^{n}e^{i\chi(g_l)}\,\xi_l\right|^2 d\Phi(\chi)\geqq 0,$$

so that $\varphi(g)$ is a positive-definite function. Finally, from the regularity of $\Phi(\mathrm{P})$ there follows the existence for every $\varepsilon>0$ of a compact set $\mathrm{K}\subset\mathrm{X}$ such that $\Phi(\mathrm{X}\setminus\mathrm{K})<\varepsilon$; moreover, since by Lemma 2 of § 23 there exists a neighborhood U of the zero element of G such that

$$|e^{i\chi(h)}-1|<\varepsilon \qquad \text{for all} \qquad h\in U \text{ and } \chi\in K,$$

we have, for all $g\in G$ and $h\in U$,

$$|\varphi(g+h)-\varphi(g)|\leqq\int_{\mathrm{K}}|e^{i\chi(h)}-1|\,d\Phi(\chi)+2\Phi(\mathrm{X}\setminus\mathrm{K})<\varepsilon\,[\varphi(0)+2],$$

so that $\varphi(g)$ is continuous. //

Note: In proving that the function $\varphi(g)$ is positive definite we have actually used the fact that *every multiplicative character $e^{i\chi(g)}$ is a positive-definite function.*

THEOREM 2: *Every positive linear functional on $V(G)$ can be represented in the form*

$$f(\mathfrak{z}) = f(\lambda e + x) = \lambda\varrho + J_\varphi(\mathfrak{z}), \tag{9}$$

where φ is a uniquely determined continuous positive-definite function on G and $\varrho \geqq 0$. Moreover, $f = J_\varphi$, i.e., $\varrho = 0$, if and only if $f(e) = \lim_{U \to 0} f(e_U)$, where $\{e_U\}$ is a δ-sequence contracting to the zero element of G.

Proof: By Theorem 5′ of § 8, we have for every $\mathfrak{z} = \lambda e + x \in V(G)$

$$f(\mathfrak{z}) = \int_{\mathfrak{M}} \mathfrak{z}(M)\, d\Phi(M) = \int_{\mathfrak{M}} [\lambda + x(M)]\, d\Phi(M),$$

where Φ is a regular completely additive non-negative real set function on the field of Borel sets of the space $\mathfrak{M}$ of maximal ideals of $V(G)$. Taking Theorem 2 of § 21 into account and applying Fubini's Theorem, we obtain

$$\begin{aligned} f(\mathfrak{z}) &= \lambda\Phi(\mathfrak{M}) + \int_{\mathrm{X}} x^{\sim}(\chi)\, d\Phi(\chi) = \\ &= \lambda\Phi(\mathfrak{M}) + \int_{\mathrm{X}} \left(\int_G e^{i\chi(g)} x(g)\, dg \right) d\Phi(\chi) = \\ &= \lambda\Phi(\mathfrak{M}) + \int_G \left(\int_{\mathrm{X}} e^{i\chi(g)}\, d\Phi(\chi) \right) x(g)\, dg = \\ &= \lambda\Phi(M_\infty) + \lambda\Phi(\mathrm{X}) + \int \varphi(-g)\, x(g)\, dg, \end{aligned}$$

where $\varphi(g) = \int_{\mathrm{X}} e^{-i\chi(g)} d\Phi(\chi)$, by Lemma 1, is a continuous positive-definite function on G. Furthermore, since $\Phi(\mathrm{X}) = \varphi(0)$, we have arrived at the formula (9), where $\varrho = \Phi(M_\infty) \geqq 0$. By taking $\mathfrak{z} = e_U$, where $\{e_U\}$ is a δ-sequence contracting to the zero element of G, and by passing to the limit for $U \to 0$, we obtain

$$\lim_{U \to 0} f(e_U) = \varphi(0).$$

On the other hand, for $\mathfrak{z} = e$ we have

$$f(e) = \Phi(M_\infty) + \varphi(0).$$

Hence $\Phi(M_\infty) = 0$, i.e., $f(\mathfrak{z}) = J_\varphi(\mathfrak{z})$ if and only if $f(e) = \lim_{U \to 0} f(e_U)$. Finally, that φ is uniquely determined follows from the fact that if

$$\int \varphi(-g)x(g)dg = \int \psi(-g)x(g)dg \qquad \text{for all} \qquad x \in L^1(G)$$

and φ and ψ are continuous, then $\varphi(g) \equiv \psi(g)$. //

***3.** In the course of proving Theorem 2 we have actually obtained at the same time the converse to Lemma 1; this converse is a generalization of the well-known theorems of Herglotz [27] and Bochner [7] on the representation of positive-definite functions on the additive group of integers and the additive topological group of real numbers, respectively:

THEOREM 3 (Generalized Theorem of Bochner): *Every continuous positive-definite function $\varphi(g)$ given on a locally compact commutative group G can be represented in a unique way in the form*

$$\varphi(g) = \int_{\mathrm{X}} e^{-i\chi(g)} d\Phi(\chi), \tag{8}$$

where Φ is a regular completely additive non-negative real set function on the field of Borel sets of the space X *of characters of G.*

For we only have to apply the proof of Theorem 2 to the case where f is the functional J_φ (which is positive, by Theorem 1). However, Theorem 1 can also be proved in a more direct way: according to Theorem 1 of the present section and Theorem 5′ of § 8,

$$\int \varphi(-g)\, x(g)dg = \int_{\mathrm{X}} x^{\sim}(\chi)d\Phi(\chi)$$

(because $x(M_\infty) = 0$). Taking $x = e_{g_0+U}$, where $\{e_{g_0+U}\}$ is a δ-sequence contracting to the element $g_0 \in G$, passing to the limit $U \to 0$, and taking Theorem 1 of § 21 into account, we obtain

$$\varphi(-g_0) = \int \exp\,(i\chi(g_0))d\Phi(\chi),$$

i.e., after changing g_0 to $-g$, we have formula (8).

***4.** In conclusion, let us show that the inversion formula that establishes Theorem 1 of § 24 is true for the class P(G) of all absolutely integrable continuous positive-definite functions on G.

LEMMA 2: *If $\varphi \in \mathrm{P}(G)$, then $\varphi^{\sim}(\chi) \geqq 0$ for all $\chi \in \mathrm{X}$.*

Proof: Since $\varphi = \varphi^*$, $\varphi^{\sim}(\chi)$ is real. Suppose, contrary to the statement of the lemma, that $\varphi^{\sim}(\chi_0) < 0$. Being continuous, $\varphi^{\sim}(\chi)$ is negative in a certain neighborhood V of χ_0. By what has been established in part 4. of the proof of Theorem 1′ of § 7, there exists a function $f_0 \in L_+(\mathrm{X})$ for which $\Delta_{f_0} \subset V$ and $f_0(\chi_0) = 1$. Then $f_0(\chi)/\varphi^\sim(\chi) \leqq 0$ everywhere on X and, since I_φ is a positive functional, we have $I_\varphi(f/\varphi^\sim) \leqq 0$. But on the other hand, by Theorem 1 the construction of the invariant integral on $L(X)$ could have been carried out by using not the class P, as was done in § 24, but the wider class $\mathrm{P}(G)$. Then, by 4. in Theorem 1 of § 24,

$$I_\varphi(f_0/\varphi^\sim) = I(f_0) > 0.$$

This contradiction proves the lemma. //

LEMMA 3: *If $\varphi \in \mathrm{P}(G)$, then for every collection of elements $h_1, \ldots, h_m \in G$ and numbers $\eta_1, \ldots, \eta_m$ we also have*

$$\psi(g) = \sum_{i,j=1}^{m} \varphi(g + h_i - h_j)\bar{\eta}_i \eta_j \in \mathrm{P}(G).$$

Proof: For every collection of elements $g_1, \ldots, g_n \in G$ and numbers $\xi_1, \ldots, \xi_n$, by putting $g_k{}^i = g_k + h_i$ and $\xi_k{}^i = \xi_k \eta_i$ and bearing in mind that the function $\varphi(g)$ is positive definite, we have

$$\sum_{k,l=1}^{n} \psi(g_k - g_l)\bar{\xi}_k \xi_l = \sum_{i,j=1}^{m} \sum_{k,l=1}^{n} \varphi(g_k^i - g_l^j)\bar{\xi}_k^i \xi_l^j \geqq 0.$$

The continuity and the absolute integrability of the function $\psi(g)$ are obvious. //

THEOREM 4: *If $\varphi \in \mathrm{P}(G)$, then $\varphi^\sim \in L^1(X)$ and*

$$\varphi(g) = \int e^{-i\chi(g)} \left(\int e^{i\chi(h)} \varphi(h)\, dh \right) d\chi.$$

The proof of this theorem is an almost verbatim repetition of the proof of Theorem 1 of § 25, the only difference being that now $\varphi^\sim(\chi) \geqq 0$ by Lemma 2, and the function

$$\omega(g) = (1 + |\lambda|^2)\varphi(g) + \lambda\varphi(g + h) + \bar{\lambda}\varphi(g - h)$$

belongs to the class $\mathrm{P}(G)$ (which replaces the class P of Theorem 1 of § 25) by Lemma 3 (with $m = 2$, $h_1 = h$, $h_2 = 0$, $\eta_1 = \lambda$ and $\eta_2 = 0$). //

CHAPTER V

THE RING OF FUNCTIONS OF BOUNDED VARIATION ON A LINE

§ 28. Functions of Bounded Variation on a Line

1. Let $f(t)$ be a function of bounded variation on the line $-\infty < t < \infty$. As is well known, $f(t)$ can be represented as the difference of two non-decreasing functions of bounded variation. Therefore at every point t there exists the right-hand limit $f(t+0)$ and the left-hand limit $f(t-0)$; the limit values $f(-\infty)$ and $f(+\infty)$ also exist. The function $f(t+0)$ is obviously continuous on the right and is also of bounded variation. Since $f(t)$ can have at most a countable set of points of discontinuity, $f(t+0)$ coincides with $f(t)$ everywhere, except at most in a countable set of points. The function $f(t+0)$ has no removable points of discontinuity. All its points of discontinuity are of the first kind. From now on, when we speak of functions of bounded variation, we shall always have in mind functions that are continuous on the right, i.e., that satisfy the condition $f(t) = f(t+0)$.

2. We denote by $V^{(b)}$ the linear space of all complex functions $f(t)$ that are of bounded variation on $-\infty < t < \infty$, satisfy the condition $f(-\infty) = 0$, and are continuous on the right, with the norm

$$\| f \| = \operatorname{Var} f.$$

The space $V^{(b)}$ is complete. We can introduce in this space an operation of multiplication of its elements by defining the product of two elements f_1 and f_2 as their 'convolution,' given by the formula

$$(f_1 * f_2)(t) = \int_{-\infty}^{\infty} f_1(t-\tau)\, df_2(\tau).$$

In fact, it is not difficult to verify that *if* $f_1, f_2 \in V^{(b)}$, *then we have* $f_1 * f_2 \in V^{(b)}$ *also and* $\| f_1 * f_2 \| \leq \| f_1 \| \, \| f_2 \|$. By means of Fubini's Theorem it can be shown that *the convolution is associative and commutative.* Obviously, the function

$$\varepsilon(t) = \begin{cases} 0 & \text{for} \quad t < 0, \\ 1 & \text{for} \quad t \geqq 0 \end{cases}$$

is the unit element with respect to convolution and $\| \varepsilon \| = 1$. Thus, $V^{(b)}$ *is a commutative normed ring with a unit element.*

Let $f(t) \in V^{(b)}$. The function

$$h(t) = \sum_{\lambda \leqq t} \omega(\lambda),$$

where we have for brevity put $f(\lambda) - f(\lambda - 0) = \omega(\lambda)$ and where the sum is extended over all the points of discontinuity λ of $f(t)$ that are not to the right of t, is called the *jump function*, or the *discrete part*, of the function $f(t)$.[1] We have

$$\operatorname{Var} h = \sum_{\lambda = -\infty}^{\infty} | f(\lambda) - f(\lambda - 0) | \leqq \operatorname{Var} f < \infty.$$

The difference $f(t) - h(t) = c(t)$ is called the *continuous part* of $f(t)$. We can separate out the *singular part* of $c(t)$, i.e., a continuous summand $s(t)$ whose total variation is concentrated in a set of measure 0 such that the difference $c(t) - s(t) = g(t)$ is even *absolutely continuous*. This means that for every $\varepsilon > 0$ there exists a $\delta > 0$ such that $\sum | g(t_{k+1}) - g(t_k) | < \varepsilon$ for every finite system of non-overlapping intervals (t_k, t_{k+1}) for which the sum of the lengths is less than δ; $s(t)$ is a continuous function of bounded variation having almost everywhere a derivative equal to zero. Thus, every function $f(t) \in V^{(b)}$ can be represented, in one and only one way, in the form of a sum

$$f(t) = g(t) + h(t) + s(t),$$

where $g(t)$ is absolutely continuous, $h(t)$ is a jump function, and $s(t)$ is a singular function. Moreover,

$$\| f \| = \| g \| + \| h \| + \| s \|.$$

Obviously, one or more of these components may vanish, i.e., they may in fact be absent.

3. The absolutely continuous function $g(t) \in V^{(b)}$ is characterized by the following properties: It has a derivative $g'(t)$ almost everywhere; $g'(t)$ is

[1] If $f(t)$ is continuous, so that the set of terms of this sum is empty, then, by a generally accepted convention, the sum must be taken to be equal to zero, i.e., the jump function of a continuous function of bounded variation is identically equal to zero.

absolutely integrable and $g(t)$ is its integral, i.e., $g(t) = \int_{-\infty}^{t} g'(\tau)d\tau$. It follows from this that $\operatorname{Var} g = \int_{-\infty}^{\infty} | g'(t) | dt$. Further, to the convolution of absolutely continuous functions there corresponds the convolution of their derivatives as functions of V, i.e., if $g(t) = \int_{-\infty}^{\infty} g_1(t-\tau)dg_2(\tau)$, then $g'(t) = \int_{-\infty}^{\infty} g_1'(t-\tau)g_2'(\tau)d\tau$. All this shows that by assigning to every element $\lambda e + x(t) \in V$ the function $\lambda\varepsilon(t) + \int_{-\infty}^{t} x(\tau)d\tau$ we obtain an isometric and isomorphic embedding of the ring V into the ring $V^{(b)}$; in other words, *V is a subring of $V^{(b)}$.*

§ 29. The Ring of Jump Functions

1. Just as absolutely continuous set functions have a natural connection with absolutely integrable functions, so jump functions are connected with 'absolutely summable' functions.[2]

A complex function $x(\lambda)$ defined on an infinite set Λ is called *absolutely summable* if it is different from zero on an at most countable point set and

$$\| x \| = \sum_{\lambda \in \Lambda} | x(\lambda) | < \infty. \tag{1}$$

The aggregate of all absolutely summable functions on Λ forms a vector space under the usual addition and multiplication by complex numbers, and the formula (1) defines a norm on it that turns it into a Banach space; we shall denote this space by $l^1(\Lambda)$.

We now consider $l^1(\Gamma)$, where Γ is the discrete additive group of all real numbers. There is a natural way of introducing in $l^1(\Gamma)$ an operation of multiplication of elements, where the product of two elements x and y is defined as their 'convolution,' given by the formula

$$(x * y)(\lambda) = \sum_{\mu} x(\lambda - \mu)y(\mu). \tag{2}$$

The series on the right-hand side is majorized by $\sum_{\mu} \| x \| \, | y(\mu) |$ and is

[2] See § 20. In order to make the present chapter independent of the preceding one, we again present information on the group ring of the discrete additive group of real numbers that is contained as a special case among the results obtained in Chapter IV for an arbitrary discrete commutative group.

therefore absolutely convergent, so that the function $z(\lambda)=(x*y)(\lambda)$ is defined for all $\lambda \in \Gamma$. Further, the sum of the series (2) can differ from zero only at points λ of the form $\lambda=\lambda_1+\lambda_2$, where $x(\lambda_1)\neq 0$ and $y(\lambda_2)\neq 0$ (because otherwise all the terms of the series are equal to zero); thus, $z(\lambda)$ differs from zero at not more than a countable set of points. Finally,

$$\sum_{\lambda}|z(\lambda)|=\sum_{\lambda}\left|\sum_{\mu}x(\lambda-\mu)\,y(\mu)\right|\leqq\sum_{\lambda}\sum_{\mu}|x(\lambda-\mu)|\,|y(\mu)|=$$

$$=\sum_{\mu}\Big(\sum_{\lambda}|x(\lambda-\mu)|\Big)|y(\mu)|=\|x\|\,\|y\|;$$

therefore, for arbitrary $x,\ y\in l^1(\Gamma)$ we also have $x*y\in l^1(\Gamma)$ and

$$\|x*y\|\leqq\|x\|\,\|y\|.$$

Obviously, the function

$$e(\lambda)=\begin{cases}1 & \text{for} \quad \lambda=0,\\ 0 & \text{for} \quad \lambda\neq 0\end{cases}$$

is the unit element for the multiplication defined by the formula (2); we have $\|e\|=1$. It is easy to verify that the multiplication (2) is associative and commutative. Thus, $l^1(\Gamma)$ is a commutative normed ring with unit element for this multiplication.

2. Let $h(\lambda)\in V^{(b)}$ be a jump function. We put

$$x_h(\lambda)=h(\lambda)-h(\lambda-0).$$

Obviously, $x_h(\lambda)\in l^1(\Gamma)$ and $\operatorname{Var} h=\|x_h\|$. In this way, to every function $x(\lambda)\in l^1(\Gamma)$ there corresponds a uniquely defined jump function $h(\lambda)$ such that $x(\lambda)=x_h(\lambda)$, namely the function $h(\lambda)=\sum_{\mu\leqq\lambda}x(\mu)$. Finally, for two arbitrary jump functions h_1 and h_2 we have

$$(h_1*h_2)(\lambda)=\int_{-\infty}^{\infty}h_1(\lambda-\mu)\,dh_2(\mu)=\sum_{\mu}h_1(\lambda-\mu)\,x_{h_2}(\mu),$$

from which it follows that h_1*h_2 is also a jump function and that

$$x_{h_1*h_2}=x_{h_1}*x_{h_2}.$$

Thus, we see that *the set H of all jump functions forms a subring of the normed ring $V^{(b)}$ that is isometrically isomorphic to the ring $l^1(\Gamma)$ of absolutely summable functions on the discrete additive group of real numbers.*

Because of this, in what follows we shall often identify the rings H and

$l^1(\Gamma)$, always implying that the identification is effected by the correspondence $h \to x_h$.

The ring $l^1(\Gamma)$ has a natural involution (§ 8)[3]

$$x(\lambda) \to x^*(\lambda) = \overline{x(-\lambda)}. \tag{3}$$

We consider the functional $f(x) = x(0)$ on $l^1(\Gamma)$. Since

$$(x * x^*)(\lambda) = \sum_{\mu} x(\lambda - \mu)\, \overline{x(-\mu)},$$

we have

$$f(x * x^*) = \sum_{\mu} | x(\mu) |^2,$$

so that $f(x * x^*) > 0$ for all non-zero $x \in l^1(\Gamma)$, i.e., f is a positive linear functional on $l^1(\Gamma)$ and the involution (3) is essential. Applying Theorem 4 of § 8, we conclude that $l^1(\Gamma)$, and hence H also, is a *ring without radical.*

3. With every function $x(\lambda) \in l^1(\Gamma)$ we can associate the *Fourier series* $\sum_{\lambda} x(\lambda) e^{i\lambda s}$, which obviously is absolutely and uniformly convergent on the entire real line $-\infty < s < \infty$. The correspondence

$$x(\lambda) \to \sum_{\lambda} x(\lambda)\, e^{i\lambda s} \tag{4}$$

is linear. It is not difficult to verify that it is one to one, i.e., (taking linearity into account) that *if* $x(\lambda) \in l^1(\Gamma)$ *and* $f(s) = \sum_{\lambda} x(\lambda) e^{i\lambda s} \equiv 0$, *then* $x = 0$. Indeed, this follows immediately from the formula

$$\lim_{T \to \infty} \frac{1}{2T} \int_{-T}^{T} |f(s)|^2\, ds = \sum_{\lambda} | x(\lambda) |^2, \tag{5}$$

whose validity can simply be established by a direct calculation (independent of the general theory of almost periodic functions).

The convolution of absolutely summable functions corresponds to the multiplication of their Fourier series. For

$$\sum_{\lambda} \Big(\sum_{\mu} x(\lambda - \mu)\, y(\mu) \Big) e^{i\lambda s} = \sum_{\mu} \Big(\sum_{\lambda} x(\lambda - \mu)\, e^{i\lambda s} \Big) y(\mu) =$$

$$= \sum_{\mu} \Big(\sum_{\lambda} x(\lambda - \mu)\, e^{i(\lambda - \mu)s} \Big) e^{i\mu s} y(\mu) = \sum_{\lambda} x(\lambda)\, e^{i\lambda s} \sum_{\mu} y(\mu)\, e^{i\mu s}. \tag{6}$$

[3] Induced by the involution $f(t) \to f^*(t) = \bar{f}(\infty) - \bar{f}(-t-0)$ of $V^{(b)}$.

Thus, *the set* $B^{(a)}$ *of all the functions on the line that are almost periodic in the sense of Bohr and that can be expanded in absolutely convergent Fourier series, furnished with the norm*

$$\left\| \sum_{\lambda} x(\lambda) e^{i\lambda s} \right\| = \sum_{\lambda} |x(\lambda)|,$$

forms a normed ring under the usual algebraic operations on functions, and this ring is isometrically isomorphic to the ring H.

At the same time, equation (6) shows that the correspondence (4), for every fixed value of s, is a homomorphism of the ring $H = l^1(\Gamma)$ into the field of complex numbers and also that every real number s determines a maximal ideal of H. These maximal ideals M_s, however, by no means exhaust the space $\mathfrak{M}(H)$ of maximal ideals of H.

4. A *character* of Γ is any homomorphic mapping of Γ into the multiplicative group of complex numbers of absolute value one, i.e., it is a complex function $\chi(\lambda)$ of the real variable λ such that

$$|\chi(\lambda)| = 1 \text{ and } \chi(\lambda + \mu) = \chi(\lambda)\chi(\mu) \text{ for all real } \lambda \text{ and } \mu.$$

Taking $\mu = 0$, we see that $\chi(0) = 1$. Now taking $\mu = -\lambda$, we find that $\chi(-\lambda) = 1/\chi(\lambda) = \overline{\chi(\lambda)}$. Obviously, the function $\chi(\lambda) = e^{is\lambda}$, for every fixed real s, is a character of Γ. It is not difficult to show that these characters exhaust all the characters of Γ that are continuous in the usual topology of the real line. On the other hand, with the help of Zermelo's axiom it is easy to construct discontinuous characters [24]; they are all non-measurable.

If χ_1 and χ_2 are characters of Γ, then $\chi_1\chi_2^{-1}$ is also a character of Γ. Thus, the characters of Γ form a group under the usual multiplication. This group X, furnished with the topology in which all possible sets of the form

$$\{\chi \in X\colon \ |\chi(\lambda_k) - \chi_0(\lambda_k)| < \varepsilon \qquad (k = 1, \ldots, n)\}$$

form a fundamental system of neighborhoods of the point $\chi_0 \in X$, is called the *group of characters of* Γ.

5. THEOREM 1: *The space $\mathfrak{M}(H)$ of maximal ideals of the ring $H = l^1(\Gamma)$ and the group X of characters of Γ can be put into a one-to-one correspondence under which the element $x \in l^1(\Gamma)$ assumes, on the maximal ideal M_χ corresponding to the character χ, the value*

$$x(M_\chi) = \sum_{\lambda} x(\lambda)\chi(\lambda)$$

or, what is the same, the element $h \in H$ assumes the value

$$h(M_\chi) = \int_{-\infty}^{\infty} \chi(t) dh(t).$$

This correspondence is a homeomorphism of the spaces $\mathfrak{M}(H)$ *and* X *such that the group* X *is compact.*

Proof: Let M be a maximal ideal of $l^1(\Gamma)$. We put $e_\lambda(t) = e(t-\lambda)$. Then

$$\chi_M(\lambda) = e_\lambda(M) \tag{7}$$

is a character of Γ. For since $e_\lambda * e_\mu = e_{\lambda+\mu}$, we have

$$\chi_M(\lambda+\mu) = e_{\lambda+\mu}(M) = e_\lambda(M)\, e_\mu(M) = \chi_M(\lambda)\, \chi_M(\mu); \tag{8}$$

further, $|\chi_M(\lambda)| \leqq \| e_\lambda \| = 1$ and $\chi_M(0) = e(M) = 1$, from which it follows by (8) that $|\chi_M(\lambda)| \equiv 1$. Since every element $x \in l^1(\Gamma)$ can be represented in the form $x = \sum_\lambda x(\lambda) e_\lambda$, where the series converges in the norm of $l^1(\Gamma)$, it follows from (7) that

$$x(M) = \sum_\lambda x(\lambda) \chi_M(\lambda). \tag{9}$$

On the other hand, by assigning to every element $x \in l^1(\Gamma)$ the number $\sum_\lambda x(\lambda)\chi(\lambda)$, where χ is a given character of Γ, we obtain a homomorphic mapping of $l^1(\Gamma)$ into the field of complex numbers. Indeed, all that needs to be verified is that the product of elements of the ring goes over into the product of the numbers corresponding to them; but for this purpose we only have to replace $e^{i\lambda s}$ in the formulas (6) by the arbitrary character $\chi(\lambda)$. The kernel M_χ of this homomorphic mapping (which is not zero, because e goes over into 1) is a maximal ideal of $l^1(\Gamma)$. Moreover

$$x(M_\chi) = \sum_\lambda x(\lambda)\chi(\lambda)$$

and, in particular,

$$e_\lambda(M_\chi) = \chi(\lambda). \tag{10}$$

A comparison of the formulas (7) and (10) shows that the maximal ideal M_χ generated by the character χ in turn generates this character: $\chi = \chi_{M_\chi}$. From (9) it follows that distinct maximal ideals generate distinct characters: if $\chi_{M_1}(\lambda) \equiv \chi_{M_2}(\lambda)$, then by (9) $x(M_1) = x(M_2)$ for all $x \in l^1(\Gamma)$, i.e., M_1 and M_2 coincide. From (10) it follows that distinct characters generate distinct maximal ideals: if $M_{\chi_1} = M_{\chi_2}$ then, by (10), $\chi_1(\lambda) = \chi_2(\lambda)$ for all $\lambda \in \Gamma$, i.e., $\chi_1 = \chi_2$.

Finally, the functions e_λ $(\lambda \neq 0)$ are generators of the ring $l^1(\Gamma)$ and therefore, according to Theorem 3 of § 5, the sets of the type

$$\{M \in \mathfrak{M}(H): \quad |e_{\lambda_k}(M) - e_{\lambda_k}(M^0)| = \\ = |\chi_M(\lambda_k) - \chi_{M^0}(\lambda)| < \varepsilon \quad (k = 1, \ldots, n)\} \tag{11}$$

form a fundamental system of neighborhoods of the point $M^0 \in \mathfrak{M}(H)$, so that in virtue of the correspondence that has been established between the maximal ideals of H and the characters of Γ the spaces $\mathfrak{M}(H)$ and X are homeomorphic. //

COROLLARY: *The ring $H = l^1(\Gamma)$ is symmetric.*

For, by formula (9),

$$x^*(M) = \sum_\lambda \overline{x(-\lambda)}\, \chi_M(\lambda) = \sum_\lambda \overline{x(\lambda)}\, \chi_M(-\lambda) = \\ = \overline{\sum_\lambda x(\lambda)\, \chi_M(\lambda)} = \overline{x(M)}.$$

6. THEOREM 2: *The set of maximal ideals M_s of the ring $H = l^1(\Gamma)$ corresponding to the continuous characters $\chi(\lambda) = e^{is\lambda}$ $(-\infty < s < \infty)$ is everywhere dense in the space $\mathfrak{M}(H)$.*

Proof: Let M^0 be an arbitrary maximal ideal of the ring $H = l^1(\Gamma)$ and $U(M^0)$ an arbitrary neighborhood of M^0; we have to show that there exists a maximal ideal $M_{s_0} \in U(M^0)$. By what has been shown at the end of the proof of Theorem 1, $U(M^0)$ contains a neighborhood $U'(M^0)$ of the form (11). By mathematical induction we can select from the numbers $\lambda_1, \ldots, \lambda_n$ numbers $\lambda_{k_1}, \ldots, \lambda_{k_m}$ such that (a) the λ_{k_j} $(j = 1, \ldots, \mu)$ are linearly independent over the field of rational numbers, i.e. that the equation $\sum_{j=1}^{m} r_j \lambda_{k_j} = 0$ in which the r_j are rational numbers implies that all the $r_j = 0$, and (b) every λ_k can be expressed as a linear combination of the λ_{k_j} with rational coefficients. Let A be the least common multiple of the denominators of the coefficients of all these expressions. The numbers λ_k can now be expressed in terms of the numbers $\mu_j = \lambda_{k_j}/A$ in the form of linear combinations with integer coefficients. The numbers μ_j, like the λ_{k_j}, are linearly independent. Since the functions e_{λ_k} are products of the functions e_{μ_j}, the neighborhood $U'(M^0)$ contains a neighborhood $U''(M^0)$ of the form

$$\{M \in \mathfrak{M}(H): \quad |e_{\mu_j}(M) - e_{\mu_j}(M^0)| = \\ = |\chi_M(\mu_j) - \chi_{M^0}(\mu_j)| < \delta \quad (j = 1, \ldots, m)\}.$$

Let $\chi_{M^0}(\mu_j) = e^{2\pi i a_j}$ $(j = 1, \ldots, m)$ and let $\mu > 0$ be such that

$$|e^{it_1}-e^{it_2}|<\delta \text{ for } |t_1-t_2|<\eta.$$

By a well-known theorem of Kronecker there exists a real t_0 and integers $p_1, \ldots, p_m$ such that

$$|a_j-t_0\mu_j-p_j|<\frac{\eta}{2\pi}, \qquad \text{i.e.,} \qquad |(2\pi a_j-2\pi p_j)-2\pi t_0\mu_j|<\eta$$

for all $j=1, \ldots, m$. Then

$$|e^{2\pi i t_0\mu_j}-e^{2\pi i a_j}|=|e^{2\pi i a_j-2\pi i p_j}-e^{2\pi i t_0\mu_j}|<\delta \ (j=1, \ldots, m),$$

i.e.,

$$|e^{2\pi i t_0\mu_j}-\chi_{M^0}(\mu_j)|<\delta \qquad (j=1, \ldots, m).$$

But this means that $U''(M^0)$, and hence also $U(M^0)$, contains the maximal ideal M_{s_0}, where $s_0=2\pi t_0$. //

***7.** Let us indicate a proof that does not depend on the theorem of Kronecker.

Proof: We shall show that

$$\left|\sum_\lambda x(\lambda)\chi(\lambda)\right| \leqq \sup_s\left|\sum_\lambda x(\lambda)e^{i\lambda s}\right| \tag{12}$$

for all $x\in l^1(\Gamma)$ and $\chi\in X$.

Suppose, to begin with, that x is 'finite,' i.e., that $x(\lambda)$ differs from zero only at a finite number of points $\lambda=\lambda_k$ ($k=1, \ldots, n$), so that

$$\sum_\lambda x(\lambda)\chi(\lambda)=\sum_{k=1}^n x(\lambda_k)\chi(\lambda_k)=\sum_{k=1}^n c_k\chi(\lambda_k).$$

By Cauchy's Inequality and formula (5),

$$\left|\sum_{k=1}^n c_k\chi(\lambda_k)\right|^2 \leqq \sum_{k=1}^n |c_k|^2 \sum_{k=1}^n |\chi(\lambda_k)|^2 = n\sum_{k=1}^n |c_k|^2 =$$
$$= n \lim_{T\to\infty}\frac{1}{2T}\int_{-T}^{T}\left|\sum_{k=1}^n c_k e^{i\lambda_k s}\right|^2 ds,$$

and hence

$$\left|\sum_{k=1}^n c_k\chi(\lambda_k)\right| \leqq \sqrt{n}\sup_s\left|\sum_{k=1}^n c_k e^{i\lambda_k s}\right|.$$

We now apply this inequality, which is valid for every finite function of $l^1(\Gamma)$, to the N-th iterate x^N of our function x. Since x is different from zero at n points, a simple calculation shows that x^N can differ from zero at not more than $\binom{N+n-1}{N}$ points (this bound is attained when the numbers λ_k are linearly independent). Therefore

$$\left|\left(\sum_{k=1}^{n} c_k \chi(\lambda_k)\right)^N\right| \leqq \sqrt{\binom{N+n-1}{N}} \sup_s \left|\sum_{k=1}^{n} c_k e^{i\lambda_k s}\right|^N,$$

i.e.,

$$\left|\sum_{k=1}^{n} c_k \chi(\lambda_k)\right| \leqq \binom{N+n-1}{N}^{\frac{1}{2N}} \sup_s \left|\sum_{k=1}^{n} c_k e^{i\lambda_k s}\right|. \tag{13}$$

But $\binom{N+n-1}{N}^{\frac{1}{2N}} \to 1$ as $N \to \infty$. Thus, by passing to the limit $N \to \infty$ in (13), we find that the inequality (12) holds for all finite functions of $l^1(\Gamma)$. And since the latter are dense in $l^1(\Gamma)$ and the functions on both sides of the inequality (12) are continuous, we conclude that (12) holds for all $x \in l^1(\Gamma)$.

The validity of Theorem 2 now follows easily from (12), in virtue of the symmetry of the ring $l^1(\Gamma)$. For if there were to exist a maximal ideal $M^0 \in \mathfrak{M}(H)$ having a neighborhood $U(M^0)$ that does not intersect the set $\{M_s\}$, then by what has been established under 4. in the proof of Theorem 1′ of § 7, $\mathfrak{M}(H)$ would contain a continuous function $\varphi(M)$ that is equal to 1 at M^0 and to 0 on the whole of $\{M_s\}$ and $l^1(\Gamma)$ would contain an element x such that $|x(M) - \varphi(M)| < 1/2$ for all $M \in \mathfrak{M}(H)$, so that

$$|x(M_s)| = \left|\sum_{\lambda} x(\lambda) e^{i\lambda s}\right| < \frac{1}{2}$$

for all s and

$$|x(M^0)| = \left|\sum_{\lambda} x(\lambda) \chi_{M^0}(\lambda)\right| > \frac{1}{2}.$$

But this would contradict the inequality (12) applied to $\chi = \chi_{M^0}$. //

***8.** We observe that the topology induced on the line $-\infty < s < \infty$ by the space $\mathfrak{M}(H)$ by identification of the maximal ideals M_s with the corresponding points s is different from the usual topology of the line. Indeed, the defining neighborhoods have the form

$$\left\{s\colon \left|e^{i\lambda_k s} - e^{i\lambda_k s_0}\right| < \varepsilon \qquad (k = 1, \ldots, n)\right\};$$

each of these inequalities distinguishes on the line a certain periodic system of intervals, and the intersections of n such systems gives an 'almost periodic' system of intervals consisting of an infinite set of intervals of different lengths that are spread over every sufficiently large segment of the line and the ratio of the total length of these 'almost periodic' intervals within the segment to the length of the segment is almost the same for every such sufficiently large segment.

9. COROLLARY 1: *If $f(s)$ is an almost periodic function with an absolutely convergent Fourier series and* $\inf_{-\infty<s<\infty} |f(s)| > 0$, *then $1/f(s)$ is also an almost periodic function with an absolutely convergent Fourier series.*

In virtue of the isomorphism established above between the rings $B^{(a)}$ and $l^1(\Gamma)$, the proof of the statement follows immediately from the fact that a continuous function the lower bound of the absolute values of which on an everywhere dense set is positive cannot vanish anywhere. //

COROLLARY 2: *If $f(s)$ is an almost periodic function with an absolutely convergent Fourier series and $F(\xi)$ is an analytic function, regular on the closure of the set of values of $f(s)$, then $F(f(s))$ is also an almost periodic function with an absolutely convergent Fourier series.*

***10.** The next theorem also follows from the results of this section.

THEOREM 3: *The ring B of all continuous almost periodic functions on the line is isometric and isomorphic to the ring $C(\mathrm{X})$ of all continuous functions on the group X of characters of the discrete additive group Γ of real numbers.*

Proof: Let $f(\chi) \in C(\mathrm{X})$ and let $f(s)$ be the restriction of $f(\chi)$ to the set of continuous characters $\{e^{is\lambda}\}$, identified with the corresponding points s of the real line. By the Corollary to Theorem 1, the ring $l^1(\Gamma)$ is symmetric; therefore, by Theorem 1 of § 7 and Theorem 1 of the present section, the functions (9) are dense in $C(\mathrm{X})$. Hence $f(s)$ is the uniform limit of their restrictions to the real line—i.e., of almost periodic functions with absolutely convergent Fourier series—and thus $f(s) \in B$. Since the real line, by Theorem 2, is dense in the space X, the correspondence $f(\chi) \to f(s)$ is one to one. Further, every function $f(s) \in B$ is the restriction of some function $f(\chi) \in C(\mathrm{X})$. For, as is well known, $f(s)$ is the limit of a sequence of trigonometric polynomials $P_n(s) = \sum_{k=1}^{k_n} c_k^{(n)} \exp(i\lambda_k s)$, uniformly convergent on the whole real line; but then, by Theorem 2, the corresponding polynomials

$$P_n(\chi) = \sum_{k=1}^{k_n} c_k^{(n)} \chi(\lambda_k)$$

converge uniformly on X to a function $f(\chi) \in C(\mathrm{X})$ whose restriction to the real line is $f(s)$. Finally, again by Theorem 2,

$$\sup_{-\infty<s<\infty} |f(s)| = \max_{\chi\in\mathrm{X}} |f(\chi)|,$$

so that the correspondence $f(s) \to f(\chi)$ is isometric. //

§ 30. Absolutely Continuous and Discrete Maximal Ideals of the Ring $V^{(b)}$

1. We now proceed to the discussion of maximal ideals of the ring $V^{(b)}$. In this section we shall consider two very simple classes of such ideals.

Let $f(t) \in V^{(b)}$. The integral

$$F(s) = \int_{-\infty}^{\infty} e^{ist}\, df(t),$$

which obviously exists for all real values of s, is called the *Fourier-Stieltjes transform* of the function $f(t)$. It is clear that the addition of functions belonging to $V^{(b)}$ and the multiplication of such functions by scalars correspond to the same operations on their Fourier-Stieltjes transforms. The convolution of functions of $V^{(b)}$ corresponds to the multiplication of their Fourier-Stieltjes transforms. In fact,

$$\int_{-\infty}^{\infty} e^{ist} d\int_{-\infty}^{\infty} f_1(t-u)\, df_2(u) = \int_{-\infty}^{\infty}\left(\int_{-\infty}^{\infty} e^{ist}\, df_1(t-u)\right) df_2(u) =$$

$$= \int_{-\infty}^{\infty}\left(\int_{-\infty}^{\infty} e^{is(t-u)}\, df_1(t-u)\right) e^{isu}\, df_2(u) = \int_{-\infty}^{\infty} e^{ist}\, df_1(t) \int_{-\infty}^{\infty} e^{isu}\, df_2(u).$$

Thus, by assigning to every function $f(t) \in V^{(b)}$ the value of its Fourier-Stieltjes transform at an arbitrary fixed point s_0, we obtain a homomorphic mapping of the ring $V^{(b)}$ into the field of complex numbers. The maximal ideal generated by this mapping will be denoted by M_{s_0}. Thus,

$$f(M_{s_0}) = \int_{-\infty}^{\infty} e^{is_0t}\, df(t) = F(s_0).$$

If $s_1 \neq s_2$, then $M_{s_1} \neq M_{s_2}$, because we can find a value $t = t_0$ such that $e^{is_1t_0} \neq e^{is_2t_0}$, and then $f(M_{s_1}) \neq f(M_{s_2})$ for $f(t) = \varepsilon(t - t_0)$.

We shall call the maximal ideals M_s $(-\infty < s < \infty)$ *absolutely continuous.*

Let us show that *every function $f(t) \in V^{(b)}$ is uniquely determined by its values on the absolutely continuous maximal ideals, i.e., by its Fourier-Stieltjes transform $F(s)$.* It obviously follows from this, in particular, that *$V^{(b)}$ is a ring without radical.*

LEMMA: *The set G of all absolutely continuous functions $g(t) \in V^{(b)}$ is an ideal of $V^{(b)}$.*

Proof: Let $g(t) \in G$ and let $f(t)$ be an arbitrary function of $V^{(b)}$, different from zero. We have to show that $f * g \in G$. Choose any $\varepsilon > 0$ and let $\delta > 0$ be such that

$$\sum |g(t_{k+1}) - g(t_k)| < \frac{\varepsilon}{\|f\|}$$

for every finite system of disjoint intervals (t_k, t_{k+1}) the sum of the lengths of which is less than δ. Then

$$\sum |(f*g)(t_{k+1}) - (f*g)(t_k)| =$$

$$= \sum \left| \int_{-\infty}^{\infty} [g(t_{k+1} - \tau) - g(t_k - \tau)] \, df(\tau) \right| \leqq$$

$$\leqq \int_{-\infty}^{\infty} \sum |g(t_{k+1} - \tau) - g(t_k - \tau)| \, |df(\tau)| < \frac{\varepsilon}{\|f\|} \operatorname{Var} f = \varepsilon,$$

and this proves the lemma. //

Now let $f(t) \in V^{(b)}$ and $f(M_s) \equiv F(s) = 0$ for all s. We put

$$g_h(t) = \begin{cases} 0 & \text{for} \quad t < -h, \\ 1 + t/h & \text{for} \quad -h \leqq t \leqq 0, \\ 1 & \text{for} \quad t > 0; \end{cases}$$

$g_h(t)$ is absolutely continuous. Therefore, by the lemma, the function

$$(f * g_h)(t) = \frac{1}{h} \int_0^h f(t + \tau) \, d\tau$$

is also absolutely continuous and hence belongs to V. Furthermore, by assumption, $(f * g_h)(M_s) = f(M_s) g_h(M_s) \equiv 0$. Bearing in mind the uniqueness theorem for the Fourier transform of an absolutely integrable function,[4] we conclude from this that $(f * g_h)(t) \equiv 0$. But since $f(t)$ is continuous on the right, we have $(f * g_h)(t) \to f(t)$ as $h \to 0$. Therefore $f(t) \equiv 0$. //

COROLLARY: *The only maximal ideals of $V^{(b)}$ that do not contain G are the absolutely continuous maximal ideals.*

For let M be a maximal ideal of $V^{(b)}$ that does not contain the entire ideal G, i.e., such that there exists an absolutely continuous function

---

[4] See, for example, E. T. Titchmarsh, *Introduction to the Theory of Fourier Integrals* Oxford, 1937, Theorem 120. (The uniqueness theorem was proved in Chapter IV (§ 22) for functions given on arbitrary commutative locally compact groups.)

$g_0(t) \in V^{(b)}$ for which $g_0(M) \neq 0$. By assigning to every element $\lambda\varepsilon(t) + g(t)$ of V the number $\lambda + g(M)$ we obtain a homomorphic mapping of this ring into the field of complex numbers, which, because $g_0(t)$ goes over into a number different from zero, is non-trivial. According to Theorem 1 of § 17, every absolutely continuous function $g(t) \in V^{(b)}$ is mapped by this correspondence into the number

$$\int_{-\infty}^{\infty} e^{is_0t}\, dg(t) = \int_{-\infty}^{\infty} e^{is_0t} g'(t)\, dt.$$

Let $f(t)$ be an arbitrary function of $V^{(b)}$. By the lemma, $g(t) = (g_0 * f)(t)$ is an absolutely continuous function. Therefore,

$$(g_0 * f)(M) = \int_{-\infty}^{\infty} e^{is_0t} d\,(g_0 * f)(t) =$$

$$= g_0(M) f(M) = \int_{-\infty}^{\infty} e^{is_0t}\, dg_0(t) \cdot f(M).$$

But, on the other hand,

$$(g_0 * f)(M) = \int_{-\infty}^{\infty} e^{is_0t}\, dg(t) = \int_{-\infty}^{\infty} e^{is_0t}\, dg_0(t) \int_{-\infty}^{\infty} e^{is_0t}\, df(t).$$

Since the first factor on the right-hand side is, by assumption, different from zero, we obtain

$$f(M) = \int_{-\infty}^{\infty} e^{is_0t}\, df(t)$$

for every $f(t) \in V^{(b)}$. But this means that $M = M_{s_0}$. //

2. Thus, *all the absolutely continuous functions of* $V^{(b)}$ *vanish on the maximal ideals that are not absolutely continuous.* Let us indicate a very simple class of such maximal ideals.

First of all, we observe that *the set C of all continuous functions $c(t) \in V^{(b)}$ is an ideal of* $V^{(b)}$. For suppose that $c(t) \in C$ and that $f(t)$ is an arbitrary function of $V^{(b)}$ different from zero. Let ε be an arbitrary fixed positive number. There exists an interval $[-A, A]$ such that the total variation of $c(t)$ outside this interval is less than $\varepsilon/2\,\|f\|$. Let $\delta > 0$ be such that $|c(t') - c(t)| < \varepsilon/2\,\|f\|$ for all points t, t' of the interval $[-A, A]$ satisfying the inequality $|t' - t| < \delta$. Then $|c(t') - c(t)| < \varepsilon/\|f\|$ for arbitrary points t, t' of the real line satisfying the same inequality. Therefore

$$|(c*f)(t') - (c*f)(t)| = \left| \int_{-\infty}^{\infty} [c(t'-\tau) - c(t-\tau)]\, df(\tau) \right| \leqq$$

$$\leqq \int_{-\infty}^{\infty} |c(t'-\tau) - c(t-\tau)|\, |df(\tau| \leqq \varepsilon \operatorname{Var} f / \| f \| = \varepsilon,$$

i.e., $(c*f)(t)$ is continuous, and this proves the above statement. Hence it follows that *the jump function of the convolution of two functions of $V^{(b)}$ is equal to the convolution of their jump functions.* For suppose that

$$f_1(t) = c_1(t) + h_1(t),\ f_2(t) = c_2(t) + h_2(t) \quad (c_1, c_2 \in C;\ h_1, h_2 \in H).$$

Then

$$(f_1 * f_2)(t) = [(c_1 * c_2)(t) + (c_1 * h_2)(t) + (c_2 * h_1)(t)] + (h_1 * h_2)(t),$$

and, by what has been proved, the expression in brackets on the right-hand side is a continuous function, whereas the last term on the right-hand side, as was proved in the preceding section, is a jump function.

Now let M be an arbitrary maximal ideal of H. By Theorem 1 of § 29, it is generated by a character χ of Γ. Let us assign to every function $f(t) \in V^{(b)}$ the number $h(M) = \sum_{\lambda} x_h(\lambda)\chi(\lambda)$, where $h(t)$ is the jump function of $f(t)$ and $x_h(\lambda) = h(\lambda) - h(\lambda - 0) = f(\lambda) - f(\lambda - 0)$. From what has just been proved it follows that we obtain a homomorphic mapping of $V^{(b)}$ into the field of complex numbers. We denote by $M_{\chi\mathcal{N}}$ the maximal ideal generated by this mapping.[5]

Obviously $c(M_{\chi\mathcal{N}}) = 0$ for every continuous function $c(t) \in V^{(b)}$ and every maximal ideal $M_{\chi\mathcal{N}}$. It is not difficult to see that, conversely, if $c(M) = 0$ for every continuous function $c(t) \in V^{(b)}$, then $M = M_{\chi\mathcal{N}}$. For the maximal ideal M in $V^{(b)}$ generates a maximal ideal of H; therefore, by Theorem 1 of § 29, there exists a character χ of Γ such that $h(M) = \sum_{\lambda} x_h(\lambda)\chi(\lambda)$ for every function $h(t) \in H$. Since for every function $f(t) = c(t) + h(t) \in V^{(b)}$ we have, by assumption, $f(M) = c(M) + h(M) = h(M)$, we see that $M = M_{\chi\mathcal{N}}$.

We shall call the maximal ideals $M_{\chi\mathcal{N}}$ the *discrete maximal ideals* of $V^{(b)}$.

For every absolutely continuous maximal ideal M_s there exists a continuous function $g(t) \in V^{(b)}$ such that $g(M_s) \neq 0$. Thus, the discrete maximal ideals differ from the absolutely continuous ones.

[5] The significance of this notation will be explained in the next section.

§ 31. Singular Maximal Ideals of the Ring $V^{(b)}$

1. The class of absolutely continuous maximal ideals and that of discrete maximal ideals by no means exhaust the set of all maximal ideals of $V^{(b)}$. We shall now describe a construction that gives both these classes (as two extreme special cases), as well as an extensive family of further classes of maximal ideals of $V^{(b)}$. But for this purpose we have to go over to a treatment of functions of bounded variation on a line as set functions.

We begin by discussing *non-decreasing* real functions $f(t) \in V^{(b)}$. Let $f(t)$ be such a function and let F be an arbitrary closed bounded set on the line $-\infty < t < \infty$. We put

$$f(F) = \inf \{[f(t_1') - f(t_1)] + [f(t_2') - f(t_2)] + \ldots + [f(t_n') - f(t_n)]\},$$

where the lower bound is taken with respect to all finite systems of non-overlapping intervals $(t_1, t_1'), (t_2, t_2'), \ldots, (t_n, t_n')$ covering the set F. For an arbitrary Borel set E of points of the line $-\infty < t < \infty$ we then put

$$f(E) = \sup f(F^E),$$

where the upper bound is taken with respect to all closed bounded sets F^E contained in E. It can be shown that $f(E)$ is a completely additive function, i.e., for every finite or countable system of pairwise non-overlapping Borel sets $E_1, E_2, \ldots$, we have the inequality

$$f(E_1 \cup E_2 \cup \ldots) = f(E_1) + f(E_2) + \ldots.$$

For the sets $E_t = (-\infty, t]$ we have $f(E_t) = f(t+0) - f(-\infty) = f(t)$.

We shall denote the total variation of the function $f(t)$ on the interval $(-\infty, t]$ (as a function of t) by $(\mathrm{Var}\, f)(t)$. Let $f(t)$ be an arbitrary function of $V^{(b)}$. Then its real and imaginary parts $\Re f(t)$ and $\Im f(t)$ are also functions of $V^{(b)}$. But every real function $\varphi(t) \in V^{(b)}$ can be represented in the form of a difference $\varphi(t) = \varphi^+(t) - \varphi^-(t)$, where $\varphi^+(t)$ and $\varphi^-(t)$ are non-decreasing functions of bounded variation, uniquely determined by $\varphi(t)$:

$$\varphi^+(t) = [(\mathrm{Var}\, f)(t) + \varphi(t)]/2,$$
$$\varphi^-(t) = [(\mathrm{Var}\, f)(t) - \varphi(t)]/2.$$

For every function $f(t) \in V^{(b)}$ we now put

$$f(E) = (\Re f)^+(E) - (\Re f)^-(E) + i(\Im f)^+(t) - i(\Im f)^-(t).$$

The equation $f(E_t) = f(t)$ shows that the function $f(t) \in V^{(b)}$ is uniquely determined by the set function $f(E)$ that it generates. The function $f(t) \in V^{(b)}$ is continuous if and only if $f(E) = 0$ for every single-point set E. $f(t) \in V^{(b)}$ is absolutely continuous if and only if $f(E) = 0$ for every Borel set E of Lebesgue measure zero. $f(t) \in V^{(b)}$ is singular if and only if it is continuous and there exists a Borel set E_0 of Lebesgue measure 0 such that $f(R) = 0$ for every Borel set E that does not intersect E_0.

The set functions $f(E)$ generated by non-negative real functions $f(t) \in V^{(b)}$ will be called *measures*.

2. To the convolution of functions of $V^{(b)}$ there corresponds the following operation on the appropriate set functions:

$$(f_1 * f_2)(E) = \int_{-\infty}^{\infty} f_1(E - u) df_2(u),$$

where the integral is to be understood in the Lebesgue-Stieltjes sense. This operation is closely connected with the operation of the arithmetical summation of sets.

By *the arithmetical sum of the sets A and B* we shall mean the set formed by all the points of the form $a + b$, where $a \in A$ and $b \in B$. To denote the arithmetical sum of sets we shall use the ordinary sign $+$. Obviously, if A and B contain the point 0, then $A \subset A + B$ and $B \subset A + B$. It is not difficult to verify that if A and B are closed bounded sets, then $A + B$ is also closed. If A and B are analytical sets, then $A + B$ is also an analytical set [69].[6] We denote the sum $A + A + \ldots + A$, where the summand A is taken n times, by $(n)A$.

LEMMA: *If A has positive measure, then $(2)A$ contains an interval.*

Proof: Obviously, it is sufficient to consider the case where the measure $m(A)$ of A is finite. Let $f_A(t)$ be the characteristic function of A; we form the function

$$\varphi(t) = \int_{-\infty}^{\infty} f_A(t - u) f_A(u) du.$$

This function is not identically zero, because

$$\int_{-\infty}^{\infty} \varphi(t) dt = \left(\int_{-\infty}^{\infty} f_A(u) du \right)^2 = [m(A)]^2,$$

and $m(A)$—the measure of A—is by assumption different from zero. Since $f_A(u)$ is bounded, $\varphi(t)$ is continuous (see, for example, the lemma of § 16). Thus, there exists an interval on which $\varphi(t)$ does not vanish. But $\varphi(t)$ can differ from zero only at the points of the set $(2)A$, as follows immediately from the formula defining it. Consequently, $(2)A$ contains an interval. //

On the basis of this lemma we shall show that there exist perfect sets F for which $(n)F$ is a set of measure 0 for every n. As an example of such a

[6] If A and B are Lebesgue-measurable, but not analytical sets, then $A + B$ may be non-measurable; see [69].

set F we can take the collection of all proper fractions (including 0) for which the digit 1 in its representation as a finite or infinite binary fraction occurs only at the a_n-th place, where $\varliminf_{n\to\infty} (a_{n+1} - a_n) = \infty$. For since the digit 1 at the $(a_n + 1)$-st place can occur only as the result of the addition of digits at the places $a_{n+1}, a_{n+2}, \ldots$, a fraction containing the digit 1 at the $(a_n + 1)$-st place can only be represented as a sum of fractions of F containing at least $a_{n+1} - a_n$ terms. Since the difference $a_{n+1} - a_n$ can be arbitrarily large, we conclude from this that the fraction whose a_N-th remainder is equal to $\sum_{n=N}^{\infty} 2^{-a_n - 1}$ cannot be contained in any of the sets $(n)F$. But the fractions with remainders of this type are everywhere dense on the line. Therefore $(n)F$ cannot contain a whole interval for any n. But if $(n)F$ were to have positive measure for a certain n, then by the lemma, the set $(2n)F$ would contain an interval and, as we have seen, this is impossible. Thus, all the $(n)F$ have measure zero, and our statement is proved.

3. Now let $\mathfrak{F}$ denote a non-empty class of sets of the type F_σ (i.e., sets that are finite or countable unions of closed bounded sets) having the following properties:

1. If $A \in \mathfrak{F}$ and A' is an arbitrary subset of A of the type F_σ, then A' is also contained in $\mathfrak{F}$.
2. If $A_1, A_2, \ldots$ is a finite or countable collection of sets contained in $\mathfrak{F}$, then $A_1 \cup A_2 \cup \ldots$ is also contained in $\mathfrak{F}$;
3. If $A \in \mathfrak{F}$ then $A - t$ is also contained in $\mathfrak{F}$ for all t $(-\infty < t < \infty)$;
4. If $A \in \mathfrak{F}$ then $(2)A$ is also contained in $\mathfrak{F}$.

Obviously, when we take the system of all subsets of type F_σ of an arbitrary set A and carry out the operations 1.-4. in all possible ways on these subsets and then, similarly, on all the new sets so obtained, we arrive at a class $\mathfrak{F}_A$ having all the properties listed.

We shall say that the function $f(E)$ is *concentrated outside* $\mathfrak{F}$ if $f(A) = 0$ for all $A \in \mathfrak{F}$; we shall say that a function f of $V^{(b)}$ is *concentrated in* $\mathfrak{F}$ if

$$(\operatorname{Var} f)(E) = \sup_{\substack{A \subset E \\ A \in \mathfrak{F}}} (\operatorname{Var} f)(A)$$

for every Borel set E.[7] If f is concentrated in a set A of $\mathfrak{F}$, i.e., $f(E) = f(E \cap A)$ for every Borel set E, then obviously f is concentrated in $\mathfrak{F}$. But the converse also holds: *if f is concentrated in $\mathfrak{F}$, then it is concentrated in a set A of F.*

[7] Since, by assumption, $\mathfrak{F}$ is not empty, it contains all the single-point sets, by the properties 1. and 3.; therefore there exists sets A of $\mathfrak{F}$ contained in E.

For if f is concentrated in $\mathfrak{F}$, then for every n there exists a $A_n \in \mathfrak{F}$ such that $(\operatorname{Var} f)(A_n) > \operatorname{Var} f - 1/n$. For $A = \bigcup_{n=1}^{\infty} A_n$ we then obtain $(\operatorname{Var} f)(A) = \operatorname{Var} f$; here $A \in \mathfrak{F}$ (property 2.). Obviously, $f(E) = f(E \cap A)$ for every Borel set E.

If f_1 is concentrated in $\mathfrak{F}$ and f_2 outside $\mathfrak{F}$, then $\operatorname{Var}(f_1 + f_2) = \operatorname{Var} f_1 + \operatorname{Var} f_2$. For by what has already been proved, f_1 is concentrated in a certain $A \in \mathfrak{F}$; obviously, f_2 is concentrated in CA. Now we have

$$\operatorname{Var}(f_1 + f_2) = (\operatorname{Var}(f_1 + f_2))(A) + (\operatorname{Var}(f_1 + f_2))(CA) =$$
$$= (\operatorname{Var} f_1)(A) + (\operatorname{Var} f_2)(CA) = \operatorname{Var} f_1 + \operatorname{Var} f_2.$$

If f_1 and f_2 are concentrated in $\mathfrak{F}$, then every linear combination $\lambda_1 f_1 + \lambda_2 f_2$ of f_1 and f_2 is also concentrated in $\mathfrak{F}$. For by what has already been proved, f_1 is concentrated in a set $A_1 \in \mathfrak{F}$ and f_2, likewise, in a set $A_2 \in \mathfrak{F}$; but then both these functions, and hence every linear combination of them, is concentrated in $A = A_1 \cup A_2$ and therefore in $\mathfrak{F}$. Obviously, *if f_1 and f_2 are concentrated outside $\mathfrak{F}$, then every linear combination $\lambda_1 f_1 + \lambda_2 f_2$ of f_1 and f_2 is also concentrated outside $\mathfrak{F}$.*

Every function f of $V^{(b)}$ can be decomposed uniquely into the sum of two terms, also in $V^{(b)}$, one of which is concentrated in $\mathfrak{F}$ and the other outside $\mathfrak{F}$. By what has just been proved, it is sufficient to show the possibility of a decomposition of this form for measures, because every function $f(E)$ in $V^{(b)}$ is a linear combination of measures. We put

$$f_{\mathfrak{F}}(E) = \sup_{\substack{A \subset E \\ A \in \mathfrak{F}}} f(A), \qquad f_{C\mathfrak{F}}(E) = f(E) - f_{\mathfrak{F}}(E).$$

It is easy to verify, using the properties 1. and 2., that $f_{\mathfrak{F}}(E)$ (and hence also $f_{C\mathfrak{F}}(E)$) is completely additive. Obviously $f(A) = f_{\mathfrak{F}}(A)$, and consequently $f_{C\mathfrak{F}}(A) = 0$ for every $A \in \mathfrak{F}$. Thus, $f_{C\mathfrak{F}}$ is concentrated outside $\mathfrak{F}$. Furthermore, $f_{\mathfrak{F}}(E) = \sup_{\substack{A \subset E \\ A \in \mathfrak{F}}} f(A) = \sup_{\substack{A \subset E \\ A \in \mathfrak{F}}} f_{\mathfrak{F}}(A)$, i.e., $f_{\mathfrak{F}}$ is concentrated in $\mathfrak{F}$. Suppose now that $f = f_1 + f_2 = f_3 + f_4$, where f_1 and f_3 are concentrated in $\mathfrak{F}$ and f_2 and f_4 outside $\mathfrak{F}$. Then

$$0 = (f_1 - f_3) + (f_2 - f_4),$$

and since $f_1 - f_3$ is concentrated in $\mathfrak{F}$ and $f_2 - f_4$ outside $\mathfrak{F}$, we have, by what was proved earlier,

$$0 = \operatorname{Var}(f_1 - f_3) + \operatorname{Var}(f_2 - f_4),$$

from which it follows that $f_3 = f_1$ and $f_4 = f_2$, i.e., that the decomposition of f is unique.

4. *The set $V_{C\mathfrak{F}}$ of all functions of $V^{(b)}$ concentrated outside $\mathfrak{F}$ is an ideal of $V^{(b)}$.* For, as we have shown, $V_{C\mathfrak{F}}$ is a linear system. Suppose now that $f \in V_{C\mathfrak{F}}$ and A is an arbitrary set of $\mathfrak{F}$; then $A - t$ also belongs to $\mathfrak{F}$ for every t (property 3.); therefore $f(A-t) \equiv 0$ and hence

$$\int_{-\infty}^{\infty} f(A-t)df_1(t) = 0,$$

i.e., $f * f_1 \in V_{C\mathfrak{F}}$ for an arbitrary function $f_1 \in V^{(b)}$.

The set $V_{\mathfrak{F}}$ of all functions of $V^{(b)}$ concentrated in $\mathfrak{F}$ is a subring of $V^{(b)}$. For, as we have shown, $V_{\mathfrak{F}}$ is a linear system. It remains to show that if f_1 and $f_2 \in V_{\mathfrak{F}}$, then $f_1 * f_2 \in V_{\mathfrak{F}}$. Obviously, it is sufficient for this purpose to confine ourselves to the case where $f_1(E)$ and $f_2(E)$ are measures. As we have seen above in the proof of the linearity of the system $V_{\mathfrak{F}}$, the functions f_1 and f_2 are concentrated on a set $A \in \mathfrak{F}$; we can obviously assume here that A contains the point 0 and that therefore $\lim\limits_{n \to \infty} (2^n)A = X$ exists. X is the sum of a countable number of closed sets and, by the properties 4. and 2., belongs to $\mathfrak{F}$. We shall show that the function $f_1 * f_2$ is concentrated in X. We have

$$\int_{-\infty}^{\infty} f_1(E-t)df_2(t) = \int_X f_1(E-t)df_2(t) =$$
$$= \int_X f_1((E \cap X)-t)df_2(t) + \int_X f_1((E \cap CX)-t)df_2(t).$$

But $X = (2)X$; therefore the set $(E \cap CX) - t$ is entirely contained in CX for $t \in X$, because otherwise the sets $E \cap CX \subset CX$ and $X + t \subset X$ would intersect, which is impossible. Consequently the last integral on the right-hand side is equal to zero, because $|f_1((E \cap CX)-t)| \leqq \operatorname{Var} f_1(CX) = 0$, and we obtain

$$\int_{-\infty}^{\infty} f_1(E-t)df_2(t) = \int_X f_1((E \cap X)-t)df_2(t) =$$
$$= \int_{-\infty}^{\infty} f_1((E \cap X)-t)df_2(t),$$

i.e., $f_1 * f_2$ is concentrated in X. And since $X \in \mathfrak{F}$, we have $f_1 * f_2 \in V_{\mathfrak{F}}$.

5. Now let Γ be the discrete additive group of all real numbers and $X_{\mathfrak{F}}$ the set of all characters $\chi(t)$ of Γ that are measurable with respect to every measure $f(E) \in V^{(b)}$ concentrated in $\mathfrak{F}$. $X_{\mathfrak{F}}$ in any case contains all the continuous characters e^{ist}. For every function $f \in V^{(b)}$ we form the integral $\int_{-\infty}^{\infty} \chi(t)df_{\mathfrak{F}}(t)$, where $\chi(t)$ is an arbitrary fixed character of $X_{\mathfrak{F}}$. We obtain in this way a homomorphic mapping of $V^{(b)}$ into the field of complex numbers. For $f_{1\mathfrak{F}}+f_{2\mathfrak{F}}\in V_{\mathfrak{F}}$, $f_{1C\mathfrak{F}}+f_{2C\mathfrak{F}}\in V_{C\mathfrak{F}}$, and consequently, $(f_1+f_2)_{\mathfrak{F}}=f_{1\mathfrak{F}}+f_{2\mathfrak{F}}$, and hence a sum of functions corresponds to the sum of the appropriate integrals. Furthermore,

$$f_1 * f_2 = f_{1\mathfrak{F}} * f_{2\mathfrak{F}} + (f_{1\mathfrak{F}} * f_{2C\mathfrak{F}} + f_{1C\mathfrak{F}} * f_{2\mathfrak{F}} + f_{1C\mathfrak{F}} * f_{2C\mathfrak{F}});$$

the first term on the right-hand side is a function of $V_{\mathfrak{F}}$, because $V_{\mathfrak{F}}$ is a ring; the expression in parentheses on the right-hand side is a function of $V_{C\mathfrak{F}}$, because $V_{C\mathfrak{F}}$ is an ideal of $V^{(b)}$; therefore $(f_1*f_2)_{\mathfrak{F}}=f_{1\mathfrak{F}}*f_{2\mathfrak{F}}$. But for every character $\chi(t)$ of $X_{\mathfrak{F}}$

$$\int_{-\infty}^{\infty}\chi(t)\,d\big(f_{1\mathfrak{F}}*f_{2\mathfrak{F}}\big)(t) \text{ exists}$$

$$\text{and is equal to } \int_{-\infty}^{\infty}\chi(t)\,df_{1\mathfrak{F}}(t)\int_{-\infty}^{\infty}\chi(t)\,df_{2\mathfrak{F}}(t). \tag{1}$$

It is sufficient to prove this for the case where $f_{1\mathfrak{F}}$ and $f_{2\mathfrak{F}}$ are measures. As we have seen above in the proof of the fact that $V_{\mathfrak{F}}$ is a ring, $f_{1\mathfrak{F}}$ and $f_{2\mathfrak{F}}$ are concentrated on a set X of $\mathfrak{F}$ such that $(2)X=X$. Since $\chi(t)$ is measurable on X with respect to the measures $f_{1\mathfrak{F}}$ and $f_{2\mathfrak{F}}$, the function $\chi(t+\tau)=\chi(t)\chi(\tau)$ is measurable in the topology of the square $X\times X$ with respect to the product of these measures. Furthermore,

$$\int\int_{X\times X}\chi(t+\tau)\,df_{1\mathfrak{F}}(t)\,df_{2\mathfrak{F}}(\tau)=$$

$$=\int\int_{X\times X}\chi(t)\chi(\tau)\,df_{1\mathfrak{F}}(t)\,df_{2\mathfrak{F}}(\tau)=$$

$$=\int_X\chi(t)\,df_{1\mathfrak{F}}(t)\int_X\chi(\tau)\,df_{2\mathfrak{F}}(\tau)=$$

$$=\int_{-\infty}^{\infty}\chi(t)\,df_{1\mathfrak{F}}(t)\int_{-\infty}^{\infty}\chi(\tau)\,df_{2\mathfrak{F}}(\tau). \tag{2}$$

But on the other hand, by applying Fubini's Theorem and bearing in mind that $X\subset X-\tau$ for $\tau\in X$, we obtain

$$\int\int_{X\times X}\chi(t+\tau)\,df_{1\mathfrak{F}}(t)\,df_{2\mathfrak{F}}(\tau)=$$

$$=\int_X\left(\int_X\chi(t+\tau)\,df_{1\mathfrak{F}}(t)\right)df_{2\mathfrak{F}}(\tau)=.$$

$$= \int\limits_X \Big(\int\limits_{X-\tau} \chi(t+\tau)\, df_{1\mathfrak{F}}(t) \Big) df_{2\mathfrak{F}}(\tau) =$$

$$= \int\limits_X \Big(\int\limits_X \chi(t)\, df_{1\mathfrak{F}}(t-\tau) \Big) df_{2\mathfrak{F}}(\tau) =$$

$$= \int\limits_X \chi(t) d \int\limits_X f_{1\mathfrak{F}}(t-\tau)\, df_{2\mathfrak{F}}(\tau) =$$

$$= \int\limits_{-\infty}^{\infty} \chi(t) d \int\limits_{-\infty}^{\infty} f_{1\mathfrak{F}}(t-\tau)\, df_{2\mathfrak{F}}(\tau). \qquad (3)$$

By comparing (2) and (3), we obtain the required equation (1). Thus, we have shown that the convolution of functions corresponds to the product of the appropriate integrals; the proof of our statement is now complete.

Let $M_{\chi\mathfrak{F}}$ denote the maximal ideal generated by the mapping in question. Thus,

$$f(M_{\chi\mathfrak{F}}) = \int\limits_{-\infty}^{\infty} \chi(t) df_{\mathfrak{F}}(t);$$

it is assumed here that the character $\chi(t)$ is measurable with respect to *all* the measures of $V^{(b)}$ that are concentrated in $\mathfrak{F}$.

If the class $\mathfrak{F}$ contains at least one set of positive measure, then it coincides with the collection of all sets F_σ. For suppose that $\mathfrak{F}$ contains the set A of positive measures, then $\mathfrak{F}$ also contains $(2)A$ (property 4.). But, by the lemma, $(2)A$ contains an entire interval; therefore $\mathfrak{F}$ contains an interval (property 1.). And then, by properties 1., 2., and 3., $\mathfrak{F}$ contains every set F_σ.

But in that case $f_{\mathfrak{F}}(t)$ is simply $f(t)$; the only characters of Γ that are measurable with respect to all the measures of $V^{(b)}$ concentrated in $\mathfrak{F}$ are the continuous characters e^{ist}. Thus, we obtain here as maximal ideals $M_{\chi\mathfrak{F}}$ the absolutely continuous maximal ideals M_s that we know already.

6. A contrasting case—a much narrower class $\mathfrak{F}$—is given by the class $\mathcal{N}$ of all countable sets (which obviously satisfies all the requirements we have imposed and is a subclass of every class $\mathfrak{F}$). For this class, $f_{\mathcal{N}}(t)$ is the jump function of $f(t)$; every character $\chi(t)$ of Γ is measurable on all the sets of $\mathcal{N}$. Thus, we have here as maximal ideals $M_{\chi\mathfrak{F}}$ the discrete maximal ideals we found in § 30.2 above (and that is why we denoted them by $M_{\chi\mathcal{N}}$).

7. Let us now discuss the case where all the sets of $\mathfrak{F}$ have measure 0 and $\mathfrak{F}$ also contains uncountable sets. Such a class, for example, is the class $\mathfrak{F}_F$ constructed in the manner indicated above by starting out from an arbitrary perfect set F for which $(n)F$ has measure 0 for every n. We shall show that the maximal ideals $M_{\chi\mathfrak{F}}$ here are different from M_s and $M_{\chi\mathcal{N}}$. In fact, for every absolutely continuous function $g(t)$ we obviously have $g_{\mathfrak{F}}(E) \equiv 0$, because, by assumption, all the sets of $\mathfrak{F}$ are of measure zero. Therefore $g(M_{\chi\mathfrak{F}}) = 0$ for every $M_{\chi\mathfrak{F}}$; and hence $M_{\chi\mathfrak{F}}$ is different from all M_s, because for every M_s there exists an absolutely continuous function $g(t)$ such that $g(M_s) \neq 0$. Now let E be an arbitrary non-countable set of $\mathfrak{F}$. This contains a certain perfect subset F. Let $f(t)$ be an arbitrary non-zero singular function whose total variation is concentrated on F; such a function can be constructed by the same method as the one that is used in constructing a singular function that maps Cantor's perfect set onto an interval. Since $f(E)$ is concentrated on F and $F \in \mathfrak{F}$, we have $f \in V_{\mathfrak{F}}$. By the uniqueness theorem for the Fourier-Stieltjes integral, there exists a point s_0 such that $\int_{-\infty}^{\infty} e^{is_0t} df(t) \neq 0$. Now let $\chi(t)$ be an arbitrary character of $X_{\mathfrak{F}}$. Then $e^{is_0t}\overline{\chi(t)}$ also belong to $X_{\mathfrak{F}}$. Therefore $\varphi(t) = \int_{-\infty}^{t} e^{is_0t}\overline{\chi(t)}df(t)$ exists. Obviously, $\varphi(t)$ is a function of bounded variation concentrated on F. But

$$\int_{-\infty}^{\infty} \chi(t)\, d\varphi_{\mathfrak{F}}(t) = \int_{-\infty}^{\infty} \chi(t)\, d\varphi(t) =$$

$$= \int_{-\infty}^{\infty} \chi(t)\, d \int_{-\infty}^{t} e^{is_0\tau}\overline{\chi(t)}\, df(\tau) = \int_{-\infty}^{\infty} e^{is_0t}\, df(t) \neq 0.$$

Thus we have shown that for every maximal ideal $M_{\chi\mathfrak{F}}$ there exists a continuous function $\varphi \in V^{(b)}$ such that $\varphi(M_{\chi\mathfrak{F}}) \neq 0$. But it follows from this that the $M_{\chi\mathfrak{F}}$ are also different from all the $M_{\chi\mathcal{N}}$, because $\varphi(M_{\chi\mathcal{N}}) = 0$ for every continuous function $\varphi \in V^{(b)}$.

The construction just described gives us not only the maximal ideals M_s and $M_{\chi\mathcal{N}}$, which we know already, but also new maximal ideals, which we shall call *singular*.

§ 32. Perfect Sets with Linearly Independent Points. The Asymmetry of the Ring $V^{(b)}$

1. The method of constructing maximal ideals in $V^{(b)}$ explained in the preceding section enables us to obtain some results of a negative character concerning this ring. In this section we shall show that $V^{(b)}$ is unsymmetric and that the absolutely continuous maximal ideals are not everywhere dense in the space $\mathfrak{M}(V^{(b)})$. The proof of both facts will be based on the existence of perfect sets with linearly independent points.

A *set with linearly independent points* is a set on the real line every finite system $(\lambda_1, \ldots, \lambda_k)$ of pairwise distinct points of which is linearly independent over the field of rational numbers, i.e., an equation $n_1\lambda_1 + \ldots + n_k\lambda_k = 0$, where $n_1, \ldots, n_k$ are integers, implies that $n_1 = \ldots = n_k = 0$.

LEMMA 1: *If the system $(\lambda_1, \ldots, \lambda_k)$ is linearly independent, then every non-countable set on the real line contains a number λ for which the system $(\lambda_1, \ldots, \lambda_k, \lambda)$ is also linearly independent.*

For, the numbers λ for which the system $(\lambda_1, \ldots, \lambda_k, \lambda)$ is linearly depend ent form only a countable set. //

We shall say that the system $(\lambda_1, \ldots, \lambda_k)$ is *linearly independent of rank n* if an equation $n_1\lambda_1 + \ldots + n_k\lambda_k = 0$, where $n_1, \ldots, n_k$ are integers of absolute value not exceeding n, implies that $n_1 = \ldots = n_k = 0$. In the contrary case, we shall call the system $(\lambda_1, \ldots, \lambda_k)$ *linearly dependent of rank n.*

LEMMA 2: *If the system $(\lambda_1, \ldots, \lambda_k)$ is linearly independent, then for every $n > 0$ there exists an $\varepsilon > 0$ such that every system $(\lambda_1', \ldots, \lambda_k')$ satisfying the inequalities $|\lambda_1' - \lambda_1| < \varepsilon, \ldots, |\lambda_k' - \lambda_k| < \varepsilon$ is linearly independent of rank n.*

For the set of points of k-dimensional space whose coordinate system is linearly dependent of rank n is closed, as the union of a finite number of hyperplanes $n_1\mu_1 + \ldots + n_k\mu_k = 0$ $(|n_1| \leqq n, \ldots, |n_k| \leqq n)$, so that a point $(\lambda_1, \ldots, \lambda_k)$ that does not occur in this set is separated from it by a sufficiently small neighborhood. //

THEOREM 1: *Every perfect set P on the real line contains a perfect subset Π with linearly independent points.*

Proof: Let P^0 be the set obtained from P by removing the end-points of all intervals contiguous with P. Since P^0 is not countable, it contains, by Lemma 1, a pair of linearly independent points. Let Δ be the interval having these points as end-points. In consequence of Lemmas 1 and 2, we can remove from this interval an interior interval δ whose end-points belong to P^0, so that the system of four points formed by the end-points of the remaining intervals Δ_0 and Δ_1 'of the first rank' is linearly independent, the lengths of these intervals are less than 1, and every pair of points, one each from each of the intervals, is linearly independent of rank 1. Next, by the same Lemmas 1 and 2, we can remove from the intervals Δ_0 and Δ_1 interior intervals δ_0 and δ_1 with end-points belonging to P^0 such that the system formed by the eight end-points of the remaining four intervals 'of the second rank' $\Delta_{00}, \Delta_{01}, \Delta_{10}, \Delta_{11}$ is linearly independent, the length of each of these segments is less than $1/2$, and every quadruple of points, one each from every segment, is linearly independent of rank 2. Continuing this construction indefinitely, we obtain at the n-th step a set Π_n that is the union of 2^n

pairwise disjoint intervals of rank n and of length $< 1/n$ whose end-points belong to P^0 and form a linearly independent system and are such that every choice of points one from each interval forms a linearly independent system of rank n. Let $\Pi = \bigcap_{n=1}^{\infty} \Pi_n$. By construction, Π is a perfect set. Since each of its points is a limit point for the set of end-points of the contiguous intervals and the latter are chosen by construction from P, we have $\Pi \subset P$. We claim that every finite system $(\lambda_1, \ldots, \lambda_k)$ of points of Π is linearly independent. For let d be the least of the distances between points of this system. Then for $n > 1/d$ no two of its points can belong to the same interval of rank n. Therefore it is linearly independent of rank n, because otherwise, by adding arbitrarily one point each from all the $2^n - k$ intervals of rank n that do not intersect it and by giving to each of them the coefficient zero, we would obtain a system of points, one from each interval of rank n, that is linearly dependent of rank n, in contradiction to the construction of these intervals. Since n is arbitrarily large, we conclude that the system $(\lambda_1, \ldots, \lambda_k)$ is linearly independent. //

***2.** *Note*: In exactly the same way, one can prove the following more general theorem.

THEOREM 1′: *Every perfect set P on the real line contains a perfect subset Π with algebraically independent points.*

3. LEMMA 3: *If Π is a perfect set with linearly independent points and $\mathfrak{F}$ the minimal class of sets F_σ containing it and satisfying the conditions* 1.-4. *of the preceding section, then the set $-\Pi$ intersects every set of $\mathfrak{F}$ in not more than a countable set of points.*

Proof: $\mathfrak{F}$ consists of subsets of type F_σ of all possible finite or countable unions of sets of the type $(n)\Pi - t$. Therefore it is sufficient to prove that $-\Pi$ intersects every set $(n)\Pi - t$ in not more than $n + 1$ points. But if $-\Pi$ were to intersect $(n)\Pi - t$ in $n + 2$ distinct points, i.e., if we had $n + 2$ equations of the form

$$x_1^{(i)} + \cdots + x_n^{(i)} - t = -x_i \qquad (i = 1, \ldots, n+2),$$

where all the $x_j^{(i)}$, $x_i \in \Pi$ and the points x_i are pairwise distinct, then by eliminating t we would obtain $n + 1$ equations of the form

$$x_1^{(i)} + \cdots + x_n^{(i)} + x_i = x_1^{(1)} + \cdots + x_n^{(1)} + x_1$$
$$(i = 2, \ldots, n+2),$$

but since the points of the set Π are linearly independent and all the x_i are distinct from x_1, this would imply that for every i in question there would exist an index j_i for which $x_i = x_{i_j}^{(1)}$. But there are $n+1$ points x_i here, whereas the number of indices j_i is not more than n. //

THEOREM 2: *The ring $V^{(b)}$ is unsymmetric.*

Proof: $V^{(b)}$, regarded as a ring of set functions on the line, admits the natural involution $f \to f^*$, where

$$f^*(E) = \overline{f(-E)} \tag{1}$$

for every Borel set E. From the formula (1) it follows that

$$\int_{-\infty}^{\infty} \exp(is\lambda)\,df^*(\lambda) = \overline{\int_{-\infty}^{\infty} \exp(is\lambda)\,df(\lambda)} \tag{2}$$

for all real values of s. But in § 30 we saw that every element of $V^{(b)}$ is uniquely determined by its Fourier-Stieltjes transform, i.e., by its values on the absolutely continuous maximal ideals. Therefore, if $V^{(b)}$ were symmetric, the equation (2) would imply that the element conjugate to $f \in V^{(b)}$ in the sense of the Definition 1 of § 8 has to be precisely the element f^* that is defined by formula (1), with the equation (2) extending to all maximal ideals, i.e.,

$$f^*(M) = \overline{f(M)} \tag{3}$$

for all $M \in \mathfrak{M}(V^{(b)})$. But this is not true, because there exist $f \in V^{(b)}$ and $M \in \mathfrak{M}(V^{(b)})$ such that $f(M) = 1$ and $f^*(M) = 0$. For let Π be a perfect set with linearly independent points and $\mathfrak{F}$ the class of sets of Lemma 3 determined by it. We choose an arbitrary monotonic singular function $f(t)$ of variation 1, concentrated in Π, and let $M = M_{\chi_0 \mathfrak{F}}$, where $\chi_0(\lambda) \equiv 1$. Clearly, $f(E) \in V_{\mathfrak{F}}$ and $f(M_{\chi_0 \mathfrak{F}}) = \int df_{\mathfrak{F}} = \int df = 1$. On the other hand, since the total variation of f is concentrated in Π, the total variation of f^* is concentrated in $-\Pi$. But then it follows from Lemma 3 that $f^* \in V_{C\mathfrak{F}}$, and hence $f^*(M_{\chi_0 \mathfrak{F}}) = 0$. //

COROLLARY: *The set of absolutely continuous maximal ideals is not everywhere dense in the space of maximal ideals of $V^{(b)}$.*

For whatever $f \in V^{(b)}$ we choose, by formula (2) the equation (3) holds for all absolutely continuous maximal ideals M. But if the set of these maximal ideals were dense in $\mathfrak{M}(V^{(b)})$, then the equation (3) would extend to all the $M \in \mathfrak{M}(V^{(b)})$ by the continuity of the functions $f(M)$ that generate the elements $f \in V^{(b)}$. But this would mean that $V^{(b)}$ is symmetric, in contradiction to Theorem 2. //

*4. With the help of the construction of singular maximal ideals explained in the preceding section one can show [67] that the set of absolutely continuous maximal ideals is not dense relative to the boundary of the space $\mathfrak{M}(V^{(b)})$.

5. THEOREM 3: *There exists a function* $\Phi(s)$ $(-\infty < s < \infty)$ *that is the Fourier-Stieltjes transform of a function of bounded variation and satisfies the condition* $\inf_{-\infty < s < \infty} |\Phi(s)| > 0$ *such that* $1/\Phi(s)$ *is not the Fourier-Stieltjes transform of any function of bounded variation.*

Proof: An example of a function that has these properties is the Fourier-Stieltjes transform of the function $\varphi(t) = f(t) - f^*(t) - \varepsilon(t) \in V^{(b)}$, where $f(t)$ is the function considered in the proof of Theorem 2 and $\varepsilon(t)$ is the unit element of $V^{(b)}$. For by (2),

$$\Phi(s) = 2i\mathfrak{F}\left(\int_{-\infty}^{\infty} \exp(is\lambda) df(\lambda)\right) - 1,$$

and therefore $\inf_{-\infty < s < \infty} |\Phi(s)| \geqq 1$. On the other hand, since $f(M_{\chi_0 \mathfrak{F}}) = 1$ and $f^*(M_{\chi_0 \mathfrak{F}}) = 0$, we have $\varphi(M_{\chi_0 \mathfrak{F}}) = 0$; consequently the element $\varphi \in V^{(b)}$ does not have an inverse in $V^{(b)}$ and therefore $1/\Phi(s)$ cannot be the Fourier-Stieltjes transform of a function in $V^{(b)}$. //

THEOREM 4: *Let* $F(s)$ *be the Fourier-Stieltjes transform of the function* $f \in V^{(b)}$ *and* $f(t) = g(t) + h(t) + s(t)$ *its decomposition into the absolutely continuous, discrete, and singular parts. If*

1) $$\inf_{-\infty < s < \infty} |F(s)| > 0$$

and

2) $$\| s \| < \inf_{-\infty < s < \infty} \left| \int_{-\infty}^{\infty} \exp(is\lambda) dh(\lambda) \right|,$$

then $1/F(s)$ *is also the Fourier-Stieltjes transform of a function of* $V^{(b)}$.

Proof: We have to show that

$$f(M) = g(M) + h(M) + s(M) \neq 0$$

for all maximal ideals M of $V^{(b)}$. For the absolutely continuous maximal ideals this is true in consequence of the condition 1). By the Corollary to the lemma of § 30, for all the remaining maximal ideals $g(M) = 0$, so that $f(M) = h(M) + s(M)$. The homomorphic mapping $V^{(b)} \to V^{(b)}/M$ induces a homomorphic mapping of H into the field of complex numbers. By

Theorem 2 of § 29, under this mapping $h(t)$ goes over into a number belonging to the closure of the set of values of $h(M_s)$. Therefore, by condition 2), $|h(M)| > \|s\| \geqq |s(M)|$ and consequently $f(M) = h(M) + s(M) \neq 0$. //

§ 33. The General Form of Maximal Ideals of the Ring $V^{(b)}$

***1.** All the maximal ideals of $V^{(b)}$ considered thus far are obtainable by the construction explained in § 31. But this does not give the entire space $\mathfrak{M}(V^{(b)})$. The general form of the maximal ideals of $V^{(b)}$ was found by Shreider [67]. In this section we shall give an account, essentially without proof, of the results obtained by him.

We shall treat the elements of $V^{(b)}$, as before, as set functions and we shall call non-negative real set functions of $V^{(b)}$ measures. Two measures are called *mutually singular* if on every Borel set on which one is positive, the other is equal to zero. Every function $f \in V^{(b)}$ can be uniquely represented in the form

$$f(E) = f_1(E) - f_2(E) + if_3(E) - if_4(E),$$

where f_1, f_2, f_3, f_4 are measures and where f_1 and f_2, and also f_3 and f_4, are mutually singular. The *variation* $(\mathrm{Var}\, f)(E)$ of the function $f \in V^{(b)}$ on a set E is defined as the sum of the variations of its real and imaginary parts. We say that *the function* $f \in V^{(b)}$ *is absolutely continuous with respect to the function* $g \in V^{(b)}$ if $(\mathrm{Var}\, f)(E) = 0$ whenever $(\mathrm{Var}\, g)(E) = 0$. We shall denote this by writing $f \prec g$. We say that a certain statement is true almost everywhere relative to $f \in V^{(b)}$ if it can be false only on a set E for which $(\mathrm{Var}\, f)(E) = 0$.

To find the general form of the maximal ideals M of $V^{(b)}$ means this, that we have to find the general form of their generating linear functionals $M(f) = f(M)$ (M fixed, f variable) that can be characterized as *multiplicative* linear functionals on $V^{(b)}$, i.e., non-zero linear functionals satisfying the additional condition

$$M(f * g) = M(f)M(g)$$

for all

$$f, g \in V^{(b)}.$$

The starting point for this is the representation of arbitrary linear functionals on $V^{(b)}$ to be given below.

***2.** A function $\varphi_f(t)$ of the real variable t and the variable element f of $V^{(b)}$ will be called a *generalized function* if it satisfies the following conditions: 1) $\varphi_f(t)$, for every fixed $f \in V^{(b)}$, is a function of t that is measurable with respect to f, i.e., with respect to the measure $(\operatorname{Var} f)(E)$; 2) if $f \dashv g$, then $\varphi_f(t) = \varphi_g(t)$ almost everywhere relative to f.

THEOREM 1: *$L(f)$ is a linear functional on $V^{(b)}$ if and only if it can be represented in the form*

$$L(f) = \int_{-\infty}^{\infty} \varphi_f(t)\,df(t), \tag{1}$$

where $\varphi_f(t)$ is a generalized function for which

$$\| \varphi_f \| \; (= \| L \|) = \sup_f \operatorname{ess\,sup}_t | \varphi_f(t) | < \infty .$$

What must a generalized function $\varphi_f(t)$ look like so that the linear functional (1) generated by it is multiplicative, i.e., coincides with the linear functional $M(f)$ generated by some maximal ideal?

A *generalized character* is a generalized function $\chi_f(t)$ satisfying the condition

$$\chi_f(s+t) = \chi_f(s)\chi_f(t)$$

for almost all pairs (s, t) relative to the product of the measures $\operatorname{Var} f \times \operatorname{Var} f$ and, in addition, the condition

$$\| \chi_f \| = 1.$$

THEOREM 2: *$f \to M(f)$ is a homomorphism of $V^{(b)}$ into the field of complex numbers, i.e., M is a multiplicative linear functional on $V^{(b)}$, if and only if*

$$M(f) = \int_{-\infty}^{\infty} \chi_f(t)\,df(t),$$

where $\chi_f(t)$ is a generalized character.

Suppose, in particular, that $M_{\chi\mathfrak{F}}$ is a maximal ideal of $V^{(b)}$ determined by a class $\mathfrak{F}$ of sets F_σ satisfying the conditions 1.-4. of § 31 and that $\chi(t)$ is a character of Γ that is measurable with respect to all the measures that are concentrated in $\mathfrak{F}$; thus,

$$M_{\chi\mathfrak{F}}(f) = f(M_{\chi\mathfrak{F}}) = \int_{-\infty}^{\infty} \chi(t)\,df_{\mathfrak{F}}(t),$$

where $f_{\mathfrak{F}}$ is that part of f that is concentrated in $\mathfrak{F}$. As was shown in § 31, the total variation of the function $f_{\mathfrak{F}}$ is concentrated in a set X such that $X + X = X$. We put

$$\chi_f(t) = \begin{cases} \chi(t) & \text{for all } t \in X, \\ 0 & \text{for all } t \notin X. \end{cases}$$

It is not difficult to verify that $\chi_f(t)$ is a generalized character and that the maximal ideal M determined by it coincides with $M_{\chi\mathfrak{F}}$. Let $V^{(b)} = V_{\mathfrak{F}} + V_{C\mathfrak{F}}$ be the decomposition of the ring $V^{(b)}$ into the semi-direct sum of the mutually singular subring and the ideal generated by the set $\mathfrak{F}$. Obviously, $V_{\mathfrak{F}}$ can be characterized as the set of those $f \in V^{(b)}$ for which $\chi_f(t) \neq 0$ almost everywhere relative to f, and $V_{C\mathfrak{F}}$ as the set of those $g \in V^{(b)}$ for which $\chi_g(t) = 0$ almost everywhere relative to g. Now it turns out that this is a general property of generalized characters.

***3.** THEOREM 3: *Let M be an arbitrary maximal ideal of $V^{(b)}$ and $\chi_f(t)$ the generalized character corresponding to it. Then the set of those $f \in V^{(b)}$ for which $\chi_f(t) \neq 0$ almost everywhere relative to f forms a subring R of $V^{(b)}$; the set of those $g \in V^{(b)}$ for which $\chi_g(t) = 0$ almost everywhere relative to g forms an ideal I of $V^{(b)}$, and*

1) *any two functions $f \in R$ and $g \in I$ are mutually singular;*

2) *every function $f \in V^{(b)}$ can be uniquely represented in the form of a sum $f = f_R + f_I$, where $f_R \in R$, $f_I \in I$.*

Thus, every maximal ideal of $V^{(b)}$ determines a decomposition of this ring into the semi-direct sum of a mutually singular subring and ideal and is obviously in turn uniquely determined by the maximal ideal that it induces in this subring. But conversely, given an arbitrary decomposition of the ring $V^{(b)}$ into the semi-direct sum of a subring R and an ideal I, every maximal ideal M' of R or, what is the same, every homomorphism $f_R \to f_R(M')$ of this subring into the field of complex numbers generates a homomorphism $f \to f_R(M')$ (where f_R is the component of f in R) of the entire ring $V^{(b)}$, i.e., generates a maximal ideal M of that ring. Thus the problem of finding the maximal ideals of $V^{(b)}$ reduces to that of finding all possible decompositions of this ring into the semi-direct sum of a mutually singular subring and ideal followed by the determination of the maximal ideals of the subring thus obtained; this subring has, in any case, the obvious maximal ideals determined by the homomorphisms $f_R \to \int_{-\infty}^{\infty} \chi(t)\, df_R(t)$, where $\chi(t)$ are arbitrary characters of Γ, measurable with respect to all functions $f_R \in R$. It was in just this way that the maximal ideals $M_{\chi\mathfrak{F}}$ were obtained in § 31. However the decompositions thus obtained of $V^{(b)}$ into the semi-direct sum of the mutually singular subring $V_{\mathfrak{F}}$ and the ideal $V_{C\mathfrak{F}}$ are not the only possible ones.

*4. THEOREM 4: *Let X be an arbitrary set of characters of* Γ, R_X *the set of all functions* $f \in V^{(b)}$ *with respect to which every character* χ *of X is measurable, and* I_X *the set of all functions* $g \in V^{(b)}$ *that are singular for every function of* R_X. *Then* R_X *is a subring and* I_X *an ideal of* $V^{(b)}$, *and* $V^{(b)}$ *is their semi-direct sum.*

THEOREM 5: *Let* φ *be a non-zero continuous measure concentrated in a perfect set* Π *with linearly independent points. Then the subring R of* $V^{(b)}$ *formed in accordance with Theorem 4 by all the functions* $f \in V^{(b)}$ *with respect to which every character is measurable, provided it is measurable with respect to* φ, *does not coincide with any of the subrings* $V_{\mathfrak{F}}$ *constructed by means of the scheme of* § 31.

Let us sketch a proof of this theorem. Suppose, in contradiction to the theorem, that $R = V_{\mathfrak{F}}$, where $\mathfrak{F}$ is some class of sets F_σ satisfying the conditions 1.-4. of § 31. Since $\varphi \in R$, φ is concentrated in a certain set $P \in \mathfrak{F}$, and we may assume that $P \subset \Pi$. Since the set P is uncountable, there is also a non-zero continuous measure ψ concentrated in it that is singular with respect to φ, so that there exists a decomposition of P into disjoint sets P_φ and P_ψ such that φ is concentrated in P_φ and ψ in P_ψ. Let $\Gamma = A \cup B$ be the decomposition of the real line into two complementary totally imperfect sets.[8] We put

$$\chi(t) = \begin{cases} 1 & \text{if} \quad t \in P_\varphi, \\ 1 & \text{if} \quad t \in P_\psi \cap A, \\ -1 & \text{if} \quad t \in P_\psi \cap B. \end{cases}$$

Since P is a set with linearly independent points, $\chi(t)$ can be extended to a character of Γ. Obviously, it is measurable with respect to φ but non-measurable with respect to ψ. Therefore ψ is not in R. But being concentrated in $P \in \mathfrak{F}$, ψ is in $V_{\mathfrak{F}}$. The contradiction proves the theorem. //

COROLLARY: *There exist maximal ideals of* $V^{(b)}$ *that cannot be obtained by the construction of* § 31.

An example of such a maximal ideal is the maximal ideal M defined by the homomorphism $f \to \int_{-\infty}^{\infty} df_R$, where R is the ring of Theorem 5. For suppose, to the contrary, that M coincides with some maximal ideal $M_{\chi\mathfrak{F}}$, so that for all $f \in V^{(b)}$ we have the equation

$$\int_{-\infty}^{\infty} df_R(t) = \int_{-\infty}^{\infty} \chi(t) df_{\mathfrak{F}}(t).$$

---

[8] I.e., sets that do not contain perfect subsets; see F. Hausdorff, *Set Theory*, New York, 1962, p. 201.

Since the measures that are concentrated at one point belong to $V_{\mathfrak{F}}$, we conclude, first of all, that $\chi(t) \equiv 1$. Thus, for all $f \in V^{(b)}$ we must have the equation

$$\int_{-\infty}^{\infty} df_R = \int_{-\infty}^{\infty} df_{\mathfrak{F}}. \tag{2}$$

But since the rings R and $V_{\mathfrak{F}}$ do not coincide, we can find a measure f that belongs to one of them but not to the other, and equation (2) obviously cannot hold for this f.

PART III

CHAPTER VI

REGULAR RINGS

1. In this chapter we shall be concerned mainly with normed rings of functions. It will become apparent that the reason that many rings of functions have theorems in common is closely connected with the fact that these rings have functions $x \neq 0$ that vanish on an arbitrary given closed set of maximal ideals. For example, Wiener's well-known Tauberian theorem, which will be discussed in § 40, appears as a corollary of just such a property of the ring of absolutely convergent Fourier integrals. Rings having this property, which will be formulated precisely in Definition 2 of § 34, are called regular (from the name of a topological space with the corresponding axiom of separability).

§ 34. Definitions, Examples, and Simplest Properties

1. Let R be a normed ring and $\mathfrak{M}$, the space of its maximal ideals. According to § 4, to every element x of R there corresponds a continuous function $x(M)$ defined on $\mathfrak{M}$ such that the mapping $x \to x(M)$ is an isomorphism, provided R has no radical, so that R can be identified in this case with the ring $\hat{R}$ of functions $x(M)$.

DEFINITION 1: The *skeleton* of an ideal I of a normed ring R is the set F of all points of the space $\mathfrak{M} = \mathfrak{M}(R)$ at which all the functions $x(M)$ corresponding to elements $x \in I$ vanish.

By Theorem 1 of § 2, every proper ideal $I \subset R$ is contained in some maximal ideal M_0. But in accordance with § 4, the relation $x_0 \in M_0$ means that $x_0(M_0) = 0$. This shows that the skeleton of every proper ideal $I \subset R$ is non-empty. On the other hand, the skeleton of the improper ideal $I = R$ is obviously the empty set. Since every continuous function that is zero on a set S is also zero on the closure of that set, the *skeleton of every ideal* $I \subset R$ *is a closed subset of the space* $\mathfrak{M}$.

***2.** Thus, associated with every ideal I of R is a closed set $F \subset \mathfrak{M}$—the skeleton of the ideal. The converse statement, that every closed set of $\mathfrak{M}$

is the skeleton of an ideal $I \subset R$ is not true in general. For example, in the ring A of functions that are analytic in the circle $|\zeta| < 1$ and continuous in the circle $|\zeta| \leqq 1$ every function that is equal to zero on a closed set F having a limit point in the interior of the circle is identically equal to zero; therefore such a set F, if it does not coincide with the whole circle $|\zeta| \leqq 1$, cannot be the skeleton of any ideal of A.

3. DEFINITION 2: A normed ring R without radical is called *regular* if for every closed set $F \subset \mathfrak{M} = \mathfrak{M}(R)$ and every point M_0 not belonging to it there exists an element $x \in R$ such that the function $x(M)$ vanishes on F and is different from zero at $M = M_0$.

Since a regular ring is by definition a ring without radical, we can regard it as a ring of functions on M and as a rule we shall do so.

***4.** Examples of regular rings are the ring $C(S)$ of all continuous complex functions on an arbitrary compactum S (this follows from the normality of S by the well-known theorem of Uryson) and the ring of all functions having continuous derivatives up to a given order (on an interval, or in a closed domain of n-dimensional space). The ring A (§ 1.2) and other rings of analytic functions are, of course, not regular. The ring W of absolutely convergent trigonometric series is regular, because it contains every function with a continuous derivative.

Let us prove the regularity of the ring V of absolutely integrable functions on the line with convolution as multiplication and with a unit element e adjoined (§ 16).

As we have seen in § 17, the space of maximal ideals of V can be identified with the real line $-\infty < s < \infty$, supplemented by the point at infinity and furnished with the topology of the projective line; here for $\mathfrak{z} = \lambda e + x(t) \in V$ we have

$$\mathfrak{z}(M) = \begin{cases} \lambda + x^{\sim}(s) = \lambda + \displaystyle\int_{-\infty}^{\infty} x(t) \exp(ist)\,dt & \text{for} \quad M = M_s, \\ \lambda & \text{for} \quad M = M_\infty. \end{cases}$$

We shall show that for every point $M^0 \in \mathfrak{M}(V)$ and every neighborhood $U(M^0)$ of this point there exists an element $\mathfrak{z} = \lambda e + x(t) \in V$ for which $\mathfrak{z}(M^0) \neq 0$ and $\mathfrak{z}(M) = 0$ everywhere outside $U(M^0)$. Suppose, to begin with, that $M^0 = M_{s_0}$, where $s_0 \neq \infty$. We take an arbitrary function $y(s)$ having a continuous second derivative $y''(s)$ that is different from zero at $s = s_0$ and equal to zero outside a finite interval $U(s_0)$ contained in the given neighborhood $U(M_{s_0})$. The function

$$x(t) = (2\pi)^{-1} \int_{-\infty}^{\infty} y(s) \exp(-its)\,ds$$

is obviously bounded and continuous. Integrating by parts, we obtain

$$t^2x(t) = -(2\pi)^{-1}\int_{-\infty}^{\infty} y''(s)\exp(-its)ds,$$

so that $t^2x(t)$ is also a bounded function. Hence it follows that

$$|x(t)| \leqq C/[1+t^2],$$

and therefore $x(t)$ belongs to V. By the inversion formula,

$$x^{\sim}(s) = \int_{-\infty}^{\infty} x(t)\exp(its)dt = y(s),$$

and so $\mathfrak{z} = x(t)$ has the required properties.

If $M = M_\infty$, we can construct, in the same way as above, a function $x(t)$ for which $x^{\sim}(s) = 1$ on the complement (which is contained in a finite interval) of the neighborhood $U(M_\infty)$ and, as is true of every absolutely integrable function, $x(M_\infty) = 0$. But then $\mathfrak{z} = e - x(t)$ satisfies the conditions stated.

5. In a regular ring R it is easy to construct an ideal whose skeleton is a preassigned closed set $F \subset \mathfrak{M}$. In fact, the set of all functions $x(M)$ that are equal to zero on F obviously forms an ideal I; on the one hand, its skeleton $F(I)$ contains every point of F, and on the other hand, by the condition of regularity, it does not contain any point M_0 not belonging to F. Thus, $F(I) = F$.

Every ideal I with the skeleton F will be said to *belong to the set* F, and the ideal constructed by the rule just given will be denoted by $I(F)$. Note that, by its very definition, $I(F)$ is the intersection of all maximal ideals that make up the set F and is therefore a closed ideal.

We form the residue-class ring $R/I(F)$. By definition, two functions $x(M)$ and $y(M)$ belong to the same residue class if their difference belongs to $I(F)$; in other words, $x(M)$ and $y(M)$ are in the same class if and only if these two functions coincide on F. Consequently, we may identify the ring $R/I(F)$ with the ring R_F formed by the restrictions of the functions $x(M)$ of R to the set F, i.e., with the ring of all functions defined on F and extendable from this set to functions of R on the entire space $\mathfrak{M}$, where the norm of every such function is given by the general formula (2) of § 2, which in our case assumes the form

$$\| x \|_{R_F} = \| x \|_{R/I(F)} = \inf \| y \|_R,$$

the lower bound being taken over all functions $y(M) \in R$ that coincide with $x(M)$ on F.

We shall now show that the space of maximal ideals of $R/I(F)$ can be identified with F as regards both the elements and the topology. Since R_F is a ring of functions defined on F, every point $M \in F$ determines a maximal ideal of R_F; and distinct points $M_1, M_2 \in F$ determine distinct maximal ideals, because they even differ by some element of R. Conversely, if M' is a maximal ideal of $R/I(F)$, then, in accordance with Note 2 of § 2, its complete inverse image M in R is a maximal ideal of R containing I and consequently belongs to F. Thus, F is the set of all maximal ideals of the ring $R_F = R/I(F)$. And since the functions of R coincide on F with the corresponding functions of R_F, it is clear that the topology of F as the space of maximal ideals of R_F also coincides with its topology as a subspace of the space $\mathfrak{M}$ of maximal ideals of R.

Theorem 1: *Let an ideal I (closed or not) of a regular ring R have as its skeleton a set $F \subset \mathfrak{M} = \mathfrak{M}(R)$. For every closed set $\Phi \subset \mathfrak{M}$ having no points in common with F there exists a function $y(M) \in I$ that is equal to* 1 *at all the points of Φ.*

Proof: We consider the residue-class ring $R/I(\Phi)$ and the image I' of I in this ring. I' is an ideal in $R/I(\Phi)$. If it were to belong to a maximal ideal M_0' of $R/I(\Phi)$, this would mean that all the functions $x(M') \in I'$ vanish at M_0'. But the functions $x(M') \in I'$ are images of functions $x(M) \in I$, and so we would find a point $M_0 \in \Phi$ at which all the functions $x(M) \in I$ vanish; but this is impossible, because I belongs to the set F, which has no points in common with Φ. Therefore the ideal I' does not belong to any maximal ideal of $R/I(\Phi)$; but then it coincides with this entire ring and, in particular, contains the unit element of this ring. We examine the function $y(M) \in I$ that is carried under the homomorphism $R \to R/I(\Phi)$ into the unit element of $R/I(\Phi)$; this function assumes the value 1 at all the points of Φ and therefore satisfies the requirements of the theorem. //

Corollary: *For any two disjoint closed sets F_1 and F_2 in the space of maximal ideals of a regular ring R there exists a function $e(M) \in R$ that is* 0 *on F_1 and* 1 *on F_2.*

For the proof, it is sufficient to apply Theorem 1 to the ideal $I = I(F_1)$ and the set $\Phi = F_2$. //

§ 35. The Local Theorem

1. Theorem 1 of § 34 and its corollary enable us to establish 'local' theorems of the Wiener type for regular rings. Norbert Wiener proved the following theorem in 1933: *If $f(t)$ coincides in a neighborhood of every point t_0 $(-\infty < t_0 < \infty)$ with a function $\varphi(t; t_0)$ that can be expanded in an absolutely convergent Fourier series, then $f(t)$ can itself be expanded in an* *absolutely convergent Fourier series.* It turns out that an analogous proposition is true for every regular ring:

THEOREM 1 (The 'Local Theorem'): *Let $f(M)$ be a function defined on the space $\mathfrak{M}$ of maximal ideals of a regular ring R. We assume that for every point $M_0 \in \mathfrak{M}$ there is a neighborhood $U(M_0)$ in which $f(M)$ coincides with a function $x(M; M_0)$ of R. Then $f(M)$ is itself an element of R.*

The proof is based on the following lemma (which we shall need in Chapter VII also).

LEMMA (On The Decomposition of the Unit Element): *For every covering of the space $\mathfrak{M}$ of maximal ideals of the regular ring R by non-empty open sets $U_1, \ldots, U_n$ there exist elements $h_1, \ldots, h_n \in R$ such that*

1) $h_j(M) = 0$ *exterior to* U_j $(j = 1, \ldots, n)$;

2) $\sum_{j=1}^{n} h_j(M) \equiv 1.$

The proof of the lemma will be by induction on n. Let $n = 2$; this means that $\mathfrak{M}$ is covered by two sets U_1 and U_2 and that $F_1 = \mathfrak{M} \setminus U_1$ and $F_2 = \mathfrak{M} \setminus U_2$ are closed and disjoint. By the Corollary to Theorem 1 of § 34, there is a function $h_1(M)$ in R that is equal to 0 on F_1, i.e., exterior to U_1, and to 1 on F_2. But then the function $h_2(M) = 1 - h_1(M)$ is equal to 0 on F_2, i.e., exterior to U_2, and, by construction, $h_1(M) + h_2(M) = 1$. Thus, the lemma is proved for $n = 2$.

Now let $n > 2$ and assume the lemma true for a covering of the space of maximal ideals of every regular ring by $n-1$ sets; we shall show that it is also true for the covering of R by n non-empty open sets. Let $U_1, \ldots, U_{n-1}, U_n$ be such a covering and let $F = \mathfrak{M} \setminus U_n$. F is closed and is contained in the open set $U_1 \cup \ldots \cup U_{n-1}$. Since $\mathfrak{M}$ is a normal space, there exists a non-empty open set $U \subset \mathfrak{M}$ such that

$$F \subset U \subset \bar{U} \subset U_1 \cup \ldots \cup U_{n-1}.$$

Let us examine the residue-class ring $R/I(\bar{U})$; by what has been shown in § 34, we may assume that it is formed by all the functions $x(M)$ given on $\bar{U}$ and extended to functions of R on the entire space $\mathfrak{M}$ and that it has $\bar{U}$ as its space of maximal ideals. From the regularity of R it obviously follows that $R/I(\bar{U})$ is also regular. By hypothesis, there then exist functions $h_1'(M), \ldots, h_{n-1}'(M) \in R/I(\bar{U})$ such that $h_j'(M) = 0$ on $\bar{U} \setminus (U_j \cap \bar{U})$ $(j = 1, \ldots, n-1)$ and $\sum_{j=1}^{n-1} h_j'(M) = 1$ for all $M \in \bar{U}$. On the other hand, M is covered by the two non-empty open sets U and U_n, and therefore, by what has been proved above, there exist functions $h(M)$, $h_n(M) \in R$ such that $h(M)=0$ outside U, $h_n(M)=0$ outside U_n, and $h(M)+h_n(M)\equiv 1$ for all $M\in\mathfrak{M}$. Let

$$h_j''(M) \qquad (j=1,\ldots,n-1)$$

be functions of R that map, under the homomorphism $R\to R/I(\bar{U})$, into the functions $h_j'(M)$, respectively. We put

$$h_j(M)=h_j''(M)h(M) \qquad (j=1,\ldots,n-1).$$

Then $h_j(M)=0$ outside U_j $(j=1,\ldots,n-1)$, because at the points $M\in\bar{U}\setminus U_j$ the factor $h_j''(M)$ is zero and outside $\bar{U}$ the factor $h(M)$ is zero. Furthermore, $h_n(M)=0$ outside U_n, by construction. Finally,

$$\sum_{j=1}^{n} h_j(M)=\left(\sum_{j=1}^{n-1} h_j''(M)\right)h(M)+h_n(M)=h(M)+h_n(M)\equiv 1.$$

Thus, the functions $h_1(M),\ldots,h_n(M)$ satisfy all the requirements, and the lemma is proved.

Let us now proceed to the *Proof of Theorem* 1: By assumption, every point of $\mathfrak{M}$ has a neighborhood in which $f(M)$ coincides with a function of R; obviously, this neighborhood can be taken to be open. Since $\mathfrak{M}$ is compact, its open covering formed by these neighborhoods contains a finite covering $U_1,\ldots,U_n$. Let $x_1(M),\ldots,x_n(M)$ be the corresponding functions of R, i.e., $f(M)=x_j(M)$ on U_j $(j=1,\ldots,n)$, and let $h_1(M),\ldots,h_n(M)$ $(\in R)$ be the functions of the lemma, i.e., $h_j(M)=0$ outside U_j $(j=1,\ldots,n)$ and $\sum_{j=1}^{n} h_j(M)\equiv 1$. Then

$$f(M)=\sum_{j=1}^{n} f(M)h_j(M)=\sum_{j=1}^{n} x_j(M)h_j(M),$$

and hence $f\in R$. //

***2.** Since the ring W of absolutely convergent trigonometrical series is regular, it is clear that Wiener's Theorem mentioned above is a corollary of the general theorem just proved.

3. THEOREM 2: *If a function $f(M)$ is defined on the space of maximal ideals $\mathfrak{M}$ of a regular ring R that can be represented in a neighborhood U_0 of every point $M_0\in\mathfrak{M}$ in the form of an analytic function of an element x_0 (depending on M_0 and U_0), then R contains an element x for which $x(M)\equiv f(M)$.*

Proof: Let us consider a neighborhood V_0 that, together with its closure, belongs to U_0 and the ideal I_0 of all functions $x(M) \in R$ that are zero on $\bar{V}_0$. The residue-class ring R/I_0 has the set $\bar{V}_0$ as its space of maximal ideals. The restriction of $f(M)$ to this set is, by Theorem 1 of § 6, an element of R/I_0. Since this is true for a neighborhood of every point $M_0 \in \mathfrak{M}$, it follows that $f(M)$ belongs locally to R and it remains to apply the local theorem just proved. //

***4.** Thus, in the ring W of absolutely convergent trigonometric series we can extract the m-th root of every function $x(t) \in W$ that does not vanish on the circle $0 \leqq t < 2\pi$ and has a unique m-th root (i.e., when t varies from 0 to 2π, $\arg x(t)$ changes by a multiple of $2\pi m$). The corresponding fact holds for the ring V of absolutely integrable functions. The representation of the element $\mathfrak{z} = \lambda e + x(t)$ of this ring on the maximal ideals M_s is given by the formula

$$\mathfrak{z}^{\sim}(s) = \lambda + \int_{-\infty}^{\infty} x(t) \exp(ist) dt,$$

and for extracting the m-th root of $\mathfrak{z}$ within V it is sufficient that $\mathfrak{z}^{\sim}(s)$ should not vanish and that its argument should change by a multiple of $2\pi m$ when s varies from $-\infty$ to $+\infty$.

***5.** Similar propositions also hold for infinitely-valued functions. For example, *in the ring V, the function* $\log \mathfrak{z}$ *exists for every element*

$$\mathfrak{z} = \lambda e + x(t)$$

that does not vanish on any maximal ideal and satisfies the condition

$$\int_{-\infty}^{\infty} d \arg \mathfrak{z}^{\sim}(s) = 0.$$

The latter fact lies at the basis of Krein's theory of integral equations on a half-line with a kernel depending on a difference (see [32]).

***6.** It stands to reason that the facts listed above also follow immediately from the general theorem on locally analytic functions of several elements of a ring (Theorem 1 of § 13). But the derivation given here is remarkably elementary, because it does not depend on Weil's integral representation of functions of several complex variables. On the other hand, it cannot be used to derive the corresponding facts for the ring $V[\alpha]$, which is not regular, where the application of Theorem 1 of § 13 is still the only method of proof known.

§ 36. Minimal Ideals

1. THEOREM 1: *Let R be a regular ring. Among all the ideals $I \subset R$ that belong to a closed set $F \subset \mathfrak{M}(R)$ there is a minimal ideal $J(F)$. It is formed by all the functions $x(M) \in R$ that are equal to zero on a neighborhood of F. Its closure $\overline{J(F)}$ is the minimal of all the closed ideals belonging to F. The functions $x(M) \in \overline{J(F)}$ are characterized by the following property: $x(M) \in R$ belongs to $\overline{J(F)}$ if and only if the ring R contains a sequence of functions $y_n(M)$ that converges to zero in norm and for which every function $y_n(M)$ coincides with $x(M)$ on some neighborhood of F.*

Proof: The ideal $J(F)$ formed by the functions $x(M) \in R$ each of which is equal to zero on some neighborhood of F belongs to this set, because for every point M_0 not in F we can find an open set $G \supset F$ whose closure does not contain M_0 and we can then construct a function $y(M) \in R$ that is equal to zero on G and is different from zero for $M = M_0$; this function, by assumption, occurs in the ideal $J(F)$, which can therefore only belong to F. Let I be an arbitrary ideal of R belonging to F; we shall show that $I \supset J(F)$. Suppose that $x(M) \in J(F)$, $F_1 = \{M \in \mathfrak{M} : x(M) = 0\}$, and that F_2 is the closure of the complement of F_1. The sets F and F_2 do not intersect; by Theorem 1 of § 34, the ideal I contains a function $e(M)$ that is equal to 1 on F_2. Obviously, $x(M) = x(M)e(M)$, and so $x(M) \in I$. Thus, $J(F) \subset I$ and $J(F)$ is indeed the minimal ideal of all the ideals that belong to F.

Every closed ideal belonging to F when it contains $J(F)$ also contains its closure $\overline{J(F)}$; therefore $\overline{J(F)}$ is the minimal of all the closed ideals belonging to F. Let $x(M) \in \overline{J(F)}$, i.e., $x = \lim_{n \to \infty} h_n$, where $h_n(M) \in J(F)$; then the function $x_n(M) \equiv x(M) - h_n(M)$ coincides with $x(M)$ in some neighborhood of F (namely, one in which $h_n(M) = 0$) and tends to zero in norm as $n \to \infty$. Conversely, suppose that for a given function $x(M) \in R$ there exists a sequence of functions $x_n(M) \in R$ converging in norm to zero and that each function $x_n(M)$ coincides with $x(M)$ in a neighborhood of F. Then $h_n(M) = x(M) - x_n(M) \in J(F)$ and $x = \lim h_n \in \overline{J(F)}$. //

COROLLARY: *In the ring $C(S)$ of all continuous functions $f(t)$ given on a compact Hausdorff space S, every closed ideal I is the collection $I(F)$ of all functions $f(t) \in C(S)$ that are equal to zero on a closed set $F \subset S$.*

For we know that $\mathfrak{M}(C(S)) = S$ (see §§ 2 and 5) and that the ring $C(S)$ is regular, so that all the above results are applicable to it. Let F be the skeleton of I. It is obviously sufficient to prove that the minimal closed ideal $\overline{J(F)}$ belonging to F coincides with the maximal closed ideal $I(F)$ belonging to F. Let $f(t)$ be an arbitrary function, continuous on S, that is equal to zero on F and let ε be an arbitrary positive number; we consider the closed sets

$$F_1 = \{t \in S: \ |f(t)| \leqq \varepsilon\}, \qquad F_2 = \{t \in S: \ |f(t)| \geqq 2\varepsilon\}.$$

Since F_1 and F_2 do not intersect, by Uryson's Theorem there exists a continuous function $a(t)$ that is equal to 1 on F_1 and to 0 on F_2 and such that $0 \leqq a(t) \leqq 1$ for all $t \in S$. The product $a(t)f(t)$ does not exceed 2ε in absolute value (and hence in the norm of the ring $C(S)$) and coincides with $f(t)$ on the set F_1 which is a (closed) neighborhood of F. Thus, $f(t)$ satisfies the criterion for belonging to $\overline{J(F)}$ established in Theorem 1. Since $f(t)$ is an arbitrary function of $I(F)$, we have $I(F) = \overline{J(F)}$. //

§ 37. Primary Ideals

1. Among all the ideals of a normed ring of special interest are those that are contained in only one maximal ideal. There is a number of problems in analysis whose solution depends on the structure of ideals that are contained in a given maximal ideal only: an example is the generalized Tauberian Theorem of Wiener (see § 40). As a matter of fact, as we shall see later, in many cases such ideals admit a simple purely algebraic description.

DEFINITION 1: A proper ideal I of a normed ring R is called *primary* if it is contained in only one maximal ideal of R. A normed ring is called *primary* if it has only one maximal ideal.

In a ring $\hat{R}$ of functions $x(M)$ a primary ideal can be characterized by the fact that the set on which all the functions $x(M) \in I$ vanish consists of a single point. If I is a closed primary ideal, then the residue-class ring R/I contains a unique maximal ideal. Thus, *the residue-class ring of every commutative normed ring with respect to a closed primary ideal is a primary ring.*

By Theorem 1 of § 36, for every maximal ideal M_0 of a regular ring R there exists a minimal closed primary ideal $\overline{J(M_0)}$. In this case $x(M) \in \overline{J(M_0)}$ if and only if there exists a sequence of functions $x_n(M) \in R$ tending to zero in norm and such that every function $x_n(M)$ coincides with $x(M)$ in a neighborhood of the point M_0.

***2.** As an illustration, let us find the minimal primary ideals of the ring D_m (§ 1, Example 2). The functions $x(t) \in D_m$ that vanish at $t = t_0$ together with all their derivatives up to and including the m-th form a closed primary ideal J in D_m. Let us show that it is the minimal primary ideal belonging to the point t_0.

We consider an arbitrary function $h(t) \in D_m$ that is equal to 1 in a neighborhood of t_0 and to 0 outside another neighborhood of the same point, for example, the neighborhood $|t-t_0| \leqq c$. Let $A = \sup_{\substack{0 \leqq t \leqq 1 \\ 0 \leqq k \leqq m}} |h^{(k)}(t)|$. The derivative of the functions $h_\nu(t) = h(\nu(t-t_0))$ can be estimated as follows:

$$|h_\nu{}^{(k)}(t)| \leqq \nu^k A = O(\nu^k) \qquad (k = 0, 1, \ldots, m). \tag{1}$$

On the other hand, for the derivatives of the given function $f(t) \in J$ in the domain $|t-t_0| \leqq \varrho$ we can easily obtain by integration the following estimates:

$$|f^{(r)}(t)| = o(\varrho^{m-r}) \qquad (r = 0, 1, \ldots, m). \tag{2}$$

We now estimate the norm of the product $h_\nu f$. We note that $h_\nu(t)$ vanishes for $|t-t_0| \geqq c/\nu$; we therefore put $\varrho = c/\nu$ in (2). By Leibniz' formula, we have

$$|(h_\nu f)^{(q)}| \leqq \sum_{k=0}^{q} \binom{q}{k} |h_\nu^{(k)}| |f^{(q-k)}| =$$

$$= \sum_{k=0}^{q} \binom{q}{k} O(\nu^k)\, o\left(\frac{1}{\nu^{m-q-k}}\right) = o\left(\frac{1}{\nu^{m-k}}\right).$$

Hence it follows that

$$\lim_{\nu \to \infty} \|h_\nu f\| = 0.$$

But every function $h_\nu f$ coincides with f in some neighborhood of t_0. By Theorem 1 of § 36, we obtain that f belongs to the minimal closed primary ideal corresponding to t_0. //

Let us construct the residue-class ring D_m/J, where J is the above ideal. Since the function $t-t_0$ is a generator of the ring D_m, its image X under the canonical mapping of D_m onto D_m/J is a generator of the ring D_m/J. Since the function $(t-t_0)^{m+1}$, together with its derivatives up to and including the m-th, vanishes at t_0, i.e., since it belongs to J, we have $X^{m+1} = 0$. It is also obvious that $X^m \neq 0$. Hence it follows that $a_0 + a_1 X + \ldots + a_m X^m = 0$ only when $a_0 = a_1 = \ldots = a_m = 0$. For by multiplying both sides by X^m we find that $a_0 X^m = 0$, and hence $a_0 = 0$; then, by multiplying by X^{m-1} we also obtain that $a_1 = 0$, etc. Thus, D_m/J is the ring of all polynomials of the form $a_0 + a_1 X + \ldots + a_m X^m$, i.e., *the ring D_m/J is finite-dimensional* (and isomorphic to the ring $I^{(m)}$ of Example 4 of § 1).

Note 1: It can be shown similarly that, in the ring $D_m(G)$ formed by the point functions $t = (t_1, \ldots, t_n)$ in a domain $G \subset R^n$ having continuous partial derivatives up to and including the m-th, the minimal primary ideal $\overline{J(t^0)}$ consists of precisely those functions $x(t)$ that, together with their partial derivatives up to and including those of the m-th order, vanish at t^0. The residue-class ring $D_m(G)/\overline{J(t^0)}$ is also finite-dimensional and consists of all polynomials of degree not exceeding m in the variables $X_1, X_2, \ldots, X_n$, where X_j is the image in $D_m(G)/\overline{J(t^0)}$ of the function $\varphi_j(t) = t_j - t_j^0$.

§ 38. Locally Isomorphic Rings

1. DEFINITION 1: Let R' and R'' be regular rings of functions and let M_0' and M_0'' be fixed points of the compacta $\mathfrak{M}(R')$ and $\mathfrak{M}(R'')$. The rings R' and R'' are called *locally isomorphic at the points M_0' and M_0''* if there exist neighborhoods $U(M_0')$ and $U(M_0'')$ that can be mapped homeomorphically onto each other so that the restriction to $U(M_0')$ of every function of R' goes over into the restriction to $U(M_0'')$ of a function of R'', and vice versa.

The very definition of the minimal closed primary ideal $J(M_0)$ (henceforth, we shall only discuss closed ideals and shall therefore omit the closure symbol) suggests that the structure of the residue-class ring $R/J(M_0)$ depends only on the behavior of the functions $x(M) \in R$ in a neighborhood of M_0. And in fact, the following theorem holds:

THEOREM 1: *If the rings R' and R'' are locally isomorphic at the points M_0' and M_0'', then the residue-class rings $R'/J(M_0')$ and $R''/J(M_0'')$ are isomorphic.*

Proof: Within $U(M_0')$ and $U(M_0'')$ we choose closed neighborhoods V' and V'' of the points M_0' and M_0''. There is a natural algebraic isomorphism between the residue-class rings $R'/I(V')$ and $R''/I(V'')$. But for rings of functions an algebraic isomorphism is always a topological one as well (Theorem 2 of § 9). Suppose now that $X' \in R'/J(M_0')$; we take an arbitrary function $x'(M) \in X'$ and consider the function $x''(M) \in R''$ that goes over into $x'(M)$ under the homeomorphism $U(M_0'') \to U(M_0')$. Under the homomorphism $R'' \to R''/J(M_0'')$ this function goes over into a certain residue class X'', which we now associate with X'. Let us show that this mapping is single-valued. Let $x_1'(M)$ be any other function of X' and $x_1''(M)$ the corresponding function in R''. The difference $x'(M) - x_1'(M)$ is in $J(M_0')$ and therefore there exists a sequence of functions $y_n'(M) \in R'$ that are equal to zero in a neighborhood of M_0' and for which

$$x'(M) - x_1'(M) - y_n'(M) \to 0$$

in the norm of the ring R'. This limit relation is preserved in the residue-class ring $R'/I(V')$, and since this ring is isomorphic with the residue-class ring $R''/I(V'')$, it is also preserved in the latter. Therefore we can find a sequence of functions $y_n''(M)$ $(n=1, 2, \ldots)$ in R'' that are equal to zero in a neighborhood of M_0'' and for which $x''(M)-x_1''(M)-y_n''(M)\to 0$ in the norm of R''. But this means that the functions $x''(M)$ and $x_1''(M)$ belong to the same residue class $X''\in R''/J(M_0'')$. Thus, the mapping $X'\to X''$ is single-valued. By the complete symmetry of our conditions, it is also single-valued in the opposite direction. It is easy to verify that this mapping is an isomorphism that preserves convergence. //

***2.** As an illustration, let us consider the ring $D_m(\mathrm{K})$ of all functions $f(t)$ on the circle K $(0\leqq t\leqq 2\pi)$ (where 0 and 2π are identified) whose first m derivatives are continuous. Obviously, this ring is locally isomorphic to the ring D_m (on an interval) at every point t_0' of K and every interior point t_0'' of the interval. Therefore the residue-class rings $D_m(\mathrm{K})/J(t_0')$ and $D_m/J(t_0'')$ are isomorphic. Moreover, it follows from the proof of Theorem 1 that the minimal ideal $J(t_0')\subset D_m(\mathrm{K})$ consists of functions with the same local characteristic as in the ring D_m, i.e., functions that, together with their first m derivatives, are equal to zero at $t=t_0'$.

***3.** A non-trivial example of a pair of locally isomorphic rings are W and V. The space of maximal ideals of the first is a circle K $(0\leqq t\leqq 2\pi)$ (where 0 and 2π are identified), and of the second, the line $-\infty<s<\infty$ supplemented by the point at infinity; here we regard V as a ring of functions on the maximal ideals, i.e., of the functions $y(s)=\lambda+\int_{-\infty}^{\infty} a(t)e^{ist}dt$, where $|\lambda|+\int_{-\infty}^{\infty}|a(t)|\,dt=\|y\|<\infty$ and $y(\infty)=\lambda$. We choose an arbitrary point t_0 on K and a point s_0 on the line $-\infty<s<\infty$ and examine the intervals $U'=\{t\in\mathrm{K}:\ |t-t_0|\leqq c<2\pi\}$, $U''=\{s:\ |s-s_0|\leqq c\}$. We shall show that the restriction of every function $x(t)\in W$ to U' goes over under the change of argument from $t-t_0$ to $s-s_0$ into the restriction of a function of V to U'', and vice versa. In the proof, we shall assume that $t_0=0$ and $s_0=0$, which we can do without loss of generality. We suppose that the chosen function $x(t)\in W$ admits, on U', the expansion

$$x(t)=\sum_{-\infty}^{\infty} a_n e^{int}, \qquad \text{where} \qquad \sum_{-\infty}^{\infty}|a_n|<\infty.$$

Suppose further that the function

$$h(s)=\int_{-\infty}^{\infty} e^{is\tau}a(\tau)\,d\tau\in V$$

is equal to 1 for $|s|\leqq c$ and to 0 outside some wider interval (we can take, for example, an arbitrary smooth function having the two last properties). Let us verify that the product $h(s)x(s)$ also belongs to V. Indeed,

$$h(s)\,x(s) = h(s)\sum_{-\infty}^{\infty} a_n e^{ins} = \sum_{-\infty}^{\infty} a_n h(s)\, e^{ins} =$$

$$= \sum_{-\infty}^{\infty} \int_{-\infty}^{\infty} e^{is(\tau+n)} a_n a(\tau)\, d\tau = \sum_{-\infty}^{\infty} \int_{-\infty}^{\infty} a_n e^{is\tau}\, a(\tau - n)\, d\tau =$$

$$= \int_{-\infty}^{\infty} e^{is\tau} \sum_{-\infty}^{\infty} a_n a(\tau - n)\, d\tau = \int_{-\infty}^{\infty} e^{is\tau} b(\tau)\, d\tau,$$

where

$$\int_{-\infty}^{\infty} | b(\tau) |\, d\tau \leqq \sum_{-\infty}^{\infty} \int_{-\infty}^{\infty} | a_n |\; | a(\tau - n) |\, d\tau =$$

$$= \sum_{-\infty}^{\infty} | a_n | \int_{-\infty}^{\infty} | a(\tau) |\, d\tau < \infty.$$

But on the interval $| s | \leqq c$ the function $h(s)x(s)$ coincides with $x(s)$; hence it follows that $x(s)$ belongs locally to V at the point $s = 0$.

Conversely, let us assume that the chosen function $y(s) \in V$ admits on U'' the expansion

$$y(s) = \lambda + \int_{-\infty}^{\infty} a(\tau)\, e^{is\tau}\, d\tau,$$

where

$$\int_{-\infty}^{\infty} | a(\tau) |\, d\tau < \infty.$$

We consider the ideal I formed in W by all the functions $x(t)$ that are equal to zero on the interval $| t | \leqq c < 2\pi$ and the residue-class ring W/I. The functions $e^{it\tau}$ exist in this ring for every real τ (because they can be extended to the whole circle K as differentiable periodic functions). Let us show that the norms of these functions in W/I are uniformly bounded with respect to τ. Now for τ integral, $\tau = n$, the function e^{itn} is, in the residue-class ring W/I, the image of the function e^{itn} in the ring W itself; but in W we have $\| e^{itn} \|_W = 1$, and hence it follows that the inequality $\| e^{itn} \|_{W/I} \leqq 1$ holds in W/I. For every real τ, we can write

$$\tau = n + \alpha,$$

where $0 \leqq \alpha < 1$; hence it follows that

$$\| e^{it\tau} \|_{W/I} \leqq \| e^{itn} \|_{W/I} \| e^{it\alpha} \|_{W/I} \leqq \sup_{0 \leqq \alpha < 1} \| e^{it\alpha} \|_{W/I}.$$

But the element $e^{it\alpha} \in W/I$, considered as a function of α, is continuous and therefore bounded in norm on the interval $0 \leqq \alpha \leqq 1$. Consequently, $\| e^{it\tau} \|_{W/I}$ is uniformly bounded for all τ. Since the element $e^{it\tau}$ of W/I is continuous in norm as a function of τ and bounded, the integral $\int_{-\infty}^{\infty} e^{it\tau} a(\tau) d\tau$ converges in norm and therefore $\lambda + \int_{-\infty}^{\infty} e^{it\tau} a(\tau) d\tau$ is an element of W/I. To find the value of the corresponding function on the maximal ideal t_0, $| t_0 | \leqq c$, we have to substitute t_0 for t and we obtain as a result the numerical integral

$$\lambda + \int_{-\infty}^{\infty} e^{it_0\tau} a(\tau)\, d\tau = y(t_0).$$

Thus, every function $y(t) \in V$ belongs locally to W at the point $t = 0$. Therefore the rings V and W are locally isomorphic at arbitrary points $t_0 \in K$ and $s_0 \neq \infty$. We shall describe below the structure of the residue-class rings of W and V with respect to the minimal primary ideals.

§ 39. Connection between the Residue-Class Rings of Two Rings of Functions, One Embedded in the Other

1. Let R' and R'' be regular rings of functions having the same space of maximal ideals $\mathfrak{M}$ and with $R' \subset R''$. We denote by J' and J'' the minimal closed ideals in these rings corresponding to a fixed point $M_0 \in \mathfrak{M}$. We shall show that *the residue-class ring R'/J' admits a natural homomorphism into the residue-class ring R''/J''.*

We fix a residue class $X' \in R'/J'$ and let $x' \in R'$ be an arbitrary function of this class. Since $R' \subset R''$, x' can be regarded as an element of R''; as such, it determines a residue class $X'' \in R''/J''$, which we shall associate with the given class X'. It is necessary to prove the uniqueness of this definition. Let $y' \in R'$ be another element of the same residue class X'. Then the difference $x' - y'$ belongs to J'. By Theorem 1 of § 36, R' contains a sequence of functions $h_\nu'(M)$ converging to zero in the norm of R' and such that every function $h_\nu'(M)$ coincides with $x'(M) - y'(M)$ in some neighborhood of M_0. But these functions $h_\nu'(M)$, considered as elements of R'', also tend to zero (in the norm of R''), in virtue of the relation between convergence in rings one of which is imbedded in the other (Theorem 1 of

§ 9). Thus, the function $x'-y'$ also belongs to J'', i.e., x' and y' belong to the same residue class of R''/J''.

The mapping $R'/J' \to R''/J''$ so constructed is obviously a homomorphism. We shall now verify that if R' is dense in R'', then its image R'/J' is dense in R''/J''. Let X'' be a residue class of R''/J'' and $x'' \in R''$ an arbitrary element of this class. By assumption, there is a sequence $x_1', \ldots, x_\nu', \ldots \subset R'$ converging to x'' in the metric of R''. We examine the corresponding classes $X_1', \ldots, X_\nu', \ldots$ in R'/J' and their images $X_1'', \ldots, X_\nu'', \ldots$ in R''/J''. We claim that in R''/J'' the equation $X''=\lim_{\nu\to\infty} X_\nu''$ holds in R''/J''. In fact, we have by definition

$$\|X''-X_\nu''\| = \inf_{y_\nu'' \in X''} \|x''-y_\nu''\|_{R''} \leqq \|x''-x_\nu''\|_{R''} \to 0,$$

as required. //

Particularly interesting is the case where the residue-class ring R'/J' is finite-dimensional. Then its image in R''/J'' is also finite-dimensional. If R' is dense in R'', then, by what has been shown, the image R'/J' is also dense in R''/J'', and so R''/J'' coincides with this image and is therefore also finite-dimensional. We formulate this result as follows:

THEOREM 1: *Let R' and R'' be regular rings having the same space of maximal ideals $\mathfrak{M}$; furthermore, let $R' \subset R''$ and let R' be dense in R''. If $J' \subset R'$ and $J'' \subset R''$ are the minimal primary ideals corresponding to the same point $M_0 \in \mathfrak{M}$ and the residue-class ring R'/J' is finite-dimensional, then the residue-class ring R''/J'', as a homomorphic image of R'/J', is also finite-dimensional and its dimension is not greater than that of R'/J'.*

***2.** Suppose, for example, that it is known that a certain normed ring R consists of functions defined on a closed domain G of n-dimensional space and that it contains the ring $D_m(G)$ of all functions that have continuous partial derivatives up to and including those of the m-th order in this domain as an everywhere-dense set. We already know (see §§ 37 and 38) that $D_m(G)$ has a finite-dimensional residue-class ring $D_m(G)/J(t^0)$ for every point $t^0 \in G$. By Theorem 1, the residue-class ring $R/J(t^0)$ is also finite-dimensional.

As an illustration, let us find the minimal ideals in the ring W. As we know, the set of maximal ideals of this ring is the circle K $(0 \leqq t \leqq 2\pi)$ (with the points 0 and 2π identified). It is well known that every periodic function $f(t)$ of period 2π having a continuous derivative belongs to W. Thus, $W \supset D_1(\mathrm{K})$. Further, $D_1(\mathrm{K})$ contains all the exponentials $\exp(int)$ $(n=0, \pm 1, \pm 2, \ldots)$ and is therefore contained in W as a dense set. The residue-class rings $D_1(\mathrm{K})/J(t_0)$ are known: they are isomorphic to the corresponding residue-class rings $D_1/J(t_0)$ (see § 38) and hence, as we have seen in § 37, each of them is a ring of linear polynomials $a_0 + a_1X$, where X is the image of the function $t - t_0$ and $X^2 = 0$. By Theorem 1, the residue-class ring $W/J(t_0)$ is a homomorphic image of the ring $\{a_0 + a_1X\}$ and is therefore either this same ring or the field of complex numbers. We shall show that the first alternative must be excluded. Let X' be the image of X in $W/J(t_0)$. Then in any case

$$\exp(inX') = 1 + inX' \qquad (n = 0, \pm 1, \pm 2, \ldots). \tag{1}$$

On the other hand, $\exp(inX')$ is the image under the homomorphism $W \to W/J(t_0)$ of the function $\exp(in(t - t_0)) = \exp(int) \cdot \exp(-int_0)$, which has the norm 1 in W. Since, under the canonical homomorphism, the norms in the residue-class rings do not increase, we must have

$$\| 1 + inX' \| = \| \exp(inX') \|_{W/J(t_0)} \leqq 1.$$

If $\| X' \|_{W/J(t_0)} > 0$, we have $\| 1 + inX' \| \geqq n \| X' \| - 1 \to \infty$ as $n \to \infty$, and this is a contradiction to (1). Thus, $X' = 0$, and *the residue-class ring of W with respect to the minimal ideal $J(t_0)$ corresponding to the point t_0 is the field of complex numbers.* Hence it follows that *in the ring W every minimal closed primary ideal coincides with the corresponding maximal ideal.*

As a consequence of the local isomorphism of the rings V and W, we now find that every closed primary ideal in V corresponding to a point $s \neq \infty$ coincides with the maximal ideal M_s.

***3.** Similar arguments can be applied to the ring of absolutely convergent Fourier series of functions of several variables. For example, in the case of functions of n variables $t_1, \ldots, t_n$, the set of maximal ideals of this ring W_n is the n-dimensional torus K^n $\{0 \leqq t_j \leqq 2\pi \quad (j = 1, \ldots, n)\}$ (where 0 and 2π are identified). Instead of the ring $D_1(K)$ of functions with a continuous first derivative that figured in the preceding construction we now have to take the ring $D_1(K^n)$ formed by the functions $x(t_1, \ldots, t_n)$ having continuous derivatives of the form $\dfrac{\partial^r x}{\partial t_{j_1} \ldots \partial t_{j_r}}$ $(r \leqq k)$, where the differentiation with respect to each argument is carried out not more than once. It is easy to verify that $D_1(K^n) \subset W_n$. In the ring $D_1(K^n)/J(t^0)$ generated by the images X_j of the functions $t_j - t_j^0$ we have $X_j^2 = 0$ $(j = 1, \ldots, n)$; this enables us to carry through the concluding part of the proof without change.

Thus, *in the ring W_n of absolutely convergent n-fold Fourier series every primary ideal coincides with a maximal one.* The same is true for the ring

V_n of absolutely convergent n-fold Fourier integrals, which is locally isomorphic to W_n.

§ 40. Wiener's Tauberian Theorem

***1.** In the ring V of absolutely convergent Fourier integrals there is one maximal ideal M_∞ at which, in general, the local isomorphism with W does not hold. We shall show presently that the maximal ideal M_∞ does not contain any other closed primary ideal. As we shall see below, this fact is equivalent to Wiener's Tauberian Theorem [76, 77].

Let us examine the minimal ideal J belonging to the point M_∞; it is formed by all the functions $y(s) \in V$ that vanish for all sufficiently large $|s|$. We have to show that its closure $\bar{J}$ coincides with the whole ideal M_∞. Since J is obviously invariant under translations and multiplications by $\exp(i\tau s)$ with an arbitrary τ, the corresponding subspace L' of L^1, formed by the integrable functions $x(t)$ whose Fourier transforms

$$y(s) = \int_{-\infty}^{\infty} x(t) \exp(its) dt$$

constitute J, is also invariant under translations and multiplications by $\exp(i\tau t)$. If the closure $\overline{L'}$ of this subspace L' did not coincide with the entire space L^1, then there would be a non-zero linear functional on L^1 that vanishes on $\overline{L'}$. Like every linear functional on L^1, it has the form

$$\langle x, f\rangle = \int_{-\infty}^{\infty} x(t) f(t) dt, \tag{1}$$

where $f(t)$ is a bounded measurable function. By what has been proved, we have $x(t) \exp(iht) \in L'$ for every function $x(t) \in L'$ and every real h, so that

$$\int_{-\infty}^{\infty} f(t) x(t) \exp(iht) dt = 0,$$

i.e., the Fourier transform of the integrable function $f(t)x(t)$ is equal to zero. By the uniqueness theorem, it follows from this that the product $f(t)x(t)$ is equal to zero for almost all t. But since $x(t)$ is not identically zero and can be translated freely along the t-axis without invalidating the result, $f(t)$ is almost everywhere equal to 0. But then (1) is the zero functional on L^1, in contradiction to the assumption. Hence $\overline{L'} = L^1$ and $\bar{J} = V$. //

***2.** Wiener's Theorem itself can be formulated as follows [76, 77]:

If the function $x_0(t) \in L^1$ has a Fourier transform $x_0^\sim(s)$ that does not vanish for any real value of s, and if $f(t)$ is a bounded measurable function *such that*

$$\int_{-\infty}^{\infty} x_0(t-\tau)f(t)dt \to l \int_{-\infty}^{\infty} x_0(t)dt \qquad as \qquad \tau\to\infty,$$

then

$$\int_{-\infty}^{\infty} x(t-\tau)f(t)dt \to l \int_{-\infty}^{\infty} x(t)dt \qquad as \qquad \tau\to\infty \tag{2}$$

for every function $x(t)\in L^1$.

Proof: For the proof, we argue as follows. Let I be the set of all functions $x(t)\in L^1$ satisfying the condition (2). If we denote by f_τ the functional on L^1 defined by the (bounded measurable) function $f(t+\tau)$, then the relation (2) can be written in the form

$$\lim_{\tau\to\infty} \langle x, f_\tau\rangle = \langle x, l\rangle.$$

Hence it follows that I is a closed linear subspace of L^1. It is also clear that I is invariant under the operations of translation $x(t)\to x(t-h)$. But then I is an ideal in V, because the convolution

$$(x * y)(t) = \int_{-\infty}^{\infty} x(t-h)y(h)dh$$

is the limit in the norm of V of linear combinations of translations of the function $x(t)$. This ideal contains $x_0(t)$, whose Fourier transform does not vanish for any s. But this means that it is not contained in any of the maximal ideals M_s $(s\neq\infty)$. Therefore I is a primary ideal belonging only to the maximal ideal M_∞. But then, by what has been proved above, $I=M_\infty=L^1$. //

***3.** Our proof of the fact that every closed primary ideal contained in M_∞ coincides with M_∞ carries over without change to the case of several variables (even of a locally compact group) and also to the case of the ring $V[\alpha]$, provided only that this ring is regular. With a more rapid growth function $\alpha(t)$, where the space of maximal ideals of $V[\alpha]$ becomes a strip in the s-plane (the ring $V[\alpha]$ is then necessarily not regular), the maximal ideal M_∞ always contains a continuum of distinct primary ideals (see Theorem 1 of § 45). They were completely investigated by Korenblum [30].

§ 41. Primary Ideals in Homogeneous Rings of Functions

***1.** A normed ring R of functions $x(t)$ with the circle K $(0\leq t\leq 2\pi)$ (where 0 and 2π are identified) as its space of maximal ideals is called *homogeneous* if: 1) it has $\exp(it)$ and $\exp(-it)$ as generating functions; and

2) all the translations $x(t+h)$ (where $t+h$ is taken reduced modulo 2π) belongs to R, and $\| x(t+h) \| = \| x(t) \|$.

Examples of homogeneous rings are the ring $C(\mathrm{K})$ of all continuous functions on K, the ring $D_m(\mathrm{K})$ of all functions on K with m continuous derivatives, the ring W of absolutely convergent Fourier series, the ring $W[\alpha]$ (§ 19) subject to the condition $\lim_{n\to\infty} \sqrt[n]{\alpha_n} = \lim_{n\to\infty} \sqrt[n]{\alpha_{-n}} = 1$, which ensures that the space of its maximal ideals coincides with K.

In what follows, we shall impose a further condition on R:

3) R contains[2] a ring $D_m(\mathrm{K})$ for some value of m.

When the condition 3) is satisfied, the ring R is regular. Further on we shall describe the structure of the primary ideals of this ring.

***2.** When condition 3) is satisfied, we can estimate the order of magnitude of the numbers $\alpha_n = \| \exp(int) \|$ as $n \to \pm\infty$. In fact, 3) is equivalent to the condition that the order of magnitude of these numbers is not greater than that of a power:

3′) $\alpha_n = O(|n|^p)$ for some value of p.

For if $R \supset D_m(\mathrm{K})$, then by Theorem 1 of § 9, there exists a constant C such that for every $x(t) \in D_m(\mathrm{K})$

$$\| x \|_R \leqq C \| x \|_{D_m(\mathrm{K})}.$$

In particular, putting $x(t) = \exp(int)$, we obtain

$$\alpha_n = \| \exp(int) \|_R \leqq C \| \exp(int) \|_{D_m(\mathrm{K})} = O(|n|^m),$$

so that condition 3′) is satisfied for $p = m$. Conversely, if condition 3′) is satisfied and $f(t)$ is an arbitrary function with continuous derivatives of order up to and including $m = p+2$, then we can show that $f(t)$ occurs in R.

For if this function is expanded in a Fourier series

$$f(t) = \sum_{-\infty}^{\infty} b_n \exp(int), \tag{1}$$

then the Fourier coefficients of the function $f^{(p+2)}(t)$, which are equal in absolute value to $|n|^{p+2} |b_n|$, are in any case bounded and then we have

$$\sum_{-\infty}^{\infty} |b_n| \, \| \exp(int) \| = \sum_{-\infty}^{\infty} |b_n| \alpha_n \leqq$$
$$\leqq \sum_{-\infty}^{\infty} O(|n|^{-p-2}) O(|n|^p) = \sum_{-\infty}^{\infty} O(1/n^2) < \infty,$$

---

[2] We could have replaced this condition by one that is formally weaker but in fact equivalent (see § 9): R contains the ring $D_\infty(\mathrm{K})$ of all infinitely differentiable functions on K.

so that the series (1) converges in the norm of R; and hence $R \supset D_{p+2}(K)$, as required.

By condition 1), $D_m(K)$ is contained in R as an everywhere-dense set and we can apply Theorem 1 of § 39, which states in this case that the residue-class ring of R with respect to the minimal closed primary ideal $J(t_0)$ corresponding to the point $t_0 \in K$ is a homomorphic image of the residue-class ring $D_m(K)/J(t_0)$. The latter ring, as we know (see §§ 38 and 37), is the ring of all polynomials of the form $a_0 + a_1 X + \ldots + a_m X^m$, where X is the image[3] in $D_m(K)/J(t_0)$ of the function $t - t_0$, $X^{m+1} = 0$, and $X^m \neq 0$. Therefore $R/J(t_0)$ is the ring of all polynomials of the form

$$a_0 + a_1 Z + \ldots + a_l Z^l,$$

where Z is the image of X under the homomorphism $D_m(K)/J(t_0) \rightarrow R/J(t_0)$ and the number $l \leqq m$ is such that[4] $Z^l \neq 0$, but $Z^{l+1} = 0$.

THEOREM 1: *The number l, which determines the dimension of the residue-class ring $R/J(t_0)$ can be obtained from the relations*

$$R \subset D_l(K), \qquad R \not\subset D_{l+1}(K).$$

Proof: A function $f(t) \in R$ that is analytic in a neighborhood of t_0 and in this neighborhood has the Taylor expansion

$$f(t) = f(t_0) + (t - t_0) f'(t_0) + \frac{1}{2!}(t - t_0)^2 f''(t_0) + \cdots$$

corresponds in the residue-class ring $R/J(t_0)$ to the element

$$f(t_0) + Z f'(t_0) + \frac{1}{2!} Z^2 f''(t_0) + \cdots + \frac{1}{l!} Z^l f^{(l)}(t_0). \tag{2}$$

For

$$\sum_{n=0}^{\infty} \frac{f^{(n)}(t_0)}{n!}(t - t_0)^n$$

is a series of analytic functions uniformly convergent in a domain containing the set of maximal ideals of $R/J(t_0)$ (i.e., the point t_0); therefore, by the results of § 6, it corresponds to the series

$$\sum_{n=0}^{\infty} \frac{f^{(n)}(t_0)}{n!} Z^n,$$

---

[3] By superimposing the canonical mapping $D_m(K) \rightarrow D_m(K)/J(t_0)$ on the local isomorphism $D_m \rightarrow D_m(K)$ at the point t_0.

[4] In fact, it follows from the last two conditions, just as in § 37.2, that

$$a_0 + a_1 Z + \ldots + a_l Z^l = 0$$

only when $a_0 = a_1 = \ldots = a_l = 0$.

converging in the norm of $R/J(t_0)$; but this series reduces to the polynomial (2).

Since $R/J(t_0)$ is finite-dimensional and all norms on a finite-dimensional space are equivalent, there exists a constant C with the property

$$\left\| \sum_{k=0}^{l} a_k Z^k \right\|_{R/J(t_0)} \geqq C \sum_{k=0}^{l} |a_k|$$

for every element $\sum_{k=0}^{l} a_k Z^k \in R/J(t_0)$. Since under the homomorphism $R \to R/J(t_0)$ the norms do not increase, we have

$$\| f(t) \|_R \geqq \left\| \sum_{k=0}^{l} \frac{1}{k!} f^{(k)}(t_0) Z^k \right\|_{R/J(t_0)} \geqq C \sum_{k=0}^{l} \frac{1}{k!} | f^{(k)}(t_0) |.$$

But inasmuch as the ring R and its norm are invariant under rotations of the circle (condition 2), the constant C can be taken to be independent of the choice of t_0, and then we have

$$\| f(t) \|_R \geqq C \max_{t_0 \in \mathrm{K}} \sum_{k=0}^{l} \frac{1}{k!} | f^{(k)}(t_0) | = C \| f(t) \|_{D_l(\mathrm{K})}. \tag{3}$$

The functions $f(t)$ to which these arguments are applicable are everywhere dense in R. Therefore, after a passage to the limit, the inequality (3) proves to hold for all functions $f(t) \in R$. In particular, it follows that *R is contained in the ring $D_l(\mathrm{K})$*. But it is not contained in the smaller ring $D_{l+1}(\mathrm{K})$. For in the residue-class ring $R/J(t_0)$ we have $Z^{l+1} = 0$ for the image Z of the element of R that coincides locally with $t - t_0$, and this must also hold in the residue-class ring $R'/J(t_0)$, where R' is any ring larger than R; but in $D_{l+1}(\mathrm{K})/J(t_0)$, as we know, $X^{l+1} \neq 0$. Thus, $R \subset D_l(\mathrm{K})$ and $R \not\subset D_{l+1}(\mathrm{K})$. //

***3.** For the ring $W[\alpha]$, the result obtained can be formulated directly in terms of the sequences $\{\alpha_n\}$. For *if $R = W[\alpha]$, then l is the smallest of the integers r for which $\varliminf_{|n| \to \infty} \alpha_n / |n|^{r+1} = 0$.*

For the proof, we assume to begin with that for some integer r

$$\varliminf_{|n| \to \infty} \alpha_n / |n|^{r+1} = 0,$$

and we show that $r \geqq l$. We choose a sequence $0 < n_1 < n_2 < \ldots < n_k < \ldots$ such that the series $\sum_{k=1}^{\infty} \frac{\alpha_{n_k}}{n_k^{r+1}}$ converges. Next we construct a sequence of positive numbers $p_k \to \infty$ that are subject to the condition that the series $\sum_{k=1}^{\infty} p_k \frac{\alpha_{n_k}}{n_k^{r+1}}$ also converges and we put

$$c_n = \begin{cases} \frac{p_k}{n_k^{r+1}} \text{ for } n = n_k \qquad (k = 1, 2, \ldots). \\ 0 \quad \text{for the remaining integers } n. \end{cases}$$

Obviously, the function $\varphi(t) = \sum_{-\infty}^{\infty} c_n \exp(int)$ belongs to $W[\alpha]$ and has no more than r continuous derivatives (because the series $\sum_{k=1}^{\infty} p_k \exp(in_k t)$ diverges). Since $W[\alpha] \subset D_l(\mathrm{K})$, it follows from this that $r \geqq l$.

On the other hand, the relation

$$\lim_{|n| \to \infty} \frac{\alpha_n}{|n|^{l+1}} = 0$$

is bound to hold. For otherwise we would have

$$\alpha_n \geqq C |n|^{l+1} \qquad (n = 0, \pm 1, \pm 2, \ldots; C > 0)$$

and consequently for every $f(t) = \sum_{-\infty}^{\infty} a_n \exp(int) \in W[\alpha]$ we would have the inequality $\sum_{-\infty}^{\infty} |a_n| |n|^{l+1} \leqq C^{-1} \sum_{-\infty}^{\infty} |a_n| |\alpha_n| < \infty$ and hence it would follow that $W[\alpha] \subset D_{l+1}$, in contradiction to Theorem 1. //

As a corollary, we obtain:

In the ring $W[\alpha]$, *where*

$$\alpha_n \leqq C(1 + |n|)^q, \tag{4}$$

every primary ideal coincides with a maximal ideal if and only if

$$\lim_{|n| \to \infty} \alpha_n / |n| = 0.$$

Applying the results of § 37, we can also state this corollary in another form.

Let the sequence $\{\alpha_n\}$ *satisfy the condition* (4). *Every function* $f(t) \in W[\alpha]$ *that vanishes for at least one point can be approximated in norm with an arbitrary degree of accuracy by a function* $\varphi(t) \in W[\alpha]$ *equal to zero on an entire interval including this point if and only if the lower limit of the sequence* $\alpha_n / |n|$ *is equal to zero.*

§ 42. Remarks on Arbitrary Closed Ideals. An Example of L. Schwartz

***1.** As we have seen in § 36, every closed ideal in the ring $C(S)$ is the intersection of maximal ideals. The structure of closed ideals is also known in certain types of rings besides $C(S)$. Thus, in the ring D_m, every closed ideal is the intersection of primary ideals; this was proved by Shilov in 1940 [55] for a single variable and by Whitney in 1948 for several variables [75]. There are also other examples of rings of functions in which all the closed ideals have been described ([59], [66], [22]). But even for such a comparatively simple ring as W—or V, which is locally isomorphic to W—the structure of the closed ideals is not known. Since there are no primary ideals in W other than the maximal ones (§ 39), it would be natural to expect that, just as in the ring C, every closed ideal in W is the intersection of maximal ideals. However, as Malliavin [36] has shown, for any number of variables $n \geqq 1$, *the ring W_n (or the ring V_n, which is locally isomorphic to it) has closed ideals that cannot be represented as intersections of maximal ideals.* We shall not give here the very complicated constructions of Malliavin; instead, we shall give a simple and instructive example that was found (at an earlier date) by L. Schwartz [68] and refers to the case $n=3$.

We consider the ring V_3 consisting of absolutely integrable functions $x(t_1, t_2, t_3) = x(t)$ of three variables (with a unit element adjoined). The Fourier transform $y(s_1, s_2, s_3) = y(s)$ of such a function $x(t)$ can be expressed by the formula

$$y(s_1, s_2, s_3) = \int\int\int x(t_1, t_2, t_3)\, e^{i(t_1 s_1 + t_2 s_2 + t_3 s_3)}\, dt_1\, dt_2\, dt_3 = \\ = \int\int\int x(t)\, e^{i\langle t, s\rangle}\, dt. \qquad (1)$$

If we go over from $x(t)$ to the function $x_h(t) = x(ht)$, where h denotes the operator of rotation in the space of points $t = (t_1, t_2, t_3)$, then $y(s)$ goes over into the function $y(hs)$. For after the changes of variable $ht = t'$, $t = h^{-1}t'$, $dt = dt'$, we have

$$\int\int\int x(ht)\, e^{i\langle t, s\rangle}\, dt = \int\int\int x(t')\, e^{i\langle h^{-1}t', s\rangle}\, dt' = \\ = \int\int\int x(t')\, e^{i\langle t', hs\rangle}\, dt' = y(hs).$$

Hence it follows that a spherically symmetrical function $x(t) = x_0(r)$, where $r = |t|$, goes over into a spherically symmetrical function $y(s) = y_0(\varrho)$, where $\varrho = |s|$. Let us find an explicit formula connecting the functions $x_0(r)$ and $y_0(\varrho)$. For this purpose we go over, in the equation (1), to spherical coordinates, taking as the polar axis the line along which the vector s lies and denoting the polar angles by ω and φ. We obtain

$$y_0(\rho) = \int\int\int x_0(r) e^{ir\rho\cos\omega} r^2 \sin\omega \, d\omega \, d\varphi \, dr =$$

$$= 2\pi \int_0^\infty x_0(r) r^2 \left(\int_0^\pi e^{ir\rho\cos\omega} \sin\omega \, d\omega \right) dr = \frac{4\pi}{\rho} \int_0^\infty x_0(r) r \sin r\rho \, dr. \tag{2}$$

We observe that $x_0(r)$ in this formula can be an arbitrary function for which the integral

$$\int_0^\infty |x_0(r)| r^2 \, dr$$

converges, because the expression

$$\int\int\int |x(t)| \, dt$$

reduces precisely to this when written in spherical coordinates.

We shall now show that the function $y_0(\varrho)$ defined for $\varrho > 0$ by the equation (2) has a continuous derivative. Indeed, formal differentiation with respect to ϱ leads to the integral

$$\int_0^\infty x_0(r) r^2 \cos r\rho \, dr,$$

which, by what was shown above, converges absolutely. Therefore the function $y_0(\varrho)$ is differentiable for $\varrho > 0$ and its derivative

$$y_0'(\rho) = \frac{4\pi}{\rho} \int_0^\infty x_0(r) r^2 \cos r\rho \, dr - \frac{4\pi}{\rho^2} \int_0^\infty x_0(r) r \sin r\rho \, dr$$

is a continuous function of ϱ.

We examine the functions $y_0(\varrho)$ on the interval $1/2 \leqq \varrho \leqq 3/2$. Of course, they form a ring having the metric of the residue-class ring of V_3 with respect to the ideal consisting of the functions that are equal to zero for $1/2 \leqq \varrho \leqq 3/2$. We denote this ring by R; by what has been proved, R is contained in the ring D_1 of functions with a continuous derivative on the interval $[1/2, 3/2]$. On the other hand, R contains all sufficiently smooth functions of ϱ; in particular, we can find in R (and consequently in V_3 as well) a function $y_0(\varrho)$ that vanishes, and whose first derivative vanishes, for $\varrho = 1$ and that is different from zero for $\varrho \neq 1$.

THEOREM 1: *The closed ideal I generated in V_3 by the functions $y_0(\varrho)$ does not coincide with the intersection of the maximal ideals containing it. In particular, it cannot contain a function $y_1(\varrho)$ that vanishes at $\varrho=1$ but has a derivative different from zero at that point.*

For let us assume that $y_1(\varrho)$ occurs in I, i.e., that it is the limit, as $\nu \to \infty$, of products $y_0(\varrho)y_\nu(s)$ in the metric of V_3. If the functions $y_\nu(s)$ were also spherically symmetrical, $y_\nu(s)=z_\nu(\varrho)$, then we would obtain a sequence of functions $y_0(\varrho)z_\nu(\varrho)$ on the interval $[1/2, 3/2]$ that converges to the function $y_1(\varrho)$ in the metric of R. This sequence would also converge in the metric of D_1 by the general theorem of the correspondence between convergence in rings embedded in each other (Theorem 1 of § 9). But convergence in the metric of D_1 is uniform convergence of the functions and their first derivatives; and for each of the functions $y_0(\varrho)z_\nu(\varrho)$ the derivative at $\varrho=1$ is equal to zero just as it is for $y_0(\varrho)$, whereas for the limit function $y_1(\varrho)$, the derivative is different from zero; the contradiction proves our assertion.

It remains to consider the case where the functions $y_\nu(s)$ are not spherically symmetrical. In that case, we consider the operator of spherical averaging Q that carries every function into its spherical average:

$$Qx = \int\int x(r\cos\omega,\ r\sin\omega\cos\varphi,\ r\sin\omega\sin\varphi)\sin\omega\, d\omega\, d\varphi.$$

Q is a linear operator in V_3 that carries every function $x(t)$ into a function $x_0(r)$ with the same norm; therefore the operator Q itself has norm 1. Since for the Fourier transform a rotation in the space of points t corresponds to the same rotation in the space of points s, Q can also be regarded as given on the points $y(s_1, s_2, s_3)$. We put $Qy_\nu(s)=z_\nu(\varrho)$. Now, a spherically symmetrical function can obviously be taken before the sign of the operator Q; hence by applying Q to the relation

$$y_0(\varrho)z_\nu(s) \to y_1(\varrho),$$

we obtain

$$y_0(\varrho)z_\nu(\varrho) \to Qy_1(\varrho) = y_1(\varrho).$$

Thus, the problem reduces to the case of spherically symmetrical factors $z_\nu(\varrho)$ examined above, and the theorem is proved. //

***2.** It would be very interesting to be able to give a description of the structure of closed ideals in the V (or in W). But, as we have already said, this problem is still open.

Note: This is not the first time that the ring W has turned out to be a source of counter-examples to hypotheses that appear natural at first sight. For a time, the hypothesis was made that a ring in which primary ideals coincide with maximal ones must contain the absolute value $|x(t)|$ of each of the functions $x(t)$ in the ring; this proved to be false in this very ring W [58]. Furthermore, as we know (§ 6), to every element x of a normed ring we can apply a function $f(\zeta)$ that is analytic on the spectrum of x; the proposition that for every ring R there exists a wider domain of functions that can be applied to each of its elements was refuted by giving this same ring W as a counter-example [29]. Finally, for the ring W we have that a change of independent variable $t = \varphi(\tau)$ that carries every function of W into a function also in W can only be linear [33], in spite of what one might expect.

CHAPTER VII

RINGS WITH UNIFORM CONVERGENCE

1. In this chapter we shall discuss certain problems concerning rings in which convergence in norm corresponds to uniform convergence of the functions $x(M)$ on the space of maximal ideals. In § 43 we shall give an account of applications to the general theory of topological spaces (the classification of compact extensions of a space S and, at the same time, of the symmetric subrings of the ring $C(S)$ of all bounded continuous functions on S). In § 44 we shall discuss arbitrary (not necessarily symmetric) subrings of $C(S)$; and in § 45, closed ideals in such subrings.

§ 43. Symmetric Subrings of $C(S)$ and Compact Extensions of a Space S

1. Let S be a topological space. A *compact extension* of S is an arbitrary compactum T (i.e., a compact Hausdorff space) containing an everywhere-dense subset homeomorphic to S.

Not every topological space has compact extensions. It is known that every compactum T is *completely regular,* i.e., that for every set $A \subset T$ and every point $t_0 \in T$ that is not an adherence point of A there exists a function $x(t)$ that is continuous on T, equal to 0 on A and different from 0 at t_0. Of course, all the subsets of T with topology induced from T have the same property. Therefore, in discussing problems on compact extensions we can restrict ourselves to completely regular spaces.

As Tychonov [71] showed in 1929, every completely regular space S has a compact extension. In 1937, Čech [10] constructed, for every regular space S, an extension βS that is maximal in the sense that every compact extension of S is obtained from βS by a continuous mapping in which all the points of S remain fixed.

In the present section we shall present a ring-theoretical approach to these problems.

Let S be a completely regular space and $C(S)$ the normed ring of all bounded continuous complex functions on S with the 'uniform norm'

$$\| x \| = \sup_{t \in S} | x(t) |. \tag{1}$$

We denote by $\mathfrak{M}$ the space of maximal ideals of $C(S)$. By Theorem 3 of § 8, $C(S)$ is isomorphic to the ring of all continuous functions on the compactum $\mathfrak{M}$. Every point $t_0 \in S$ determines a maximal ideal M_{t_0} of $C(S)$ (which is formed by the functions $x(t)$ that vanish at $t = t_0$), and distinct points correspond to distinct maximal ideals. Therefore S can be regarded as a subset of $\mathfrak{M}$. We shall show that *the original topology of S coincides with the topology induced in S by $\mathfrak{M}$.*

If $t_0 \in S$ is not an adherence point of the set $A \subset S$, then, in virtue of the complete regularity of S, there is a function $x_0(t) \in C(S)$ that is equal to 0 on A and to 1 at $t = t_0$. But in that case, the maximal ideal M_{t_0} corresponding to t_0 is not an adherence point of the set $\mathfrak{A}$ of maximal ideals corresponding to the points of A: the neighborhood $| x_0(M) - 1 | < 1$ separates M_{t_0} from $\mathfrak{A}$. Conversely, if the maximal ideal M_{t_0} is not an adherence point of $\mathfrak{A}$ in $\mathfrak{M}$, then, in virtue of the complete regularity of $\mathfrak{M}$, there exists a continuous function $x(M)$ that is equal to 0 on $\mathfrak{A}$ and to 1 for $M = M_{t_0}$; the corresponding function $x(t) \in S$ is equal to 0 at the points of A and to 1 at t_0 and hence t_0 is not an adherence point of A.

Finally, *the closure of S in $\mathfrak{M}$ is the entire space $\mathfrak{M}$.* For otherwise we could find a non-zero function $x(M)$ that is equal to zero on all the maximal ideals that correspond to the points of S; but according to (1), such a function would have to have the norm 0, in contradiction to the construction.

Thus, *$\mathfrak{M}$ is a compact extension of the space S.*

Let us show that *$\mathfrak{M}$ is a maximal compact extension in the sense of Čech.*

Let Q be a compact extension of the space S. We consider the ring $C(Q)$ of all continuous functions on Q. Since every continuous function on Q is also continuous on $S \subset Q$ and the upper bound of its absolute values on S coincides with its maximal absolute value on Q, $C(Q)$, regarded as a ring of functions on S, is a closed subring of $C(S)$. The space of maximal ideals of $C(Q)$, as we know, coincides with Q itself (see § 2.2). Now let M be a point of the space $\mathfrak{M}$; by selecting, in the maximal ideal M, all the functions contained in $C(Q)$, we obviously obtain a maximal ideal M' of $C(Q)$ and so a point of Q. The mapping $M \to M'$ leaves all the points of S fixed. Let us verify that it is continuous. Suppose that M' is not an adherence point of the set $A' \subset Q$; then the ring $C(Q)$ contains a function $x(t)$ that vanishes on A', and at the point M' is equal, say, to 1. By examining it in $C(S)$ we see that the inverse image M of the maximal ideal M' is not an adherence point of the inverse image A of A'. Hence it follows that the relation $M \in \overline{A}$ in $\mathfrak{M}$ implies that $M' \in \overline{A'}$ in Q, i.e., the mapping

$M \to M'$ is continuous. Since the image of $\mathfrak{M}$ under the mapping $M \to M'$ contains the entire space S and is compact (as the continuous image of a compact space), $\mathfrak{M}$ is mapped onto the entire space Q. Thus, Q is a continuous image of $\mathfrak{M}$ under which S goes over into itself, as required.

At the same time, we have connected every compact extension Q of S with a certain subring of $C(S)$, namely, with $C(Q)$ regarded as a ring of functions on S. The question naturally arises as to what the intrinsic properties are of these subrings $C(Q)$ that distinguishes them among all the subrings of $C(S)$. The answer is given by the following theorem:

THEOREM 1: *Let $C(S)$ be the ring of all bounded continuous functions on a completely regular space S, furnished with the uniform norm* (1). *Its subrings of the form $C(Q)$, where Q is an (arbitrary) compact extension of S, can be characterized as (arbitrary) closed subrings K of $C(S)$ satisfying the following two conditions:*

(*) *$x(t) \in K$ implies that $\overline{x(t)} \in K$;*

(**) *For every set $A \subset S$ and every point $t_0 \in S$ not belonging to its closure $\bar{A}$ there is a function $x(t) \in K$ that is equal to* 0 *on A and is different from* 0 *at $t = t_0$.*

Proof: If K satisfies the condition (*), then, by Theorem 3 of §8, it is isomorphic to the ring $C(Q)$ of all continuous functions on the space Q of its maximal ideals. Assuming that K also satisfies condition (**) we then find, in the same way as in the case of the ring $C(S)$, that Q is a compact extension of S, identified with the set of maximal ideals generated by its points; and from this it is also clear that the isomorphism between K and $C(Q)$ can be realized by assigning to every function of $C(Q)$ its restriction to S. Conversely, if Q is an arbitrary compact extension of S, then the subring K of $C(S)$ formed by the restrictions to S of all possible functions of $C(Q)$ satisfies the condition (*), because $C(Q)$ satisfies this condition, and it satisfies the condition (**), because Q, as a compact Hausdorff space, is completely regular. //

We can go further and characterize in topological terms all the symmetric closed subrings of $C(S)$, and we do not even have to assume that S is regular.

THEOREM 2: *Let $C(S)$ be the ring of all bounded continuous functions on a topological space S furnished with the uniform norm* (1). *Its closed subrings K satisfying the condition* (*) *of Theorem* 1 *(i.e., those that contain the complex conjugate of each of the functions belonging to them) are precisely the subrings that are isomorphic to an arbitrary ring of the form $C(Q')$, where Q' can be any compact extension of an arbitrary continuous completely regular image S' of S.*

Proof: Let S' be an arbitrary completely regular space for which there exists a continuous mapping φ of S onto S', and Q an arbitrary compact extension of this space S'. With every function $f(q')\in C(Q')$ we associate the function $f(\varphi(t))$ on S. Since φ can be regarded as a continuous mapping of S into Q', we have $f(\varphi(t))\in C(S)$, and it is not difficult to verify that these functions $f(\varphi(t))$ form a closed subring K of $C(S)$ that is isomorphic to $C(Q')$ and satisfies the condition (*). Conversely, let K be an arbitrary closed subring of $C(S)$ satisfying condition (*). By Theorem 3 of § 8, it is isomorphic to the ring $C(Q')$ of all continuous functions on the space Q' of its maximal ideals. To every point $t\in S$ there corresponds a maximal ideal of K and hence of $C(Q')$, i.e., a point $t'\in Q'$ such that, for every function $x\in K$ and the corresponding function $x'\in C(Q')$, we have

$$x(t)=x'(t'). \tag{2}$$

We put $t'=\varphi(t)$ and assume that $S'=\varphi(S)$. Since every neighborhood of the point $t_0'=\varphi(t_0)\in S'$ in the topology induced by $C(Q')$ contains a neighborhood of the form

$$\{t'\in S': |x_k'(t')-x_k'(t_0')|<\varepsilon \quad (k=1,\ldots,n)\}=$$
$$=\varphi(\{t\in S: |x_k(t)-x_k(t_0)|<\varepsilon \quad (k=1,\ldots,n)\}),$$

φ is a continuous mapping of S onto S'. Furthermore, S' is dense in $C(Q')$, because every function $x'\in C(Q')$ that is equal to zero on S' is also equal to zero on all of Q' by formula (2), and Q', as a compact Hausdorff space, is completely regular. Thus, Q' is a compact extension of S'. //

***2.** By way of illustration, let us consider as the space S the line $-\infty<t<\infty$ with its usual topology. There are many distinct compact extensions of this space. The minimal compact extension is obtained by adding to it the single point at infinity. The corresponding subring $K\subset C(S)$ consists of the functions $x(t)$ that have limits as $t\to+\infty$ and $t\to-\infty$, and

$$\lim_{t\to+\infty} x(t)=\lim_{t\to-\infty} x(t).$$

A wider subring is obtained when we consider the functions $x(t)$ for which these limits exist but do not necessarily coincide. The corresponding compact extension is obtained by adjoining two points, $-\infty$ and $+\infty$.

One of the symmetric subrings of $C(S)$ is the ring B of all continuous almost periodic functions on the line S. Since it separates the points of the line, S can be regarded as part of the space $\mathfrak{M}(B)$ of maximal ideals of this subring. But B does not satisfy the condition (**) of Theorem 1; for example, B contains no function that is equal to 0 for all $t\leqq 0$ and is different from 0 for an arbitrary $t_0 > 0$. Therefore the space $\mathfrak{M}(B)$ cannot be a compact extension of the line S in its original topology. But by Theorem 2, $\mathfrak{M}(B)$ is a compact extension of the line S furnished with a certain *weaker* topology. Since the functions e^{ist} are generators of B, this is nothing other than the topology in which the periodic systems of intervals and the intersections of all possible finite collections of such systems form a fundamental system of neighborhoods.

3. Let us examine the special case where the space S is itself compact. Then every continuous image of S is also a compact space and has no compact extension other than itself. Therefore, if S is compact, every symmetric closed subring K of $C(S)$ is isomorphic to the ring $C(Q)$, where Q is a continuous image of S. The space Q can be constructed from the subring K as follows. We call two points t' and t'' equivalent if $x(t') = x(t'')$ for every function $x(t) \in K$. This relation between points is reflexive, symmetric, and transitive, i.e., it is an equivalence relation and therefore splits S into a collection of classes $\tau = \{t\}$ of mutually equivalent points. By regarding the classes τ as new 'points' and by introducing the natural topology in the set of these points we obtain the space Q of maximal ideals of K. Note that *the mapping $S \to Q$ is not only continuous, but also open* (because S is compact).

§ 44. The Problem of Arbitrary Closed Subrings of the Ring $C(S)$

1. In this section it is assumed that S is compact; $C(S)$ denotes, as usual, the ring of all continuous functions on S furnished with a uniform norm, and K a subring of $C(S)$ that is closed under uniform convergence and separates the points of S.

If the subring K is symmetric, i.e., if the complex conjugate $\overline{x(t)}$ of every function $x(t)$ in K is also in K, then, by what has been proved at the end of the preceding section, $K = C(S)$. Thus, it makes sense to consider only asymmetric subrings K of $C(S)$ that separate the points of S.

The *anti-symmetric* subrings form, in a way, a class opposite to that of the symmetric subrings. A ring of functions K is called *anti-symmetric* if it follows from $x(t) \in K$, $\overline{x(t)} \in K$ that $x(t) = \text{const}$. A simple example is the subring A of the ring $C(S)$ of all continuous functions on the closed unit circle S of the complex plane formed by the functions of $C(S)$ that are analytic inside this circle.

We shall soon show that a description of arbitrary closed subrings of $C(S)$ reduces in principle to a description of the anti-symmetric subrings of this ring.

2. Let K be an arbitrary closed subring of $C(S)$; the functions $x(t)$ such that both $x(t)$ and $\overline{x(t)}$ are in K form its greatest symmetric subring K' (if K is symmetric, then $K'=K$; if K is anti-symmetric, then K' consists of constants only). The relation '$x(t_1)=x(t_2)$ for every $x \in K$'' between points t_1, t_2 of S is an equivalence relation that determines a splitting of S into classes $\tau=\{t\}$ of equivalent points (when K is symmetric, every individual point is a class; when K is anti-symmetric, the entire space S forms a single class). Every class τ is obviously a closed set. We denote by K_τ the set of functions $x(t)$ of K considered only on the class τ.

The following theorem shows that the whole ring K is, in a certain definite sense, 'pasted together' from the rings K_τ.

THEOREM 1: *If a function $f(t) \in C(S)$ belongs to the ring K_τ on every class τ, then $f(t)$ belongs to K.*

Proof: By what has been shown at the end of the preceding section, the space of maximal ideals of the symmetric ring K' is a compactum S', a continuous image of the compactum S having the classes τ as its points. We choose a number $\varepsilon > 0$ and consider an arbitrary class τ_0. By assumption, the function $f(t)$ coincides on this class with a function $x_0(t) \in K$. Since both $f(t)$ and $x_0(t)$ are continuous on S, we can find a neighborhood U_0 of τ_0 in S within which the inequality

$$|f(t)-x_0(t)| < \varepsilon$$

is satisfied. Because the mapping $S \to S'$ is open, there exists a neighborhood $V(\tau_0)$ of τ_0 in S' such that every class $\tau \in V(\tau_0)$ is entirely contained in U_0. These neighborhoods $V(\tau_0)$, constructed for every point $\tau_0 \in S'$, form a covering of the compactum S'. From this covering we select a finite covering $V_1, \ldots, V_n$ such that no proper part of it covers the entire space S'. The corresponding functions of K will be denoted by $x_1(t), \ldots, x_n(t)$ and the neighborhoods in S, by $U_1, \ldots, U_n$. By the lemma of § 35, the ring $K'=C(S')$ contains functions $h_1'(\tau), \ldots, h_n'(\tau)$ with the following properties:

1) $\quad 0 \leqq h_j'(\tau) \leqq 1;$

2) $\quad h_j'(\tau)=0 \quad$ outside $\quad V_j;$

3) $\quad h_1'(\tau)+\ldots+h_n'(\tau) \equiv 1.$

For every point $t \in S$ we put $h_j(t)=h_j'(\tau)$, where τ is the class containing t. We form the difference

$$f(t)-\sum_{j=1}^{n} h_j(t)x_j(t)=\sum_{j=1}^{n} h_j(t)\,[f(t)-x_j(t)].$$

For every point t only those terms of this sum can be different from zero for which $t \in U_j$ (because otherwise $h_j(t) = 0$). But for such terms we have, by construction, $|f(t) - x_j(t)| < \varepsilon$. Therefore, everywhere on S, we have

$$\left| f(t) - \sum_{j=1}^{n} h_j(t) x_j(t) \right| \leqq \sum_{j=1}^{n} h_j(t) \left| f(t) - x_j(t) \right| < \varepsilon \sum_{j=1}^{n} h_j(t) = \varepsilon.$$

Since ε is arbitrary, we see that $f(t)$ is the limit of a sequence of elements of K. Since K is closed, we infer that $f(t)$ itself belongs to K, as required. //

The next theorem shows that every ring of functions K_τ is closed with respect to uniform convergence.

THEOREM 2: *K_{τ_0} is a complete normed ring with the norm*

$$\| x \| = \max_{t \in \tau_0} | x(t) |. \tag{1}$$

Proof: We consider the ideal J_{τ_0} in K consisting of all the functions of K that are equal to zero on the class τ_0. This ideal is obviously closed and belongs to τ_0, because for an arbitrary point t_0 not in τ_0 there is a function $x(t) \in K'$ that assumes distinct values at t_0 and at a certain point $t \in \tau_0$. The residue-class ring K/J_{τ_0} can be regarded as consisting of functions of K, considered only on τ_0 (see § 34), i.e., coinciding with K_{τ_0}. But it is known that the residue-class ring of a complete normed ring with respect to any closed ideal is complete in its norm; therefore it is sufficient to verify that the norm in K/J_{τ_0} coincides with the value (1). Let ε be an arbitrarily given positive number, $\tilde{x}(t)$ an element of K/J_{τ_0}, $x(t)$ a function of K that coincides with $\tilde{x}(t)$ on τ_0, and U a neighborhood of τ_0 in S for which the following inequality holds:

$$\max_{t \in U} | x(t) | < \max_{t \in \tau_0} | x(t) | + \varepsilon.$$

We choose the neighborhood V_0 of τ_0 in S', as in the proof of Theorem 1, in such a way that every class $\tau \in V_0$ is entirely contained in U. We construct a function $h'(\tau) \in K' = C(S')$ having the properties

1) $0 \leqq h'(\tau) \leqq 1$, 2) $h'(\tau_0) = 1$, 3) $h'(\tau) = 0$ outside V_0.

We consider the product of the corresponding functions $h(t) \in K$ and $x(t)$. Since $h(t)$ vanishes outside U and assumes values between 0 and 1 on U, we have

$$\max_S | h(t) x(t) | = \max_U | h(t) x(t) | \leqq \max_U | x(t) | < \max_{t \in \tau_0} | x(t) | + \varepsilon.$$

But since $h(t) = 1$ on τ_0, $h(t)x(t)$ occurs in the same residue class of

J_{τ_0} as $x(t)$. Therefore, in accordance with the definition of a norm in a residue-class ring,

$$\| \tilde{x}(t) \|_{K/J_{\tau_0}} \leqq \| h(t)x(t) \|_K = \max_S | h(t)x(t) | < \max_{t \in \tau_0} | x(t) | + \varepsilon,$$

and hence, since ε is arbitrary,

$$\| \tilde{x}(t) \|_{K/J_{\tau_0}} \leqq \max_{t \in \tau_0} | x(t) |. \tag{2}$$

On the other hand, every element $y(t)$ of the residue class determined by the function $\tilde{x}(t)$ has the norm

$$\max_{t \in S} | y(t) | \geqq \max_{t \in \tau_0} | y(t) | = \max_{t \in \tau_0} | \tilde{x}(t) |$$

in K. Therefore

$$\| \tilde{x}(t) \|_{K/J_{\tau_0}} = \inf \| y(t) \|_K \geqq \max_{\tau \in \tau_0} | x(t) |. \tag{3}$$

Combining (2) and (3), we obtain

$$\| \tilde{x}(t) \|_{K/J_{\tau_0}} = \max_{t \in \tau_0} | x(t) |,$$

as required. //

***3.** By way of illustration, we examine the structure of a ring K that is obtained as follows: Let S be the closed unit circle of the complex plane and A, the ring formed by the functions $f(\zeta)$ that are continuous on S and analytic within S; we adjoin to A a set $\mathcal{E}$ of continuous real functions defined on S. The classes τ into which S splits under the construction described above fall into three types according to the following rule:

1. The class τ does not have interior points and does not bound any domain (in the sense that every point ζ not in τ can be joined to ∞ by a line passing outside τ).
2. The class τ has interior points, but does not bound any domain.
3. The class τ bounds a domain.

For a class τ of the first type, the ring K_τ contains all the polynomials in ζ and, by Theorem 2, is closed under uniform convergence in τ; by a well-known approximation theorem in the theory of functions of a complex variable [39] it is therefore the ring of all continuous functions on τ.

For a class τ of the second type, the ring K_τ, for the same reasons, is the ring of all continuous functions on τ that are analytic at interior points of this class.

Of the greatest interest are the rings K_τ corresponding to classes of the third type. In this case, as in the second, the ring K_τ consists of functions continuous on τ and analytic at all interior points both of the class τ and of the domain bounded by it; therefore the space of maximal ideals of K_τ is wider than τ.

By Theorem 1, the ring K consists of all continuous functions that on each class τ belong to the corresponding ring K_τ. The ring K coincides with $C(S)$ (and hence, every class consists of a single point only) if and only if all the classes τ belong to the first type.

If all the classes τ belong to the first or the second type, then the ring K consists of all continuous functions on S that are analytic at the interior points of the classes τ; the space of maximal ideals of K coincides in this case with the entire circle S. Finally, if there is a class τ_0 of the third type, then the set of maximal ideals of K is necessarily wider than S, because every maximal ideal of K_{τ_0} determines a homomorphism into the field of complex numbers and thus a maximal ideal of K. Speaking geometrically, we must paste onto S the 'film' spanning the domain bounded by τ_0, the edge of the film being attached to the boundary of the domain.

Here are two simple examples:

a) $\mathcal{E}$ is the set of all real functions on S that are equal to 0 on the circle $|\zeta|=1$. The classes τ are the individual interior points and the whole circumference $|\zeta|=1$. The latter class is of the third type. The space of maximal ideals of K is the aggregate of two circles pasted together at the circumference; this is a two-dimensional set homeomorphic to a sphere.

b) $\mathcal{E}$ is the set of all real functions on S whose values at every point depend only on its distance from the origin. The classes τ are circles with center at O (one of them being this point itself); they are all (except the point O) of the third type. The space of maximal ideals of K is obtained from S by adjoining a continuum of 'films' stretched over each of these circles. Thus, $M(K)$ in this case is three-dimensional (homeomorphic to a hemisphere described on S).

4. As we have seen, Theorems 1 and 2, in general, reduce the investigation of the ring K to that of rings K_τ given on smaller sets. If the rings K_τ are anti-symmetric, this means that the ring K in question is 'pasted together' in a definite manner from anti-symmetric rings. But if among the rings K_τ there are also some that are not anti-symmetric, then the construction described can be repeated, and K_τ is decomposed into still 'smaller' rings $K_{\tau\tau'}$. Repeating this operation transfinitely, if necessary, we must eventually arrive at a complete decomposition of K into anti-symmetric rings.

Observe that the question of an accurate description of the character of the pastings of anti-symmetric rings that are applied to the original ring K

(i.e., the analogue to Theorem 1) remains open in the general case (namely, for an infinite continuation of the process). But in any case, we have the right to restrict ourselves, in what follows, to a discussion of anti-symmetric subrings of K.

The following natural condition has the effect of making the space S the *carrier*—the space of maximal ideals—of the subring K in question. Namely, if Q is the compactum of maximal ideals of K, then for $Q \neq S$ the ring K will naturally be studied as a subring of $C(Q)$, and not of $C(S)$.

Further, the description of the structure of the subrings of $C(S)$ will naturally be given in dependence on the topological invariants of S, in the first instance on its dimension. Thus, *if the space S is zero-dimensional, then any closed subring K of $C(S)$ having S as its space of maximal ideals coincides with the whole ring $C(S)$, just as in the symmetric case.* For if S is zero-dimensional, then for any two distinct points t', t'' of this space, S can be divided into two closed sets without common points such that the first of these sets, F', contains t', and the second, F'', contains t''. But then, according to the proof of Theorem 2 of § 14 on the decomposition of a ring into a direct sum of ideals, K contains a function $e(t; t', t'')$ that assumes the value 0 on F' and the value 1 on F''. The subring of real functions generated by the functions $e(t; t', t'')$, where t' and t'' $(\neq t')$ ranges over the whole of S, satisfies the conditions of Theorem 1′ of § 7 and therefore contains all the continuous real functions on S. But then K contains all the continuous functions on S, i.e., $K = C(S)$, as required.

The case of a carrier of positive dimension has not been completely investigated. Let us assume that there exists a function $z(t) \in K$ that assumes distinct values at distinct points of S. With the help of this function we can map S homeomorphically onto a closed bounded set F of points of the complex plane. For every point λ_0 that does not belong to F the ring K contains the function $(z - \lambda_0)^{-1}$; hence it follows that K contains all the rational functions of z with poles not belonging to F. Let us assume that F is of planar measure 0. Then, by a theorem of Hartogs and Rosenthal [39, 25], every continuous function on F is the limit of a uniformly convergent sequence of rational functions with poles outside F; thus, $K = C(F) = C(S)$ in this case as well.

Recently Helson and Quigley [26] have observed that this argument can be generalized as follows. Let us assume that K contains a system of functions $z_1(t), \ldots, z_n(t), \ldots$ separating the points of S and that the set of values of each of these functions is of planar measure 0; then it turns out that in this case also, $K = C(S)$. For by the preceding argument, all the functions of the form $f(z_1(t))$, $g(z_2(t))$, etc. belong to K, where $f, g, \ldots$ are arbitrary continuous functions on the set of values of the corresponding functions $z_1(t)$, $z_2(t)$, etc. Restricting ourselves to real functions only, we see that K has an 'ample set' of them: for every pair of distinct points $t', t'' \in S$ we can find a real function of the form $f(z_j(t))$ that assumes distinct values at t' and t''. But then, by Theorem 1′ of § 7, K contains all the continuous real functions and consequently coincides with $C(S)$. The conditions of this theorem are satisfied, for example, when the set of maximal ideals of K is a differentiable curve $\{z_1(t), \ldots, z_n(t)\}$ in complex n-dimensional space.

Wermer [74] has shown that $K = C(S)$ if K has two generators and if, in the corresponding complex two-dimensional space, the set S is a curve homeomorphic to a circle that does not bound a piece of an analytic surface.

Thus, for one-dimensional carriers of the special type described, we can prove that $K = C(S)$; in the general case, the problem remains open.

It is rather likely that for a one-dimensional S we always have $K = C(S)$.

The ring $A(G)$ of functions that are analytic inside a plane domain G and continuous on $\overline{G}$ (where $\overline{G}$ is the closure of G on the Riemann sphere) constitutes an example of an anti-symmetric ring with a two-dimensional carrier $S = \overline{G}$. Every subring of this ring with the same carrier—for example, the subring of functions that have the derivative 0 at a given interior point—is obviously also anti-symmetric.

Note that the carrier $\overline{G}$ of such a ring does not always fit into the Euclidean plane. For example, if G is obtained by removing from the full complex plane (the Riemann sphere) a bounded closed set of planar measure 0 that does not divide the plane, then $S = \overline{G}$ is the whole Riemann sphere [21].

It is not known whether there exist anti-symmetric rings with two-dimensional carriers that are homeomorphic to a torus or more complicated closed surfaces.

Recently Goffman and Singer [21] have proposed the following method of constructing anti-symmetric rings with carriers of an arbitrary dimension $\geqq 2$. Let G be a plane domain and Q a compactum; we consider the ring R consisting of all continuous functions $f(\zeta, q)$ $(\zeta \in G, q \in Q)$ defined on the topological product $T = \overline{G} \times Q$ and analytic in ζ for every fixed q. The carrier of R, as is easy to verify, is the space T. Real functions occurring in R are constant in ζ for every fixed q. We now consider the subring $R_1 \subset R$ formed by the functions $f(\zeta, q) \in R$ that are constant in q for a fixed $\zeta = \zeta_1$. The carrier of R_1 is the set T in which all the points (ζ_1, q) have been identified. It is obvious that there are no real functions in R_1 except constants.

If Q is zero-dimensional, then $T = \overline{G} \times Q$ is two-dimensional, and for a sufficiently complicated structure of the compactum Q the carrier of the ring R_1 cannot even be embedded in a Riemann sphere. If Q is one-dimensional, then the carrier of R_1 is three-dimensional; for example, when G is a circle and Q an interval, R_1 is an anti-symmetric ring with a carrier that is homeomorphic to a three-dimensional sphere. By increasing the dimension of Q we can construct anti-symmetric rings with carriers of arbitrary dimension $\geqq 2$.

Another example of an anti-symmetric ring is the ring $A_2(G_2)$ of analytic functions of two complex variables defined in a two-dimensional complex (four-dimensional real) domain G_2. By pasting onto $A_2(G_2)$, as before, we can obtain a new series of anti-symmetric rings with carriers of dimensionality $\geqq 4$.

It would be interesting to ascertain whether every anti-symmetric ring can be obtained by a similar construction from the rings $A(G)$, $A_2(G_2)$, ...

§ 45. Ideals in Rings with Uniform Convergence

1. As we already know from § 36, every closed ideal in the ring $C(S)$ of all continuous functions on a compactum S is the intersection of maximal ideals. In this section we shall discuss the problem of the structure of closed ideals in other rings with uniform convergence.

Let K be an arbitrary subring of $C(S)$ that is closed under uniform convergence. We split S into classes, as in § 44; within each class τ every function $x(t)$ in K whose complex conjugate $\overline{x(t)}$ is also in K has a constant value. This splitting gives rise to a collection of sets K_τ each of which consists of functions of K considered only on the class τ. To every ideal $J \subset K$ there corresponds the ideal $J_\tau \subset K_\tau$ formed by the restrictions of the functions of J to τ. We claim, first of all, that *to a closed ideal J in $C(S)$ there corresponds an ideal J_τ that is closed with respect to uniform convergence on τ.* Let a sequence of functions $x_n(t) \in J_\tau$ be given that is uniformly convergent on τ to a certain function $x(t)$. Without loss of generality we may assume that the inequalities

$$\max_{t \in \tau} | x_{n+1}(t) - x_n(t) | < 1/2^n \qquad (n = 1, 2, \ldots)$$

are satisfied. By assumption, the functions $x_n(t)$ can be extended from τ to the entire space S so that a function of the ideal J is obtained. As in the proof of Theorem 2 of § 44, we can find a function $h_n(t)$ in K that is equal to 1 on τ and such that

$$\max_{t \in S} | h_n(t) [x_{n+1}(t) - x_n(t)] | < 1/2^{n-1}.$$

The series

$$x_1(t) + \sum_{n=1}^{\infty} h_n(t) [x_{n+1}(t) - x_n(t)]$$

converges in K and all of the terms of the series, and hence the sum, belongs to J. But on τ this sum is obviously $x(t)$; thus, $x(t)$ belongs to the ideal J_τ, as we have claimed.

Next, we claim that the closed ideal $J \subset K$ is the 'pasting together' of the ideals J_τ in the same sense as the ring K is the 'pasting together' of the rings K_τ: *every function $x(t) \in C(S)$ that belongs on each class τ to the ideal J_τ belongs to J.* To satisfy ourselves of this, we only have to turn to the proof of Theorem 1 of § 44. It shows that the function $x(t)$ is the limit of a sum of functions of the form $x_k(t)h_k(t)$, where $x_k(t)$ is an arbitrary extension to the entire space S of a function $\tilde{x}_k(t)$ defined on a class τ and belonging on this class to the ideal J_τ. But for such an extension we can always take a function occurring in J. Hence $x(t)$ is the limit of functions of J, and since J is closed, $x(t) \in J$, as required.

Thus, the problem of ideals in arbitrary rings with uniform convergence reduces to the same problem in anti-symmetric rings.

2. As an example, let us examine the anti-symmetric ring A, again letting S denote the closed unit circle in the complex plane.

Every maximal ideal of this ring is formed by all the functions $f(\zeta) \in A$ that vanish at a certain point ζ_0, $|\zeta_0| \leqq 1$. For a maximal ideal M_{ζ_0} corresponding to an interior point ζ_0 of S we can indicate all the primary ideals it contains; for every n the set of functions $f(\zeta) \in A$ that vanish, and whose first n derivates vanish, at ζ_0 is a primary ideal, and it is easy to verify that there are no other primary ideals belonging to M_{ζ_0}.[1]

The situation is more complicated with primary ideals of A corresponding to boundary points of S. It turns out that for a boundary point ζ_0 the ideals generated by the functions $\zeta - \zeta_0$, $(\zeta - \zeta_0)^2, \ldots, (\zeta - \zeta_0)^n, \ldots$ coincide with the maximal ideal M_{ζ_0}. Moreover, Carleman, in 1926, proved the following theorem [9]:

---

[1] It is sufficient to show that when all the functions $(\zeta - \zeta_0)^n \varphi_0(\zeta)$, where $\varphi_0(\zeta_0) \neq 0$, belong to a primary ideal J contained in M_{ζ_0}, then so does $(\zeta - \zeta_0)^n$. For every point $\zeta \neq \zeta_0$ of S we can find a function $\varphi \in J$ that does not vanish in some neighborhood of this point. By adjoining to these neighborhoods a neighborhood of ζ_0 in which the function $\varphi_0(\zeta)$ does not vanish, we obtain a covering of S from which we can choose a finite covering; let $\varphi_0, \varphi_1, \ldots, \varphi_m$ be the corresponding functions; by construction, with the exception of φ_0 they all belong to J. The ideal generated by these functions is not contained in any maximal ideal and therefore coincides with the whole ring A; consequently, for suitable $f_0, f_1, \ldots, f_m$ we have the equation

$$f_0\varphi_0 + f_1\varphi_1 + \ldots + f_m\varphi_m \equiv 1.$$

We multiply this equation by $(\zeta - \zeta_0)^n$. Then all the terms on the left-hand side belong to J, and on the right we have the function $(\zeta - \zeta_0)^n$, which therefore also belongs to J.

Let the function $u(\zeta) \in A$ *vanish on* S *at the point* $\zeta = 1$ *only and let it not have a logarithmic residue.*[2] *For every other function* $v(\zeta) \in A$ *that is equal to* 0 *at* $\zeta = 1$ *there exists a sequence of polynomials* $P_n(\zeta)$ *such that the functions* $u(\zeta)P_n(\zeta)$ *converge uniformly on* S *to* $v(\zeta)$.

In terms of the theory of rings, this means that *in the ring* A *the ideal generated by a function* $u(\zeta)$ *that is equal to* 0 *at* $\zeta = 1$ *and does not have a logarithmic residue* (in particular, the functions $\zeta - 1$, $(\zeta - 1)^2, \ldots$ are of this kind) *coincides with the maximal ideal corresponding to the point* $\zeta = 1$.

3. Now it turns out that with a more rapid decrease of the function $u(\zeta)$ as $\zeta \to 1$ one can discover non-trivial primary ideals generated by $u(\zeta)$. The following general theorem contains Carleman's Theorem as a special case.

THEOREM 1: *The functions* $u_\alpha(\zeta) = (\zeta - 1) \exp(\alpha/(\zeta - 1))$ $(0 \leqq \alpha < \infty)$ *belong to the ring* A *and vanish at the point* $\zeta = 1$ *only. They generate distinct closed ideals of* A. *Furthermore, every function* $u(\zeta) \in A$ *that vanishes on* S *at* $\zeta = 1$ *only generates an ideal that coincides with one of these ideals.*

Proof: For every real γ the real part of the function $\gamma/(\zeta - 1)$ assumes the constant value $-\gamma/d$ on the circle of diameter d that touches the line $\xi = 1$ from the left at the point $\zeta = 1$. Therefore $|\exp(\alpha/(\zeta - 1))|$ is bounded on the unit circle for $\gamma = \alpha > 0$:

$$\left|\exp \frac{\alpha}{\zeta - 1}\right| = \exp\left(-\frac{\alpha}{d}\right) \leqq \exp\left(-\frac{\alpha}{2}\right).$$

Since the function $\exp(\alpha/(\zeta - 1))$ is analytic everywhere except at $\zeta = 1$ and does not vanish, the product $(\zeta - 1)\exp(\alpha/(\zeta - 1)) = u_\alpha(\zeta)$ belongs to A and vanishes at $\zeta = 1$ only.

Let us suppose that the ideal generated by $u_\beta(\zeta)$ coincides with that generated by $u_\alpha(\zeta)$, where $\alpha < \beta$. In that case there exists a sequence of functions $f_n(\zeta) \in A$ such that $\max\limits_{|\zeta| \leqq 1} |f_n(\zeta)u_\beta(\zeta) - u_\alpha(\zeta)| \to 0$ as $n \to \infty$. In particular, at every point $\zeta \neq 1$

$$\lim_{n \to \infty} f_n(\zeta) = \exp \frac{\alpha - \beta}{\zeta - 1}. \tag{1}$$

For a given $\varepsilon > 0$ we can find a number N such that for $n > N$ we have the inequality

$$\left|(\zeta - 1)\, f_n(\zeta) \exp \frac{\beta}{\zeta - 1} - (\zeta - 1) \exp \frac{\alpha}{\zeta - 1}\right| < \varepsilon$$

for all $|\zeta| \leqq 1$. Hence we obtain, in succession, for $\zeta \neq 1$

[2] In other words, the harmonic function $\log |u(\zeta)|$ is determined on the circle $|\zeta| \leqq 1$ by its boundary values in accordance with Poisson's formula.

$$\left|(\zeta-1)\exp\frac{\alpha}{\zeta-1}\cdot\left[f_n(\zeta)\exp\frac{\beta-\alpha}{\zeta-1}-1\right]\right|<\varepsilon,$$

$$\left|f_n(\zeta)\exp\frac{\beta-\alpha}{\zeta-1}-1\right|<\frac{\varepsilon}{|\zeta-1|}\left|\exp\left(-\frac{\alpha}{\zeta-1}\right)\right|,$$

$$\left|f_n(\zeta)\exp\frac{\beta-\alpha}{\zeta-1}\right|<\frac{\varepsilon}{|\zeta-1|}\left|\exp\left(-\frac{\alpha}{\zeta-1}\right)\right|+1<$$

$$<\frac{2+\varepsilon\left|\exp\left(-\frac{\alpha}{\zeta-1}\right)\right|}{|\zeta-1|},$$

$$|(\zeta-1)f_n(\zeta)|<\left|\exp\frac{\alpha-\beta}{\zeta-1}\right|\left\{2+\varepsilon\left|\exp\left(-\frac{\alpha}{\zeta-1}\right)\right|\right\}.$$

If, in particular, $|\zeta|=1$, then, as we have seen,

$$\left|\exp\frac{\alpha-\beta}{\zeta-1}\right|=\exp\frac{\beta-\alpha}{2},\qquad\left|\exp\left(-\frac{\alpha}{\zeta-1}\right)\right|=\exp\frac{\alpha}{2}.$$

Therefore we obtain, for $|\zeta|=1$, $\zeta\neq1$,

$$|(\zeta-1)f_n(\zeta)|\leqq\exp\frac{\beta-\alpha}{2}\cdot\left(2+\varepsilon\exp\frac{\alpha}{2}\right)\leqq\text{const.}$$

Since the function $(\zeta-1)f_n(\zeta)$ is continuous, this inequality is also true for $\zeta=1$, and by the maximum principle, for $|\zeta|<1$ as well. Therefore we have for all ζ with $|\zeta|\leqq1$

$$|f_n(\zeta)|\leqq\text{const.}/|\zeta-1|,\qquad\overline{\lim_{n\to\infty}}|f_n(\zeta)|\leqq\text{const.}/|\zeta-1|.$$

But this last inequality contradicts the limit relation (1), and this proves the first part of the theorem.

For the proof of the second part, we consider, following Carleman, the harmonic function $f(\zeta)=-\log|u(\zeta)|$, for a given function $u(\zeta)\in A$ that is equal to 0 on S only at the point $\zeta=1$. There is a neighborhood of $\zeta=1$ in which we have $|u(\zeta)|<1$ and $f(\zeta)>0$; therefore, for sufficiently large $r<1$ and sufficiently small $\varepsilon>0$, the second term on the right-hand side of the equation

$$f(0)=\frac{1}{2\pi}\int_0^{2\pi}f(re^{i\theta})\,d\theta=\frac{1}{2\pi}\int_\varepsilon^{2\pi-\varepsilon}f(re^{i\theta})\,d\theta+\frac{1}{2\pi}\int_{-\varepsilon}^{\varepsilon}f(re^{i\theta})\,d\theta$$

is positive. In the inequality

$$\frac{1}{2\pi}\int_\varepsilon^{2\pi-\varepsilon}f(re^{i\theta})\,d\theta<f(0)$$

we can pass to the limit as $r \to 1$ and then as $\varepsilon \to 0$, and this yields the existence of the improper integral

$$\frac{1}{2\pi}\int_0^{2\pi} f(e^{i\theta})\,d\theta$$

and the inequality

$$\frac{1}{2\pi}\int_0^{2\pi} f(e^{i\theta})\,d\theta \leqq f(0).$$

Letting α denote the difference between the left-hand and right-hand sides of the inequality, we obviously have

$$\alpha = \lim_{\varepsilon\to 0}\lim_{r\to 1}\frac{1}{2\pi}\int_{-\varepsilon}^{\varepsilon} f(re^{i\theta})\,d\theta.$$

For $r < R < 1$ we have, by Poisson's Formula,

$$f(re^{i\varphi}) = \frac{1}{2\pi}\int_0^{2\pi}\frac{R^2-r^2}{R^2-2Rr\cos(\theta-\varphi)+r^2}f(Re^{i\theta})\,d\theta =$$

$$= \frac{1}{2\pi}\int_{\varepsilon}^{2\pi-\varepsilon}\frac{R^2-r^2}{R^2-2Rr\cos(\theta-\varphi)+r^2}f(Re^{i\theta})\,d\theta +$$

$$+\frac{1}{2\pi}\int_{-\varepsilon}^{\varepsilon}\frac{R^2-r^2}{R^2-2Rr\cos(\theta-\varphi)+r^2}f(Re^{i\theta})\,d\theta. \tag{2}$$

Since for fixed r and $R \to 1$ the Poisson kernel is bounded, the first term of the sum (2), and consequently the second term as well, has a limit for $R \to 1$, $\varepsilon \to 0$. It is easy to verify that the limit of the second term is equal to

$$\alpha\frac{1-r^2}{1-2r\cos\varphi+r^2}.$$

Thus, we obtain

$$f(re^{i\varphi}) = \frac{1}{2\pi}\int_0^{2\pi}\frac{1-r^2}{1-2r\cos(\theta-\varphi)+r^2}f(e^{i\theta})\,d\theta + \alpha\frac{1-r^2}{1-2r\cos\varphi+r^2}.$$

For the function $u(\zeta)$ we have, correspondingly,

$$u(\zeta)=\exp\left(-\frac{1}{2\pi}\int_0^{2\pi}\frac{e^{i\theta}+\zeta}{e^{i\theta}-\zeta}f(e^{i\theta})\,d\theta+iC\right)\exp\left(-\alpha\frac{1+\zeta}{1-\zeta}\right)=$$
$$=u_1(\zeta)\,u_2(\zeta),$$

where $u_1(\zeta)$ is a function without logarithmic residue and

$$u_2(\zeta)=\exp\left(\alpha\frac{\zeta+1}{\zeta-1}\right)=C_1\exp\frac{2\alpha}{\zeta-1}.$$

By Carleman's Theorem, there exists a sequence of polynomials $P_n(\zeta)$ such that the functions $u_1(\zeta)P_n(\zeta)$ converge uniformly to $\zeta-1$. Hence the sequence $u_1(\zeta)P_n(\zeta)u_2(\zeta)$ converges uniformly to $u_{2\alpha}(\zeta)$. Thus, the ideal generated by $u(\zeta)$ contains $u_{2\alpha}(\zeta)$. But also, conversely, in an obvious way we can form a sequence of polynomials of the form $(\zeta-1)Q_n(\zeta)$ that converges uniformly to $u_1(\zeta)$; hence the functions $Q_n(\zeta)u_{2\alpha}(\zeta)$ converge uniformly to $u_1(\zeta)u_2(\zeta)=u(\zeta)$, i.e., $u(\zeta)$ belongs to the ideal generated by $u_{2\alpha}(\zeta)$. Thus, the ideals generated by $u(\zeta)$ and $u_{2\alpha}(\zeta)$ coincide. //

This theorem gives an idea of the disposition of primary ideals of the ring A.

4. Recently, Rudin has given a complete description of the closed ideals of A. Let us quote his results.

Assume that there are given: 1) a Blaschke product

$$B(\zeta)=\zeta^m\prod_n\frac{a_n-\zeta}{1-\bar{a}_n\zeta}\frac{|a_n|}{a_n}$$
$$(0<|a_n|<1,\ \sum_n(1-|a_n|)<\infty);$$

2) a measure μ that is concentrated on a set E lying on the circle $|\zeta|=1$, including all the limit points of the sequence a_n and having Lebesgue measure zero. We put

$$M(\zeta)=B(\zeta)\exp\left\{-\int_{|z|=1}\frac{z+\zeta}{z-\zeta}\,d\mu(z)\right\};$$

$M(\zeta)$ is a bounded function whose radial limit values for $|\zeta|\to 1$ are almost everywhere 1. The set $I(B,\mu)$ of all functions $f(\zeta)\in A$ that vanish on E and have a bounded quotient $f(\zeta)/M(\zeta)$ is a closed ideal of A. It can be shown that every closed ideal of A is of the form $I(B,\mu)$.

Furthermore, Rudin has shown that every closed ideal J of A is a principal ideal (i.e., is generated by a single function $f(\zeta)$) and that the ideal $I(B,\mu)$ is the intersection of primary ideals if and only if the measure μ is discrete.

CHAPTER VIII[1]

NORMED RINGS WITH AN INVOLUTION AND THEIR REPRESENTATIONS[2]

1. In this chapter[1] we study rings (mainly *non-commutative* rings) in which an operation of involution ($*$-operation) is introduced axiomatically. The representation of such rings by operators in a Hilbert space are investigated by means of positive functionals. The results obtained are applied to the theory of representations of locally compact groups.

2. The theory of commutative normed rings, as developed in the papers [79] and [83] has proved a useful apparatus for the solution of various problems of analysis. It has also been applied with success in the theory of commutative topological groups.

For applications of analogous methods to non-commutative groups it has turned out to be necessary to develop a theory of non-commutative normed rings.

Since an operation of involution can be introduced in a group ring in a natural way, the problem that arises in the first place is the investigation of rings in which an operation of involution ($*$-operation) is given.

One class of such rings (the so-called $*$-rings) was discussed by the authors in the paper [80]. In that paper it appeared that in the theory of rings with an involution an important role is played by positive functionals that is, functionals that satisfy the condition $f(x^*x) \geqq 0$.

Positive functionals on a group ring and the positive-definite functions on the group connected with them were used by Gelfand and Raikov in [82] in order to prove the existence and completeness of the system of continuous representations of a locally compact group.

---

[1] In the original Russian text, the present chapter was called an appendix.

[2] The present chapter is written by I. M. Gelfand and M. A. Naimark; it reproduces, with some improvements of an editorial nature, the paper that appeared in *Izv. AN. USSR., ser. mat.,* 12 (1948), 445-480. A section contained in the original concerned with the generalized Schur Lemma, which is rather far removed from the basic theme of the present chapter, is omitted here.

The present chapter is devoted to the general theory of rings with an involution and their representations in conjunction with the theory of positive functionals.

Certain methods to be explained here, particularly in § 50, are essentially an extension of the methods of the paper [82] to the case of an arbitrary ring with an involution.[3]

§ 46. Rings with an Involution and their Representations

1. A set R of elements $x, y, \ldots$ is called a *normed ring* if:

1. R is a ring, i.e., operations of addition and multiplication satisfying the usual algebraic conditions are defined in R. We also assume that R has a unit element e.

2. R is a linear vector space with multiplication by complex numbers, where this multiplication is permutable with the operation of multiplication of elements in R.

3. A norm is defined in R, i.e., every element x is associated with a number $|x|$ such that

$$|x+y| \leqq |x| + |y|, \qquad |xy| \leqq |x| \cdot |y|.$$

$$|x| \geqq 0 \quad \text{and is equal to zero only for} \quad x = 0.$$

$$|\lambda x| = |\lambda|\,|x|, \qquad |e| = 1.$$

4. The ring is complete, i.e., from

$$\lim_{m,\, n \to \infty} |x_n - x_m| = 0$$

there follows the existence of an x such that

$$\lim_{n \to \infty} |x_n - x| = 0.$$

2. DEFINITION 1: A normed ring R is said to be a *ring with an involution* if an operation is defined in it that assigns to every element x an element x^* such that the following conditions are satisfied:

a) $(\lambda x + \mu y)^* = \bar{\lambda} x^* + \bar{\mu} y^*$; c) $(xy)^* = y^* x^*$;

b) $x^{**} = x$; d) $|x^*| = |x|$.

In the balance of this chapter it is to be understood, even without specific mention, that every ring under discussion is a ring with an involution.

An element x is called *Hermitian* if $x^* = x$.

[3] We should like to express our thanks to D. A. Raikov, who has read the paper and made a number of valuable critical remarks.

Every element x can be represented in the form $x = x_1 + ix_2$, where x_1, x_2 are Hermitian elements. For it is sufficient to put

$$x_1 = [x + x^*]/2,$$
$$x_2 = [x - x^*]/2i.$$

The element x^*x is always Hermitian, because

$$(x^*x)^* = x^*x^{**} = x^*x.$$

In particular, since $e^* = e^*e$, we have $e = e^*$, i.e., e is a Hermitian element.

Some results of this chapter remain valid when only the algebraic conditions of our list of conditions are retained, namely, Axioms 1. and 2. and a), b), c).

A typical example of a ring with an involution is the ring K of all bounded linear operators in a Hilbert space. Here the $*$-operation is interpreted as the transition from an operator to its Hermitian conjugate.

In this context, it is natural to study homomorphic mappings of a ring with an involution into K that preserve the $*$-operation. Such a mapping will be called a *representation of the ring*. In other words, we introduce the following definition.

3. DEFINITION 2: We shall say that *a representation of a ring R is given* if every element $a \in R$ is associated with an operation $A \in K$ in a Hilbert space $\mathfrak{H}$ (we shall denote this for brevity as $a \to A$, or $A(a)$), provided the following conditions are satisfied:

1. If $a \to A$, $b \to B$, then $ab \to AB$ and $\lambda a + \mu b \to \lambda A + \mu B$;
2. If $a \to A$, then $a^* \to A^*$;
3. $e \to E$.

DEFINITION 3: The representation is called *cyclic* if the space contains a vector ξ_0 such that the vectors $A\xi_0$ (A are the operators corresponding to the elements of R) are everywhere dense in $\mathfrak{H}$. The vector ξ_0 is also called *cyclic*.

Suppose that two representations are given, the first of which assigns to the element a the operator $A(a)$ in a space $\mathfrak{H}$ and the other, the operator $A'(a)$ in a space $\mathfrak{H}'$. We shall say that these are *equivalent representations* if a one-to-one correspondence can be set up between $\mathfrak{H}$ and $\mathfrak{H}'$ in which the operator $A(a)$ corresponds to $A'(a)$.

A subspace $\mathfrak{H}_1 \subset \mathfrak{H}$ is called *invariant* if every vector of $\mathfrak{H}_1$ is carried by all the operators $A(a)$ into vectors of $\mathfrak{H}_1$.

By regarding all the operators of the representation only as operators in $\mathfrak{H}_1$ we obtain a representation of R in the space $\mathfrak{H}_1$. We shall call this representation a *part* of the original representation in $\mathfrak{H}$.

If $\mathfrak{H}_1$ is invariant, then its orthogonal complement is also invariant. For let ξ be orthogonal to $\mathfrak{H}_1$, i.e., $(\xi, \eta) = 0$ for all $\eta \in \mathfrak{H}_1$. Then

$$(A(a)\xi, \eta) = (\xi, A^*(a)\eta) = 0,$$

because $\mathfrak{H}_1$ is invariant with respect to the operators that are images of elements of R, and $A^*(a)$ is the image of a^*.

4. If a representation of a ring R in a space $\mathfrak{H}$ is given, then the space can be decomposed into the direct sum of invariant subspaces such that the representation is cyclic in each of them. For let $\xi_0 \neq 0$ be an arbitrary fixed vector of $\mathfrak{H}$. We consider the set of all vectors $A(a)\xi_0$, where a ranges over the entire ring R. The closure of this set forms an invariant subspace $\mathfrak{H}_1$ of $\mathfrak{H}$ in which the representation is cyclic. The orthogonal complement of this subspace is also invariant. In this space we proceed as before, etc. Using transfinite induction we arrive at the required decomposition.

DEFINITION 4: A representation is called *irreducible* if $\mathfrak{H}$ has no subspaces invariant with respect to all the operators $A(a)$, other than $\mathfrak{H}$ and 0.

If a representation is irreducible, then it is clear that every vector $\xi \neq 0$ is cyclic. Obviously, the converse is also true.

THEOREM 1: *A representation is irreducible if and only if every bounded operator B that is permutable with the operators $A(a)$ is a multiple of the unit element.*

Proof of the Necessity: Let B be permutable with all the $A(a)$. To begin with, we assume that B is Hermitian. Then every function of B is also permutable with $A(a)$. In particular, the projection operators $E(\lambda)$ that give the spectral decomposition of B are also permutable with $A(a)$. But this means that the subspaces corresponding to them are invariant with respect to $A(a)$. Since the representation is irreducible, this means that each of these subspaces is either the null space or the whole space. Thus, for every λ, $E(\lambda)$ is either zero or E. Since $(E(\lambda)\xi, \xi)$ increases monotonically with increasing λ, it follows from this that there exists a λ_0 such that $E(\lambda) = E$ for $\lambda > \lambda_0$ and $E(\lambda) = 0$ for $\lambda < \lambda_0$. Hence it follows that

$$B = \int_{-\infty}^{\infty} \lambda dE(\lambda) = \lambda_0 E.$$

If B is an arbitrary bounded operator, then B^* is also permutable with all the $A(a)$. For,

$$B^*A(a) = (A^*(a)B)^* = (BA^*(a))^* = A(a)B^*.$$

Therefore the Hermitian operators $(B+B^*)/2$ and $(B-B^*)/2i$ are multiples of the unit operators, and consequently so is B.

Proof of the Sufficiency: Let us assume the contrary, i.e., that the representation is reducible. Then there exists an invariant subspace $\mathfrak{H}_1$ different from 0 and from the whole space. We denote the projection operator corresponding to this subspace by E_1. Then E_1 is permutable with all the $A(a)$ (since $\mathfrak{H}_1$ is invariant with respect to $A(a)$). But E_1 is not a multiple of E, because $\mathfrak{H}_1$ is different from 0 and from the whole space. We have thus reached a contradiction. //

§ 47. Positive Functionals and their Connection with Representations of Rings

1. DEFINITION 5: *A positive linear functional* is a function $f(x)$ that assigns to every $x \in R$ a complex number $f(x)$ such that:

1. $f(\lambda x+\mu y)=\lambda f(x)+\mu f(y)$;
2. $f(x^*x) \geqq 0$ for every x.

The function $f(x)$ is called a *real linear functional* if:

1. $f(\lambda x+\mu y)=\lambda f(x)+\mu f(y)$;
2. $f(x)$ is continuous;
3. $f(x^*)=\overline{f(x)}$ for every x.

It is obvious that if $f(x)$ is a positive functional, then $f(e) \geqq 0$, because $e^*e=e$.

Every continuous linear functional $f(x)$ can be represented in the form $f=f_1+if_2$, where f_1 and f_2 are real functionals. For this it is sufficient to put

$$f_1(x)=[f(x)+\overline{f(x^*)}]/2 \quad \text{and} \quad f_2(x)=[f(x)-\overline{f(x^*)}]/2i.$$

It is easy to verify that f_1 and f_2 are real linear functionals.

Below, we shall show that every positive functional is real and hence that every linear combination of positive functionals with real coefficients is a real linear functional. The converse is not true in general. We shall later give a counter-example.

2. In what follows, we shall often make use of the following inequality: Let $f(x)$ be a positive functional. Then for arbitrary x and y we have

$$|f(y^*x)|^2 \leqq f(y^*y)f(x^*x). \tag{1}$$

The proof of this inequality is an exact replica of the usual proof of the Schwarz inequality.[4]

---

[4] In the original text the Cauchy-Bunyakovskiĭ Inequality.

THEOREM 1: *Every positive linear functional f is real and satisfies the inequality*

$$|f(x)| \leqq f(e)\,|x|. \tag{2}$$

Proof: Let us assume, to begin with, that $|x| < 1$ and $x^* = x$. We put

$$y = (e - x)^{\frac{1}{2}} = \\ = e - \frac{1}{2}x - \frac{1}{2!}\cdot\frac{1}{2}\cdot\frac{1}{2}x^2 - \frac{1}{3!}\,\frac{1}{2}\cdot\frac{1}{2}\cdot\frac{3}{2}x^3 - \ldots;$$

this series converges, because $|x| < 1$. In virtue of condition d) of Definition 1 of § 46, an involution is continuous; therefore $y^* = y$; moreover

$$yy^* = y^2 = e - x,$$

which is easy to prove by squaring the power series. Therefore .

$$f(e - x) = f(y^*y) \geqq 0,$$

i.e., $f(x)$ is real and $f(x) \leqq f(e)$.

We obtain $f(x) \geqq -f(e)$ similarly.

It is easy to get rid of the restriction $|x| < 1$. We then obtain: if $x^* = x$, then $f(x)$ is real and

$$|f(x)| \leqq f(e)\,|x|.$$

Hence it follows that $f(x^*) = \overline{f(x)}$ for every x. For,

$$f(x) = f\left(\frac{x + x^*}{2}\right) + if\left(\frac{x - x^*}{2i}\right),$$

$$f(x^*) = f\left(\frac{x + x^*}{2}\right) - if\left(\frac{x - x^*}{2i}\right);$$

$f((x + x^*)/2)$ and $f((x - x^*)/2i)$, by what has been proved above, are real; therefore $f(x^*) = \overline{f(x)}$. It now remains to prove that $f(x)$ is a continuous function of x. For this, it is sufficient to prove the inequality (2) for every $x \in R$. We have already proved this inequality for Hermitian elements x—in particular, for elements of the form x^*x. Thus,

$$f(x^*x) \leqq f(e)\,|x^*x|,$$

and therefore

$$f(x^*x) \leqq f(e)\,|x|^2. \tag{3}$$

On the other hand, by putting $y = e$ in (1), we obtain

$$| f(x) |^2 \leqq f(e) f(x^* x),$$

and therefore, by (3),

$$| f(x) |^2 \leqq f(e)^2 | x |^2.$$

Thus, the inequality (2) is proved for all elements $x \in R$. //

3. We shall now give an example of a ring and a real functional on it that is not representable as a linear combination of positive functionals.

Let the elements of the ring R be the set of complex functions $x(z)$ that are analytic for $| z | < 1$ and continuous in the circle $| z | \leqq 1$. We put $| x | = \max_{|z| \leqq 1} | x(z) |$. We define sums and products as the sums and products of the functions; x^* is defined by the equation $x^*(z) = \overline{x(\bar{z})}$. It will be shown in § 50 that every positive functional in this ring is of the form

$$f(x) = \int_{-1}^{+1} x(t) d\sigma(t),$$

where $\sigma(t)$ is a monotonic function given on the interval $[-1, +1]$ of the real t-axis.

We now consider the following real functional:

$$f_1(x) = [x(z_0) + x(\bar{z}_0)]/2,$$

where z_0 is a fixed non-real number such that $| z_0 | \leqq 1$. Then it is not difficult to verify that there does not exist a complex function of bounded variation $\sigma_1(t)$ such that

$$[x(z_0) + x(\bar{z}_0)]/2 = \int_{-1}^{+1} x(t) d\sigma_1(t).$$

For let us suppose that there exists a $\sigma_1(t)$ such that

$$\int_{-1}^{+1} x(t) d\sigma_1(t) = [x(z_0) + x(\bar{z}_0)]/2$$

for every function that is analytic in the unit circle. We substitute $x_n(z) = (1/n) \exp(inz)$ for $x(z)$ in this equation. Then the left-hand side of the equation tends to zero as $n \to \infty$, but the absolute value of the right-hand side tends to ∞; that is, we have reached a contradiction. This means that $f_1(x)$ is not representable as a linear combination of positive functionals.

Every representation of a ring R provides us with a set of positive functionals. For let ξ_0 be any vector of the space $\mathfrak{H}$. We put

$$f(a) = (A(a)\xi_0, \xi_0). \tag{4}$$

Then $f(a)$ is a positive functional. For,

$$f(a^*a) = (A(a^*a)\xi_0, \xi_0) = \\ = (A^*(a)A(a)\xi_0, \xi_0) = (A(a)\xi_0, A(a)\xi_0) \geqq 0.$$

THEOREM 2: *Every representation of a ring R with an involution is continuous. Furthermore,* $|A| \leqq |a|$.

Proof: Applying the inequality (3) to the positive functional

$$f(a) = (A(a)\xi_0, \xi_0),$$

we obtain

$$(A(a^*a)\xi_0, \xi_0) \leqq (\xi_0, \xi_0) |a|^2,$$

i.e.,

$$|A(a)\xi_0|^2 \leqq |a|^2 |\xi_0|^2.$$

Since ξ_0 is an arbitrary vector, this inequality means that $|A| \leqq |a|$.

4. Our immediate object is to give a description of representations by means of positive functionals. It is better to do this first for cyclic representations.

Let two cyclic representations of a ring R be given, one in the space $\mathfrak{H}$ and one in the space $\mathfrak{H}'$. We denote the operators corresponding to an element a by $A(a)$ and $A'(a)$, respectively. Let ξ_0 and ξ_0' be cyclic vectors of the corresponding representations. We put

$$f(a) = (A(a)\xi_0, \xi_0) \qquad \text{and} \qquad f'(a) = (A'(a)\xi_0', \xi_0').$$

We shall show that if $f(a) = f'(a)$ *for every* a, then the representations are equivalent.

We set up a correspondence between the vectors of the spaces $\mathfrak{H}$ and $\mathfrak{H}'$ in the following way. Let $\xi = A\xi_0$. Then we associate with it the vector $\xi' = A'\xi_0'$. We shall show that this correspondence is isometric. It then follows that it is one to one. In order to prove the isometry, let us show that the scalar products of corresponding vectors coincide. Let

$$\xi_1 = A_1\xi_0, \quad \xi_1' = A_1'\xi_0',$$
$$\xi_2 = A_2\xi_0, \quad \xi_2' = A_2'\xi_0'.$$

Then

$$(\xi_1, \xi_2) = (A_1\xi_0, A_2\xi_0) = (A_2^*A_1\xi_0, \xi_0) = f(a_2^*a_1),$$
$$(\xi_1', \xi_2') = (A_1'\xi_0', A_2'\xi_0') = (A_2'^*A_1'\xi_0', \xi_0') = f'(a_2^*a_1).$$

Since $f(x) \equiv f'(x)$, we see that $(\xi_1, \xi_2) = (\xi_1', \xi_2')$, i.e., for elements of the form $A\xi_0$ (or of the form $A'\xi_0'$) the isometry is proved.

Since both representations are cyclic, the set of such elements is dense in $\mathfrak{H}$ and $\mathfrak{H}'$, respectively. We may therefore extend this correspondence by continuity to all of $\mathfrak{H}$ and $\mathfrak{H}'$.

We have thus seen that a cyclic representation is uniquely determined to within an equivalence by the positive functional (4). The problem now arises, Does there exist, for every positive functional, a representation in which this functional can be written in the form (4)? We shall show that the answer to this question is in the affirmative.

Suppose that a positive functional $f(x)$ in R is given. By making use of this functional, we introduce a scalar product in R in the following way. We put

$$(x, y) = f(y^*x).$$

We shall consider x to be equivalent to zero if

$$(x, x) = f(x^*x) = 0.$$

Two elements shall be called *equivalent* if their difference is equivalent to zero.

The set of elements that are equivalent to zero forms a left ideal in R. For suppose that $x \sim 0$ and that y is an arbitrary element. Then

$$f((yx)^*yx) = f(x^*y^*yx) = f(zx),$$

where $z = x^*y^*y$. By the inequality (1), we have

$$| f(zx) | \leqq \sqrt{f(x^*x)} \sqrt{f(z^*z)},$$

and consequently $f(zx) = 0$, i.e., $yx \sim 0$. If $x_1 \sim 0$ and $x_2 \sim 0$, then $x_1 + x_2 \sim 0$, because

$$f((x_1 + x_2)^*(x_1 + x_2)) = \\ = f(x_1^*x_1) + f(x_2^*x_1) + f(x_1^*x_2) + f(x_2^*x_2).$$

The first and fourth terms on the right-hand side are equal to zero, by definition; and that the second and third terms are zero can again be deduced from (1).

Let us verify that the axioms for a scalar product are satisfied.

1. $(x, y) = \overline{(y, x)}$. For in virtue of the fact that a positive functional is real, we have

$$f(y^*x) = \overline{f((y^*x)^*)} = \overline{f(x^*y)}.$$

2. $(\lambda x + \mu y, z) = \lambda(x, z) + \mu(y, z)$. This is obvious.

3. $(x, x) \geqq 0$. This is obvious. The fact that $(x, x) = 0$ only for $x \sim 0$ follows from the definition of equivalence to zero.

We denote the space so obtained by $\tilde{\mathfrak{H}}$. In general, it is not complete. We denote its completion by $\mathfrak{H}$. We shall now construct a representation in this space in the following way: With every element a we associate an operator A in $\mathfrak{H}$ by the formula $Ax = ax$. We have only to verify that if $x_1 \sim x_2$, then $Ax_1 \sim Ax_2$. This is clear, because if $x_1 - x_2 \sim 0$, then $a(x_1 - x_2) \sim 0$, in virtue of the fact that the set of elements equivalent to zero forms an ideal.

Let us show that the operator A is bounded and that

$$|A| \leqq |a|. \tag{5}$$

By definition, $(Ax, Ax) = f(x^*a^*ax)$. We put

$$f_1(y) = f(x^*yx),$$

keeping x fixed. $f_1(y)$ is also a positive functional. For,

$$f_1(y^*y) = f(x^*y^*yx) = f((yx)^*yx) \geqq 0.$$

Therefore, by (2),

$$f_1(a^*a) \leqq f_1(e) \, |a^*a| \leqq |f_1(e)| \, |a|^2,$$

i.e.,

$$f(x^*a^*ax) \leqq f(x^*x) \, |a|^2,$$

or

$$(Ax, Ax) \leqq (x, x) \, |a|^2.$$

Thus,

$$|A|^2 = \sup_{(x, x) = 1} (Ax, Ax) \leqq |a|^2,$$

i.e., the inequality (5) is proved. We have shown that the operator A, defined on $\tilde{\mathfrak{H}}$, is bounded; therefore we can extend the definition to the closure $\mathfrak{H}$ of $\tilde{\mathfrak{H}}$. After this extension, the norm of A, as before, does not exceed $|a|$.

We shall now show that the mapping $a \rightarrow A$ is a representation of the ring R. First, it is obvious that if $a \rightarrow A$, $b \rightarrow B$, then $\lambda a + \mu b \rightarrow \lambda A + \mu B$ and $ab \rightarrow AB$.

Let us show that if $a \rightarrow A$, then a^* goes over into A^*—in other words, that $(ax, y) = (x, a^*y)$. But this is in fact the case, because $(ax, y) = f(y^*ax)$ and

$$(x, a^*y) = f((a^*y)^*x) = f(y^*ax) = (ax, y).$$

Let us show that the representation so obtained is cyclic. As our vector ξ_0, we choose the element $x=e$. Then the set of vectors $A\xi_0$ is in our case the set of all a, i.e., the entire space $\tilde{\mathfrak{H}}$, and it is therefore dense in $\mathfrak{H}$. This proves the cyclicity.

Next, it is obvious that for this choice of ξ_0 we have, by definition of the scalar product,

$$(A\xi_0, \xi_0) = (a, e) = f(e^*a) = f(a).$$

Thus, we have constructed a representation of the ring by means of the preassigned positive linear functional f. Our results can be combined in the form of a theorem:

THEOREM 3: *To every cyclic representation of a ring R with a cyclic vector ξ_0 there corresponds a positive linear functional*

$$f(a) = (A(a)\xi_0, \xi_0), \tag{6}$$

where $A(a)$ is the operator corresponding to the element a. The representation is uniquely determined by the functional $f(x)$ to within equivalence.

Conversely, to every positive linear functional $f(a)$ there corresponds a cyclic representation such that $f(a)$ is defined by the formula (6).

5. If a representation $a \to A$ in a space $\mathfrak{H}$ is given, then to every element $\xi \in \mathfrak{H}$ there corresponds the positive functional $f(a) = (A(a)\xi, \xi)$.

When we replace the vector ξ by $\lambda\xi$, with $|\lambda| = 1$, then $f(a)$ remains unchanged.

In general, vectors that are not proportional may yield the same functional $f(a)$. However, we have the following theorem.

THEOREM 4: *Let there be given an irreducible representation of a ring R. We put $(A\xi_1, \xi_1) = f_1(a)$ and $(A\xi_2, \xi_2) = f_2(a)$.*

If $f_1(a) \equiv f_2(a)$, then $\xi_1 = \lambda\xi_2$, where $|\lambda| = 1$.

Proof: Since the representation is irreducible and $\xi_1 \neq 0$, the set of vectors $A\xi_1$ is everywhere dense in $\mathfrak{H}$.

We define an operator U in the following way: if $\xi = A\xi_1$, then we put $U\xi = A\xi_2$. Let us show that the operator so constructed preserves the length of vectors. We have

$$(U\xi, U\xi) = (A\xi_2, A\xi_2) = (A^*A\xi_2, \xi_2) =$$
$$= (A^*A\xi_1, \xi_1) = (A\xi_1, A\xi_1) = (\xi, \xi),$$

and hence it follows that the operator U is uniquely determined. For if $A'\xi_1 = A''\xi_1$, then $\xi = (A' - A'')\xi_1 = 0$, and therefore $U\xi = 0$, i.e., $A'\xi_2 = A''\xi_2$. We extend the bounded operator U by continuity to all of $\mathfrak{H}$.

Let us show that the operator U is permutable with all the operators A

of the representation. For let A_0 be an operator of the representation and let the vector ξ have the form $\xi = A\xi_1$. Then $A_0\xi = A_0 A\xi_1$, i.e., by definition of U, we have $UA_0\xi = A_0 A\xi_2$. But

$$A_0 U\xi = A_0 UA\xi_1 = A_0 A\xi_2.$$

Thus, for vectors of the form $\xi = A\xi_1$, we have $A_0 U\xi = UA_0\xi$. In virtue of the fact that these elements are everywhere dense, we obtain $A_0 U = UA_0$. Thus, U is permutable with all the operators of an irreducible representation and therefore, by Theorem 1 of § 46, $U = \lambda E$. But this means that

$$\xi_2 = U\xi_1 = \lambda\xi_1.$$

§ 48. Embedding of a Ring with an Involution in a Ring of Operators

1. Let a ring R with an involution be given. It may happen that the ring can be simplified, but simplified in such a way that the set of positive functionals—or, what is the same, of representations—remains the same.

For example, in § 47 we mentioned the ring of analytic functions given in a circle. Every positive functional in this ring is given by a formula

$$f(x) = \int_{-1}^{+1} x(t)\,d\sigma(t),$$

where $\sigma(t)$ is a monotonic function. But the same set of positive functionals belongs to the ring of all continuous functions on the interval $[-1, +1]$.

More accurately, we pose the problem in the following way.

To replace the norm $|x|$ in R by a norm $|x|_1$ as small as possible, but so that the set of positive functionals in R (i.e., the set of cyclic representations) remains the same. The new norm must be such that

$$\left.\begin{array}{lll} |e|_1 = 1, & |x+y|_1 \leqq |x|_1 + |y|_1, & |\lambda x|_1 = |\lambda|\,|x|_1, \\ |xy|_1 \leqq |x|_1 |y|_1, & |x^*|_1 = |x|_1, & |x|_1 \geqq 0. \end{array}\right\} \quad (1)$$

In the new norm there may be elements $x \neq 0$ for which $|x|_1 = 0$. In that case, we declare them to be equivalent to zero and thus go over to a new ring.

Furthermore, it may happen that in the new norm the ring R is not complete. Then we complete it. We also see that after introduction of the new norm the ring R becomes isomorphic to a ring of operators in a Hilbert space and the norm goes over into the operator norm in the Hilbert space.

2. LEMMA 1: *Let a set of norms $|x|_\alpha$ be introduced in R, each of them satisfying the condition* (1). *Let each of these norms be such that R can be mapped isomorphically, and with preservation of the involution and the norm, into a ring of operators in a Hilbert space. Lastly, for every x let* $\sup_\alpha |x|_\alpha < \infty$. *Then in the norm* $|x|_1 = \sup_\alpha |x|_\alpha$ *the ring R can be mapped isomorphically, and with preservation of the norm and the involution, into a ring of operators in a Hilbert space.*

Proof: Let us assume that for the norm $|x|_\alpha$ the ring is realized as a ring of operators in the space $\mathfrak{H}_\alpha$.

We consider the space $\mathfrak{H}$ that is the orthogonal direct sum of the spaces $\mathfrak{H}_\alpha$. If we denote the vectors in $\mathfrak{H}_\alpha$ by ξ^α, then the vectors in $\mathfrak{H}$ are $\xi = \{\xi^\alpha\}$, where $\{\xi^\alpha\}$ is different from zero for not more than a countable set of values α and

$$\sum_\alpha |\xi_\alpha|^2 = |\xi|^2 < \infty.$$

The scalar product is defined by the formula

$$(\xi, \eta) = \sum_\alpha (\xi_\alpha, \eta_\alpha).$$

To every a there corresponds, by definition, an operator $X_a{}^{(\alpha)}$ in $\mathfrak{H}_\alpha$, with $|X_a{}^{(\alpha)}| = |a|_\alpha$. In the space $\mathfrak{H}$, to the element a there corresponds the operator X_a, defined as follows:

$$X_a\{\xi_\alpha\} = \{X_a^{(\alpha)}\xi_\alpha\}.$$

The operator X_a is bounded. For

$$|X_a\xi|^2 = \sum_\alpha |X_a^{(\alpha)}\xi_\alpha|^2 \leqq \sum |X_a^{(\alpha)}|^2 |\xi_\alpha|^2 \leqq \sup_\alpha |X_a^{(\alpha)}|^2 \sum |\xi_\alpha|^2.$$

It is easy to see that $\sup_\alpha |X_a{}^{(\alpha)}| = |X_a| = |a|_1$. Thus we have realized the ring R with the norm $|a|_1$ in the form of a ring of operators in a Hilbert space. //

LEMMA 2: *In the ring R let a norm* $|x|_0$ *be introduced satisfying the condition* (1), *and let a representation* $a \to X_a$ *of R be given that is continuous in this norm.*[5] *Then* $|X_a| \leqq |a|_0$.

Proof: Since this representation is continuous, we can extend it to the completion R_0 of R. We then obtain a representation of the complete

[5] When we say that the representation $a \to X_a$ is continuous in the norm $|x|_0$ satisfying the conditions (1), then we also require that $|a|_0 = 0$ should imply $X_a = 0$, i.e., that elements equivalent to zero in this norm go over into 0.

ring R_0. For this, the inequality $|X_a| \leqq |a|_0$ holds, according to Theorem 2 of § 47. //

THEOREM 1: *In the ring R we can introduce a norm* $|x|_1$ *satisfying the following conditions:*

1. $|x|_1$ *satisfies* (1).
2. *Every representation of R is continuous in the norm* $|x|_1$.
3. *If any other norm* $|x|_2$ *also satisfies the conditions* 1. *and* 2., *then* $|x|_1 \leqq |x|_2$ *for every x.*
4. *The ring R with the norm* $|x|_1$ *can be mapped isomorphically, and with preservation of the involution and the norm, into a ring of operators in a Hilbert space.*

It is clear that the norm $|x|_1$ is uniquely determined by the conditions 1., 2., 3.

Proof: We consider the set of all cyclic representations of the given ring[6] and denote each cyclic representation by a symbol α. In this representation let the element a correspond to the operator $X_a^{(\alpha)}$. By Theorem 2 of § 47, $|X_a^{(\alpha)}| \leqq |a|$. We put

$$|a|_\alpha = |X_a^{(\alpha)}|;$$

$|a|_\alpha$ is a norm satisfying the conditions (1) of this section. Now we put $|a|_1 = \sup_\alpha |a|_\alpha$. Then we have $|a|_1 \leqq |a|$, since $|a|_\alpha \leqq |a|$.

By Lemma 1, the ring R with the norm $|a|_1$ satisfies the conditions (1) and can be realized as a ring of operators in a certain Hilbert space $\mathfrak{H}$.

Every cyclic representation of R is continuous in the norm $|a|_1$. For let a cyclic representation $a \to X_a^{(\alpha)}$ be given. Then

$$|X_a^{(\alpha)}| = |a|_\alpha \leqq \sup_\alpha |a|_\alpha = |a|_1.$$

Every representation splits into a direct sum of cyclic representations and is therefore also continuous in this norm.

Thus, we have shown that $|a|_1$ satisfies the conditions 1., 2., and 4. Let us show that the condition 3. is also satisfied. Let a norm $|x|_2$ satisfying the conditions (1) be given in which all the representations of R are continuous. Then we have, by Lemma 2,

$$|X_a^{(\alpha)}| \leqq |a|_2,$$

i.e., $|a|_\alpha \leqq |a|_2$. Therefore,

[6] We only take a cyclic representation, because every representation breaks up into cyclic ones.

$$|a|_1 = \sup_\alpha |a|_\alpha \leqq |a|_2.$$

This completes the proof of the theorem. //

THEOREM 2: *The equation*

$$|x|_1 = \sup \sqrt{f(x^*x)} \tag{2}$$

holds, where sup *is extended over all positive functionals f for which* $f(e) = 1$.

Proof: Every cyclic representation can be described (§ 47) by a positive linear functional f. Let us find the norm of the operator X_a corresponding to the element a in this representation. We have

$$(X_a x, X_a x) = f((ax)^*ax) = f(x^*a^*ax), \qquad (x, x) = f(x^*x).$$

We introduce the functional $f_x(y) = f(x^*yx)$, keeping x fixed. $f_x(y)$ is a positive functional and

$$f_x(e) = f(x^*x), \qquad |X_a|^2 = \sup (X_a x, X_a x),$$

where sup is taken over all x for which $(x, x) = 1$; thus,

$$|X_a|^2 = \sup f_x(a^*a),$$

where sup is taken over all x for which $f_x(e) = 1$.

Therefore $|X_a|^2 \leqq \sup f(a^*a)$, where sup is taken over all the positive functionals f for which $f(e) = 1$.

If we denote this representation by the subscript α, then

$$|a|_\alpha^2 \leqq \sup f(a^*a),$$

in virtue of the conditions that $f(e) = 1$ and f is a positive linear functional. Therefore

$$|a|_1^2 = \sup_\alpha |a|_\alpha^2 \leqq \sup_{f(e)=1} f(a^*a).$$

On the other hand,

$$f(a^*a) = (X_a e, X_a e).$$

Therefore, if $f(e) = 1$, then

$$f(a^*a) \leqq |X_a|^2 = |a|_\alpha^2 \leqq \sup_\alpha |a|_\alpha^2 = |a|_1^2,$$

and consequently

$$\sup_{f(e)=1} f(a^*a) \leqq |a|_1^2.$$

This completes the proof. //

3. *Note* 1: All the arguments of this section remain valid if there is no norm at all in the original ring, i.e., if it is defined only by the axioms 1., 2. (§ 46.1) and a), b), c), d) (§ 46.2); we need only impose the additional condition that for every x

$$\sup f(x^*x) < \infty,$$

where sup is taken over all positive f for which $f(e) = 1$.

Note 2: The set I of elements for which $|x|_1 = 0$ (we call them equivalent to zero) forms a two-sided ideal. These elements x can be characterized by the fact that they go over into 0 in every representation or, to put it differently, that every positive functional vanishes on them.

Let us show that I is in fact an ideal.

Let $|x|_1 = 0$ and let y be an arbitrary element. Then

$$|xy|_1 \leqq |x|_1 |y|_1 = 0,$$

and similarly, $|yx|_1 = 0$. Furthermore, if $|x|_1 = 0$ and $|y|_1 = 0$, then

$$|\lambda x + \mu y|_1 \leqq |\lambda| |x|_1 + |\mu| |y|_1 = 0.$$

Thus, we have shown that I is an ideal. In studying representations or positive functionals, we may replace R by the residue-class ring of this ideal. We denote this residue-class ring by R'. We shall call it a *reduced ring*.

Note 3: Every representation of R is continuous in the norm $|x|_1$ and is therefore a continuous representation of R'. But a continuous representation of R' can be extended to a representation of the completion of R'. We denote the completion of R' by $\bar{R}$.

Thus, *every representation of the ring R is also a representation of $\bar{R}$, and vice versa.*

Similarly, *every positive linear functional on R can be extended to $\bar{R}$, and vice versa.*

§ 49. Indecomposable Functionals and Irreducible Representations

1. In the finite-dimensional case, every representation splits into irreducible representations. In the general case, the existence of such representations is not clear a priori. Without touching on the problem of decomposing representations into irreducible ones, we shall show in this section that there exist irreducible representations. It is very convenient to do this in terms of positive functionals.

DEFINITION 1: We shall say that a positive functional f_1 is *subordinate to a functional* f ($f_1 \ll f$) if there exists a $\lambda > 0$ such that $\lambda f - f_1$ is a positive functional.

We construct a cyclic representation $a \to X_a$ (X_a are operators in a space $\mathfrak{H}$) corresponding to the functional f:

$$f(a) = (X_a \xi_0, \xi_0),$$

where ξ_0 is a cyclic vector. Let B be a bounded positive-definite operator in the Hilbert space $\mathfrak{H}$ that is permutable with all the operators of the representation. We put

$$f_1(a) = (X_a B \xi_0, \xi_0).$$

In particular, the functional f corresponds to the operator $B = E$. We claim that $f_1(a)$ *is a positive functional and is subordinate to* $f(a)$.

$f_1(a)$ is positive; for

$$f_1(a^* a) = (X_a^* X_a B \xi_0, \xi_0) = (X_a B \xi_0, X_a \xi_0) = (B X_a \xi_0, X_a \xi_0) \geqq 0,$$

since B is a positive-definite operator. Furthermore, f_1 is subordinate to f. For B is bounded. Therefore there exists a λ such that $(B\xi, \xi) \leqq \lambda(\xi, \xi)$, or

$$\lambda(\xi, \xi) - (B\xi, \xi) \geqq 0.$$

Putting $\xi = X_a \xi_0$, we obtain

$$\lambda (X_a^* X_a \xi_0, \xi_0) - (X_a^* X_a B \xi_0, \xi_0) \geqq 0,$$

i.e., $\lambda f - f_1$ is a positive functional; but this means that f_1 is subordinate to f.

Conversely, suppose that f_1 is a positive functional subordinate to f. We construct a cyclic representation by means of f (§ 47.4). *Then the functional* f_1 *corresponds to a positive-definite operator* B *that is permutable with all the operators of the representation.*

Let us prove this. We know that the space $\mathfrak{H}$ is obtained as the completion of the space $\tilde{\mathfrak{H}}$ formed from the classes of equivalent elements of R, and that the scalar product in $\tilde{\mathfrak{H}}$ is given by the formula

$$(x, y) = f(y^* x),$$

where the elements x for which $(x, x) = 0$, i.e., $f(x^* x) = 0$, are taken to be equivalent to zero.

In the space $\tilde{\mathfrak{H}}$ we consider the Hermitian form $f_1(y^* x)$. We shall show that f_1 is uniquely determined as a Hermitian bilinear form continuous in $\tilde{\mathfrak{H}}$. f_1 is subordinate to f. This means that there exists a λ such that $\lambda f - f_1$ is a positive functional. Thus, we have:

$$\lambda f(x^*x) - f_1(x^*x) \geqq 0,$$

and therefore

$$0 \leqq f_1(x^*x) \leqq \lambda f(x^*x).$$

This shows that $f(x^*x) = 0$ implies that $f_1(x^*x) = 0$. Hence we see, say by means of the inequality

$$|f_1(y^*x)|^2 \leqq f_1(x^*x) f_1(y^*y),$$

that the expression $f_1(y^*x)$ is zero.

We have thus shown that $f_1(y^*x)$ is uniquely determined, i.e., that $f_1(y^*x) = 0$ for $x \sim 0$. f_1 is a bounded bilinear form. In fact,

$$0 \leqq f_1(x^*x) \leqq \lambda f(x^*x).$$

Therefore

$$|f_1(y^*x)|^2 \leqq f_1(x^*x) f_1(y^*y) \leqq \lambda^2 f(x^*x) f(y^*y) = \lambda^2 (x, x)(y, y).$$

Being bounded, this bilinear form can be extended to the completion of $\tilde{\mathfrak{H}}$, i.e., to the space $\mathfrak{H}$. But to a bounded bilinear form in $\mathfrak{H}$ there corresponds a bounded operator B. Therefore there exists an operator B such that

$$f_1(y^*x) = (Bx, y).$$

Let us show that the operator B is permutable with all the operators X_a of the representation. For this purpose it is sufficient to show that

$$(BX_a x, y) = (Bx, X_a^* y).$$

But this is in fact so, because

$$(BX_a x, y) = (Bax, y) = f_1(y^*ax)$$

and

$$(Bx, X_a^* y) = (Bx, a^* y) = f_1((a^*y)^*x) = f_1(y^*ax).$$

Thus, we have proved the following theorem.

THEOREM 1: *Let $f(x)$ be a positive functional, $a \rightarrow X_a$ a cyclic representation corresponding to it, and ξ_0 a corresponding cyclic vector, i.e.,*

$$f(a) = (X_a \xi_0, \xi_0).$$

Then to every positive functional f_1 subordinate to f there corresponds a positive-definite operator B that is permutable with all the X_a, and

$$f_1(a) = (X_a B \xi_0, \xi_0).$$

Conversely, to every bounded positive-definite operator B that is permutable with all the operators X_a there corresponds a positive functional subordinate to f.

In particular, to $f(x)$ itself there corresponds the unit operator. Now let us consider linear combinations of positive functionals subordinate to $f(x)$ We call them functionals subordinate to the positive functional $f(x)$. They correspond to arbitrary bounded operators that are permutable with the operators X_a of the representation.

Indeed, every operator that permutes with the X_a can be represented as a linear combination of positive operators. Thus we have the following result:

In the set of functionals subordinate to the positive functional f we can introduce an operation of multiplication such that it becomes isomorphic to the ring of operators permutable with the operators X_a of the representation generated by the functional $f(x)$. Here f itself plays the role of the unit element of the ring.

We have turned the set C_f of functionals subordinate to f into a ring with an involution. Moreover:

1. To the functional f there corresponds the unit element of the ring.
2. The operation $*$ is defined as follows: $f^*(x) = \overline{f(x^*)}$.
3. Multiplication is connected with the $*$-operation by the usual condition:

$$(f_1 f_2)^* = f_2^* f_1^*.$$

2. DEFINITION 2: A positive function f is called *indecomposable* if every functional f_1 subordinate to f is a multiple of f, i.e., $f_1(x) = \lambda f(x)$.

THEOREM 2: *Let f be a positive functional. The representation corresponding to it is irreducible if and only if f is indecomposable.*

Proof: To every functional f_1 subordinate to f there corresponds an operator that is permutable with all the operators X_a of the representation. To f itself there corresponds the unit operator. The irreducibility of the representation is equivalent to the condition that every operator B that is permutable with the operators of the representation be a multiple of the unit (Theorem 1 of § 46), i.e., to the condition that every functional f_1 subordinate to f be a multiple of f. //

3. We shall now proceed to prove the existence of irreducible representations. By the theorem just proved, it is sufficient for this purpose to prove the existence of indecomposable positive functionals.

The set of positive functionals $f(x)$ such that $f(e) \leqq 1$ forms a convex set. For if $f_1(x)$ and $f_2(x)$ are positive and $f_1(e) \leqq 1$, $f_2(e) \leqq 1$, then

$$f(x) = \alpha f_1(x) + \beta f_2(x) \qquad (\alpha \geqq 0, \beta \geqq 0, \alpha + \beta = 1)$$

satisfies the same conditions. Therefore the existence of indecomposable positive functionals follows immediately from a theorem of Krein and Milman [85]. Furthermore, let x be an element of the ring such that $|x|_1 \neq 0$. Then there exists a positive functional f such that

$$f(x^*x) \neq 0, \qquad f(e) = 1.$$

On the other hand, by the same theorem of Krein and Milman the set of all positive functionals f satisfying the condition $f(e) \leqq 1$ is the least weakly closed convex set containing all the indecomposable positive functionals that satisfy the condition $f(e) \leqq 1$. Therefore there exists such an indecomposable functional f_0 satisfying the conditions

$$f_0(e) \leqq 1, \qquad f_0(x^*x) \neq 0.$$

Thus, we have proved the following theorem.

THEOREM 3: *Let x be an element of the ring $\bar{R}$ such that $|x|_1 \neq 0$. Then there exists an indecomposable positive functional f_0 satisfying the conditions*

$$f_0(e) \leqq 1, \qquad f_0(x^*x) \neq 0. \tag{1}$$

By Theorem 2 of the present section and Theorem 3 of § 47, this theorem can also be stated in the following way:

THEOREM 4: *Let x_0 be an element of a ring $\bar{R}$ such that $|x_0|_1 \neq 0$. Then there exists an irreducible representation $a \to X_a$ of the ring $\bar{R}$ such that the operator X_{x_0} corresponding to x_0 in this representation is different from zero.*

In fact, condition (1) can be rewritten in the form

$$|\xi_0| \leqq 1, \qquad |X_{x_0}\xi_0|^2 \neq 0.$$

The latter inequality means that $X_{x_0} \neq 0$.

§ 50. The Case of Commutative Rings

1. The whole picture becomes particularly simple when R is a commutative ring.

LEMMA 1: *If R is commutative, then the ring $\bar{R}$ is isomorphic to the ring of all continuous functions $x(M)$ on a compact space. Furthermore,*

$$x^*(M) = \overline{x(M)}.$$

Proof: $\bar{R}$ is a $*$-ring, i.e.,

$$|x^*x|_1 = |x|_1 |x^*|_1. \tag{1}$$

For $|x|_1 = \sup_f \sqrt{f(x^*x)}$, where f is a positive functional and $f(e) = 1$. By the inequality (2) of § 47, we have

$$\sup_f \sqrt{f(x^*x)} \leqq \sqrt{|xx^*|_1};$$

thus, $|x|_1^2 \leqq |xx^*|_1$. On the other hand, we always have

$$|xx^*|_1 \leqq |x|_1 |x^*|_1 = |x|_1^2,$$

so that

$$|xx^*|_1 = |x|_1^2 = |x|_1 |x^*|_1.$$

By Lemma 1 in [80], a commutative ring in which an involution and a norm are introduced satisfying the condition (1) is isomorphic to the ring of all continuous functions on a compact set $\mathfrak{M}_1$. This $\mathfrak{M}_1$ is the set of maximal ideals of $\bar{R}$.

2. DEFINITION: A maximal ideal M of a ring R is called *symmetric* if for every $x \in R$ we have $x^*(M) = \overline{x(M)}$. If M is a maximal ideal, then we denote by M^* the maximal ideal for which $x^*(M^*) = \overline{x(M)}$. It is easy to show that this M^* exists for every M. A symmetric maximal ideal is one for which $M^* = M$.

It can be shown that the set of symmetric maximal ideals of a ring R forms a closed subset in the set of all maximal ideals of R.

THEOREM 1: *$\bar{R}$ is isomorphic to the ring of all the continuous functions on the set $\mathfrak{M}_1$ of symmetric maximal ideals of R.*

Proof: To prove the theorem it is sufficient, by Lemma 1, to show that the set of maximal ideals of $\bar{R}$ is homeomorphic to the set of symmetric maximal ideals of R.

To every maximal ideal of $\bar{R}$ there corresponds a symmetric maximal ideal of R.

For let a homomorphism of $\bar{R}$ into the field of complex numbers be given. This homomorphism is at the same time a homomorphism of the ring R', which is part of $\bar{R}$. But R' is a residue-class ring of R. Therefore this homomorphism is at the same time a homomorphism of R itself into the field of complex numbers.

A maximal ideal of R is thus determined. This maximal ideal is symmetric, because all the maximal ideals of $\bar{R}$ are symmetric. The continuity of the correspondence between ideals follows immediately from the definition of the topology in the set of maximal ideals (see [79], § 7).

Conversely, let M be a symmetric maximal ideal of R. We consider the functional

$$f(x) = x(M).$$

This functional is positive. In fact,

$$f(x^*x) = x^*(M)x(M) = |x(M)|^2 \geqq 0, \qquad f(e) = e(M) = 1.$$

Therefore, it is equal to zero for elements for which $|x|_1 = 0$. Moreover, it is continuous in the norm $|x|_1$ and can therefore be extended to $\bar{R}$.

Thus, we can establish a one-to-one continuous correspondence between the symmetric maximal ideals M of R and the maximal ideals $\bar{M}$ of $\bar{R}$. If we denote an element of R and the corresponding element of $\bar{R}$ by the same letter x, then we have Theorem 2:

THEOREM 2: *Every positive linear functional $f(x)$ on a commutative ring R can be represented in a unique way in the form*

$$f(x) = \int x(M) d\sigma(\Delta), \tag{2}$$

where $\sigma(\Delta)$ is a positive completely additive set function on the set $\mathfrak{M}_1$ of symmetric maximal ideals of R.

Proof: Every positive functional on R can be extended to $\bar{R}$ (§ 48, Note 3). It then turns into a positive functional on the set of continuous functions on the compact set $\mathfrak{M}_1$. Such a functional is described by the formula (2), where $\sigma(\Delta)$ is a positive completely additive set function. The set function $\sigma(\Delta)$ is uniquely determined.

Conversely, if $\sigma(\Delta)$ is a positive set function on the set of symmetric maximal ideals, then $f(x)$, given by formula (2), is a positive functional. For

$$f(x^*x) = \int x^*(M)x(M) d\sigma(\Delta) = \int |x(M)|^2 d\sigma(\Delta) \geqq 0.$$

This completes the proof. //

It follows from formula (2) that every indecomposable positive functional is of the form $f(x) = x(M_0)$, where M_0 is a fixed symmetric maximal ideal.

The theorem just proved means, strictly speaking, that every functional decomposes in a unique way into indecomposable positive functionals. We have required R to be commutative. This condition is essential, as is clear from the following theorem.

THEOREM 3: *Assume that every positive functional given on a ring R decomposes in a unique way into indecomposable positive functionals. Then the reduced ring $\bar{R}$ is commutative (i.e., $xy - yx$ is an element equivalent to zero in R).*

Proof: Let us consider an arbitrary continuous representation $a \to X_a$, where X_a is an operator in the space $\mathfrak{H}$. We shall show that $\mathfrak{H}$ is one-dimensional. Let us assume the contrary. Then $\mathfrak{H}$ contains at least two linearly independent vectors ξ_1 and ξ_2. We put

$$f_1(a) = (X_a\xi_1, \xi_1), \quad f_2(a) = (X_a\xi_2, \xi_2),$$

$$f_1'(a) = \frac{1}{2}(X_a(\xi_1 + \xi_2), \quad \xi_1 + \xi_2),$$

$$f_2'(a) = \frac{1}{2}(X_a(\xi_1 - \xi_2), \quad \xi_1 - \xi_2),$$

and assume that

$$\varphi(a) = f_1(a) + f_2(a) = f_1'(a) + f_2'(a).$$

$\varphi(a)$ is a positive functional. By Theorem 2 of § 49, the functionals $f_1(a)$, $f_2(a)$, $f_1'(a)$, $f_2'(a)$ are indecomposable. Therefore $f(a)$ decomposes in two ways into indecomposable functionals. These are distinct. For the vectors ξ_1, ξ_2, $\xi_1 + \xi_2$, $\xi_1 - \xi_2$ are not proportional; but on the other hand, by Theorem 4 of § 47, in the case of an irreducible representation by functionals of the form $(X_a\xi, \xi)$, the vector ξ is uniquely determined to within a factor.

Thus, every irreducible representation of R is one-dimensional and hence commutative. Since every element that is carried into zero by all continuous representations is equivalent to zero, $xy - yx$ is equivalent to zero, i.e., the reduced ring is commutative. //

3. *Example*: We denote by R_0' the set of functions given on the half-line $0 \leqq u < \infty$ such that

$$\| f \| = \int_0^\infty | f(u) | \sinh 2u du < \infty.$$

We define a multiplication $f = f_1 \times f_2$ in R_0', where $f(u)$ is given by the formula

$$f(u) = \int_0^\infty \int_{|u-t|}^{u+t} f_1(s) f_2(t) ds dt.$$

Further, we let R_0 denote the set of all elements of the form $\lambda e + f$, where e is a formally adjoined unit element and $f \in R_0'$.

It can be verified that this turns R_0 into a normed ring.

We also define an operation $*$ by putting

$$f^*(u) = \overline{f(u)}, \qquad (\lambda e + f)^* = \bar{\lambda} e + f^*.$$

It is easy to verify that this yields a commutative ring with an involution. Let us find the maximal ideals of this ring. The arguments here are exactly analogous to those given in [83].

Just as in the other case, we come to the conclusion that the maximal ideals are determined by homomorphisms

$$f \to \int_0^\infty f(s)\psi(s)ds,$$

where $\psi(s)$ is defined by the relation

$$\psi(s)\psi(t) = \int_{|t-s|}^{t+s} \psi(u)du.$$

The solutions of this equation are the functions

$$\psi(s) = [2 \sin \varrho s]/\varrho,$$

where ϱ is an arbitrary complex number. This homomorphism is defined for all elements of R_0 if and only if $\varrho = \varrho_1 + i\varrho_2$, where $|\varrho_2| \leqq 2$.

Thus, every maximal ideal of the ring is determined by a number $\varrho = \varrho_1 + i\varrho_2$, where $|\varrho_2| \leqq 2$, and with ϱ and $-\varrho$ determining the same maximal ideal.

Let M be the maximal ideal determined by ϱ. Then the maximal ideal M^* is determined by $\bar{\varrho}$. Therefore, symmetric maximal ideals are determined by the conditions $\varrho = \bar{\varrho}$ or $\varrho = -\bar{\varrho}$; thus, symmetric maximal ideals arise when ϱ is either real or pure imaginary. The corresponding homomorphisms are given by the formulas

$$f \to \int_0^\infty f(s)[2 \sin \varrho s]/\varrho \, d\varrho,$$

where ϱ is real, and by

$$f \to \int_0^\infty f(s)[2 \sinh \varrho s]/\varrho \, d\varrho,$$

where $0 \leqq \varrho \leqq 2$.

§ 51. Group Rings

1. As a special case of the rings studied earlier we shall now discuss the so-called group rings. This enables us to obtain certain results concerning representations of groups.

Let G be a locally compact group. For simplicity of the exposition we shall assume that the left and right invariant Haar measures on G coincide.

We examine the ring R' whose elements are the absolutely integrable functions $x(g)$ on the group.

Multiplication is given by the formula

$$x_1 \times x_2 = \int x_1(gg_1{}^{-1})x_2(g_1)dg_1.$$

An involution is defined by the equation

$$x^*(g) = \overline{x(g^{-1})}.$$

The norm $|x|$ of the element x is taken to be

$$|x| = \int |x(g)| dg.$$

To this ring we adjoin a formal unit element (if the group is non-discrete), so that finally the elements of the ring can be expressed by the symbol $\lambda e + x$, where e is the formally adjoined unit element. The multiplication is naturally given by the formula

$$(\lambda_1 e + x_1)(\lambda_2 e + x_2) = \lambda_1\lambda_2 e + (\lambda_1 x_2 + \lambda_2 x_1 + \lambda_1\lambda_2 x_1 x_2),$$

and the involution and norm are extended as follows: We put

$$|\lambda e + x| = |\lambda| + |x|, \qquad (\lambda e + x)^* = \bar{\lambda} e + x^*.$$

We denote the ring so obtained by R and call it *the group ring of G.*

THEOREM 1: *To every representation $a + \lambda e \to X_a + \lambda E$ of the group ring there corresponds a continuous unitary representation $g \to T_g$ of the group. Conversely, to every measurable unitary representation of the group there corresponds a representation $a \to X_a$ of the group ring. These representations are connected by the formula*

$$X_a = \int a(g) T_g dg.$$

Let us prove the theorem for a cyclic representation. Such a representation, as we know (§ 47), can be realized in the following way: The space $\mathfrak{H}$ is obtained by completing the space $\tilde{\mathfrak{H}}$ whose vectors are the elements $x, y, \ldots$ of R (considered to within equivalence). The scalar product is given by the formula

$$f(y^*x) = (x, y),$$

where f is a positive functional. To the element a there corresponds the operator X_a given by $X_a x = ax$. We have shown that here $|X_a| \leqq |a|$.

With the element g_0 of G we associate the operator $T_{g_0}x=y$, where $y(g)=x(g_0^{-1}g)$.

Let us show that this operator is unitary. Observe that the following equation holds for elements of R:

$$(T_{g_0}y)^*T_{g_0}x=y^*x.$$

For

$$y^*x=\int\overline{y(g_1g^{-1})}x(g_1)dg.$$

Therefore under the application of T_{g_0} scalar products do not change. Since, first of all, the operator T_g maps our set of functions into itself, it is clear that the operator T_{g_0} extended to the entire space $\mathfrak{H}$ is unitary. Let us now show that the representation $g\to T_g$ of G is continuous. For this purpose, we note that the functional f (like every positive functional) is continuous; on the other hand

$$|T_{g_0}x-x|=\int|T_{g_0}x-x|dg\to 0 \quad \text{as} \quad g_0\to e,$$

and therefore

$$f((T_{g_0}x-x)^*(T_gx-x))\to 0 \quad \text{as} \quad g_0\to e.$$

This means that

$$(T_{g_0}x-x,\ T_{g_0}x-x)\to 0 \quad \text{as} \quad g_0\to e.$$

This proves the continuity at the unit element of the group, and consequently at every other point.

It now remains to show that the representation of the ring generated by $g\to T_g$ is the one with which we started.

Let $a(g)$ be an absolutely integrable function. In the representation of R it corresponds to the operator $Ax=a\times x$. Our aim is to show that

$$Ax=\int a(g)T_gxdg.$$

or, in terms of scalar products,

$$(Ax,y)=\int a(g)(T_gx,y)dg,$$

i.e.,

$$f(y^*ax)=\int a(g)f(y^*T_gx)dg.$$

Since the functional f and the operation T_g are continuous, we can rewrite the right-hand side of this equation as follows:

$$f\left(y^* \int a(g) T_g x dg\right) = f(y^* a x).$$

This proves our statement.

We have shown that to every representation of the group ring there corresponds a representation of the group. The converse is easy to show. For suppose that a unitary representation of the group is given: $g \to T_g$.

Let us assume that the function T_g is weakly continuous, i.e., that $(T_g \xi, \eta)$ is a continuous function of g for every ξ and η. We put

$$A = \int a(g) T_g dg.$$

This operator exists, provided only that the function $a(g)$ is integrable. We thus obtain a representation of the group ring. It is easy to see that, conversely, to this operator there corresponds the representation T_g, provided only that T_g is cyclic.[7]

We make one further remark concerning the proof of the theorem. In constructing T_g, the representation we have used is not that of the extended ring R, but only that of the original ring R' without the adjoined unit element. But since the space $\mathfrak{H}$ was constructed by means of a completion of the space $\tilde{\mathfrak{H}}$ formed from elements of R, we have to show that the unextended ring R' leads to the same space $\mathfrak{H}$. For this purpose, we shall show that the element e is the limit, in the sense of our scalar product, of elements x of R'.

To every neighborhood V of the unit element of G we assign a function $e_V(g)$ satisfying the following conditions:

$$e_V(g) \geqq 0; \qquad e_V(g) = 0 \quad \text{if} \quad g \notin V;$$

$$e_V(g^{-1}) = e_V(g); \qquad \int e_V(g) dg = 1.$$

We call such a system of functions a *unit system*. It is easy to show that a unit system has the following properties:

$$|x \times e_V - x| \to 0 \qquad \text{as} \qquad V \to e$$

(the limit is interpreted in the partially ordered system of neighborhoods given in the sense of their natural ordering). We have

$$f(e_V x) \to f(x) \qquad \text{for every} \qquad x \in R,$$

[7] In the case of a separable space $\mathfrak{H}$, this result can be strengthened. Specifically, it follows from the preceding arguments that if the function $(T_g \xi, \eta)$ is measurable for all ξ, $\eta \in \mathfrak{H}$, then the representation $g \to T_g$ coincides almost everywhere on G with a continuous representation.

i.e., $(e_V, x) \to (e, x)$ for every x. Moreover,

$$(e_V, e_V) = f(e_V e_V) \leqq f(e)|e_V e_V| = f(e),$$

—i.e., the lengths of the vectors e_V are bounded—so that e_V is weakly convergent. We denote this weak limit by ξ_0. Then we have

$$(\xi_0, x) = (e, x) \qquad \text{for every} \qquad x \in R.$$

In particular,

$$(e_V, \xi_0) = (e_V, e) = f(e_V),$$

and therefore $f(e_V)$ has a limit. Thus, the vector $\xi_0 - e$, which is orthogonal to all the elements x, splits off. It is not essential for the representation, because in this one-dimensional space every element x yields the null operator. By discarding this one-dimensional space we obtain the required result. For this 'parasitical' one-dimensional space to be absent it is necessary and sufficient that

$$\lim_{V \to e} f(e_V) = f(e). \ //$$

2. Applying Theorem 4 of § 49 to the group ring R and using Theorem 1 just proved, we obtain a fundamental result of Gelfand and Raikov [82] on the completeness of the system of irreducible representations of a locally compact group.

Using the expression (6) of § 47 for a positive functional and Theorem 1 of this section, we find that every positive functional in the group ring R is determined by the formula

$$f(\lambda e + a) = \lambda C + \int a(g)(T_g \xi, \xi) dg,$$

where $g \to T_g$ is a continuous unitary representation of G and the function $\varphi(g) = (T_g \xi, \xi)$ is a continuous positive-definite function on G (see [82]).

Conversely, every bounded measurable positive-definite function $\varphi(g)$ corresponds to a positive functional on the group ring defined by the formula

$$f(a) = \int a(g)\varphi(g) dg.$$

Hence it follows, in particular, that every bounded measurable positive-definite function coincides almost everywhere on G with a continuous positive-definite function (see [82]).

Positive-definite functions were the starting point in [82] for the construction of unitary representations of a locally compact group.

§ 52. Example of an Unsymmetric Group Ring

1. It is known that the group ring of a compact or a commutative group is symmetric (see [87]). However, it turns out that for a locally compact group this is in general not true.

We shall now give an example of a locally compact group whose group ring is unsymmetric.

2. Let G be the group of complex matrices of order 2 with determinant 1; let $\mathfrak{H}$ denote the subgroup of G that consists of unitary matrices.

According to a well-known definition, a double coset of $\mathfrak{H}$ in G is a set of elements of the form h_1gh_2, where g is fixed and h_1 and h_2 range over the whole subgroup $\mathfrak{H}$. Since $\mathfrak{H}$ is a compact subgroup, the set of elements of the form h_1gh_2 has finite measure whenever the set of elements g has finite measure.

We consider the set R_0' of functions $f(g)$ that are summable and constant on the double cosets of $\mathfrak{H}$ in G. The set R_0 of elements of the form $\lambda e+f$ $(f\in R_0')$ forms a subring of the group ring R. For assume that $f_1(g)$ and $f_2(g)$ belong to R_0'. We have to show that the function $f=f_1\times f_2$ also belongs to R, i.e., is constant on the double cosets of $\mathfrak{H}$ in G. But

$$f(gh)=\int f_1(ghg_1{}^{-1})f_2(g_1)dg_1=$$
$$=\int f_1(gg_2{}^{-1})f_2(g_2h)dg_2=\int f_1(gg_2{}^{-1})f_2(g_2)dg_2=f(g);$$

and similarly, $f(hg)=f(g)$ for all $h\in\mathfrak{H}$.

The ring R_0 is commutative. For the Haar measure on G is invariant under the transformation $g\to g^{-1}$ and therefore, when f_1 and $f_2\in R_0'$, we have

$$\int f_1(hg^{-1})f_2(g)dg=\int f_1(g^{-1})f_2(gh)dg=$$
$$=\int f_1(g^{-1})f_2(hg)dg=\int f_2(hg^{-1})f_1(g)dg$$

for all $h\in\mathfrak{H}$.

3. In order to prove that the group ring is unsymmetric, we shall first show that R_0 is unsymmetric.

Let us therefore examine the ring R_0 in more detail. Every matrix g can be represented in the form $g=ha$, where h is unitary and a is a positive-definite Hermitian matrix. Every Hermitian matrix can be written in the form $a = h_1 \delta h_1^{-1}$, where h_1 is unitary and δ is a diagonal matrix. Since the group is unimodular, δ is of the form

$$\delta = \begin{pmatrix} \lambda & 0 \\ 0 & \lambda^{-1} \end{pmatrix}, \qquad \lambda > 0.$$

Thus, in every coset of $\mathfrak{H}$ in G there is a diagonal matrix $\begin{pmatrix} \lambda & 0 \\ 0 & \lambda^{-1} \end{pmatrix}$, and it is easy to see that every coset containing the matrix $\begin{pmatrix} \lambda & 0 \\ 0 & \lambda^{-1} \end{pmatrix}$ also contains $\begin{pmatrix} \lambda^{-1} & 0 \\ 0 & \lambda \end{pmatrix}$ and, apart from this, no further diagonal matrices.

Thus, every double coset is characterized by a number λ; and λ and λ^{-1} correspond to the same coset. It is convenient to consider $t = \log \lambda$ in place of λ. Then t and $-t$ determine the same coset.

Thus, we can regard the functions in R_0—that is, the functions that are constant on the double cosets, as even functions of t.

4. Now let us find the rule of multiplication and the norm in R_0. For this purpose, we observe that the space of left cosets of $\mathfrak{H}$ in G is the set of positive-definite matrices with determinant 1. The transformation of left cosets reduces to the transformation of the corresponding quadratic forms. It can be shown, furthermore, that this is the group of transformations of a three-dimensional Lobachevsky space, where the points of the Lobachevsky space are in one-to-one correspondence with the left cosets. A double coset is a collection of left cosets; therefore it corresponds to a set of points in the Lobachevsky space. In order to find this set, we note the following: The elements of our group can be regarded as transformations of the left cosets that consist in multiplying each coset on the right by a given element g of the group. An element g leaves the unit coset invariant if and only if it is an element of $\mathfrak{H}$.

Thus, the subgroup $\mathfrak{H}$ consists of the motions of a Lobachevsky space that leave a fixed point of the space in place.

Since a double coset is obtained from a left coset by multiplying it on the right by all the elements of $\mathfrak{H}$, this corresponds in the Lobachevsky space to a sphere with its center at a fixed point.

Thus, to a function on the group that is constant on double cosets there corresponds a function in the Lobachevsky space that is constant on spheres with a fixed center. The integral of a function over the group differs only by a constant factor from the integral of the corresponding function in the Lobachevsky space. Therefore the norm of such a function is equal to

$\int |f(\bar{g})|\,d\bar{g}$, where $d\bar{g}$ is the element of volume in the Lobachevsky space. If we denote the radius of the sphere by t, then the norm of such a function is equal to $\int_0^\infty |f(t)|\,\varphi(t)$, where $\varphi(t)$ is the area of the surface of the sphere.

5. We now proceed to compute the rule of multiplication for the functions $f(t)$ corresponding to the involution of functions on the group. Avoiding a direct computation, we use the following argument.

The maximal ideals of a commutative ring are given by homomorphisms

$$f \to \int_0^\infty f(t)\alpha_\rho(t)\varphi(t)dt,$$

where $\alpha_\rho(t)$ is a so-called spherical function of the given group. These functions are calculated in the paper [81]. They are equal to $[2 \sin \varrho t]/[\varrho \sinh 2t]$; every ϱ corresponds to its own homomorphism, i.e., its own maximal ideal. Under the homomorphism, products of functions go over into products of numbers. Therefore, if $f = f_1 \times f_2$, we have

$$\int f(u)\,\alpha_\rho(u)\,\varphi(u)\,du = \int f_1(u)\,\alpha_\rho(u)\,du \cdot \int f_2(u)\,\alpha_\rho(u)\,du.$$

In order to express f in terms of f_1 and f_2, we put

$$\left.\begin{aligned} \tilde{f}_1(u) = f_1(u)\cdot[\varphi(u)/\sinh 2u], \qquad \tilde{f}_2(u) = f_2(u)\cdot[\varphi(u)/\sinh 2u], \\ \tilde{f}(u) = f(u)\cdot[\varphi(u)/\sinh 2u]. \end{aligned}\right\} \quad (1)$$

Then the homomorphism is given by the formula

$$f \to \int_0^\infty \tilde{f}(u)\,\frac{2 \sin \rho u}{\rho}\,du.$$

We can satisfy condition (1) by putting

$$\tilde{f}(u) = \int_0^\infty \left(\int_{|t-u|}^{t+u} \tilde{f}_1(s)\tilde{f}_2(t)\,ds \right) dt.$$

For let us denote

$$\int_0^\infty \left(\int_{|t-u|}^{t+u} \tilde{f}_1(s)\tilde{f}_2(t)\,ds \right) dt$$

by $g(u)$. Then

$$\int_0^\infty g(u)\frac{2\sin\rho u}{\rho}\,du=$$

$$=\int_0^\infty\left(\int_0^\infty\left(\int_{|t-u|}^{t+u}\tilde{f}_1(s)\tilde{f}_2(t)\,ds\right)dt\right)\frac{2\sin\rho u}{\rho}\,du=$$

$$=\int_0^\infty\left(\int_0^\infty\tilde{f}_1(s)\left(\int_{|s-t|}^{s+t}\frac{2\sin\rho u}{\rho}\,du\right)ds\right)\tilde{f}_2(t)\,dt=$$

$$=\int_0^\infty\left(\int_0^\infty\tilde{f}_1(s)\frac{4\sin\rho s\sin\rho t}{\rho^2}\,ds\right)\tilde{f}_2(t)\,dt=$$

$$=\int_0^\infty\tilde{f}_1(s)\frac{2\sin\rho s}{\rho}\,ds\int_0^\infty\tilde{f}_2(t)\frac{2\sin\rho t}{\rho}\,dt,$$

i.e.,

$$\int_0^\infty g(u)\frac{2\sin\rho u}{\rho}\,du=\int_0^\infty\tilde{f}(u)\frac{2\sin\rho u}{\rho}\,du.$$

In virtue of the completeness of the system of functions $\sin\varrho u$, it follows from this that $g(u)=\tilde{f}(u)$.

Now let us find all the maximal ideals of our ring. For this purpose, in what follows we shall denote the elements of the ring by $\tilde{f}(u)$. In terms of $\tilde{f}$, the norm is expressed by the formula

$$\int_0^\infty|\tilde{f}(u)|\sinh 2u\,du,$$

because

$$\int_0^\infty|f(u)|\varphi(u)du=\int_0^\infty[|f(u)|/\sinh 2u]\varphi(u)\cdot\sinh 2u\,du=$$

$$=\int_0^\infty|\tilde{f}(u)|\sinh 2u\,du.$$

Therefore the ring introduced here is isomorphic to the ring discussed in § 50. The maximal ideals of R_0 were examined in that same section. They were given by the formula

$$\int_0^\infty \tilde{f}(u)\,[2 \sin \varrho u]\,\varrho^{-1} du,$$

where ϱ is the complex number $\varrho = \varrho_1 + i\varrho_2$, with $|\varrho_2| \leqq 2$.

Remark: This implies, among other things, that the remaining spherical functions corresponding to the given group are given by the formula $[2 \sinh \varrho u]/[\varrho \sinh 2u]$ $(0 \leqq \varrho \leqq 2)$.

Since the transition from f to f^* corresponds to a replacement of $\tilde{f}(u)$ by $\bar{\tilde{f}}(u)$, it is clear that unsymmetric maximal ideals exist in the ring and consequently that the ring is unsymmetric.

6. Now let us show that the group ring R_0 is unsymmetric.

For this purpose, we note that if the element $f + \lambda e$ belongs to a subring of R_0 and has an inverse, then this inverse also belongs to R_0.

For let the element $\varphi + e$ of R be the inverse of the element $f + e \in R_0$. Let us show that $\varphi + e \in R_0$, i.e., that the function $\varphi(g)$ is constant on the double cosets of $\mathfrak{H}$. We have:

$$(f + e) \times (\varphi + e) = e.$$

Hence

$$\varphi = -f - f \times \varphi,$$

i.e.,

$$\varphi(g) = -f(g) - \int f(gg_1^{-1})\varphi(g_1)dg_1.$$

But then for $h \in \mathfrak{H}$

$$\varphi(hg) = -f(hg) - \int f(hgg_1^{-1})\varphi(g_1)dg_1 =$$
$$= -f(g) - \int f(gg_1^{-1})\varphi(g_1)dg_1 = \varphi(g),$$

because $f \in R_0'$. Thus, the function $\varphi(g)$ is invariant under left translation by $h \in \mathfrak{H}$. Similarly, using the equation

$$(\varphi + e) \times (f + e) = e,$$

we can show its invariance under a right translation. Therefore

$$\varphi \in R_0', \qquad \varphi + e \in R_0.$$

Assume now that $x \in R_0$ and that $(e + x^*x)^{-1}$ does not exist in R_0'. Such an element x can be found, because R_0 is unsymmetric. But then $(e + x^*x)^{-1}$ also does not exist in R. This completes the proof that the group ring R is unsymmetric.

7. In the paper [84] it is shown that Beurling's Theorem holds in commutative locally compact groups. The example we have given above shows that in non-commutative locally compact groups this theorem does not hold in general.

Let R be a normed ring with an involution. We shall say that the generalized Beurling Theorem holds in R if for every linear functional $f(x)$ in the ring R there exists an indecomposable positive functional that is a weak limit point of functionals $f(xa)$ $(a \in R)$.

THEOREM: *The generalized Beurling Theorem holds in a ring R if and only if R is a symmetric ring.*

Proof of the Necessity: Let R be an unsymmetric ring. Then there exists an element x_0 such that $(e + x_0{}^* x_0)^{-1}$ does not exist. Therefore, $e + x_0{}^* x_0$ belongs only to one maximal right ideal I_r of R. Since e does not belong to I_r, there exists a linear functional $f(x)$ such that $f(e) = 1$, $f(x) = 0$ for all $x \in I_r$. For $x \in I_r$, $a \in R$ we also have $xa \in I_r$, and therefore $f(xa) = 0$ for all $x \in I_r$. Hence every weak limit point $f_0(x)$ of the functionals $f_a(x) = f(xa)$ also vanishes on I_r. By assumption, Beurling's Theorem holds, i.e., among these limit points there is a positive normed indecomposable functional $f_0(x)$. Therefore, this functional also vanishes on I_r. In particular,

$$f_0(e + x_0{}^* x_0) = 0,$$

i.e.,

$$1 + f(x_0{}^* x_0) = 0;$$

but this is impossible, because $f(x_0{}^* x_0) \geqq 0$.

Proof of the Sufficiency: Let $f(x)$ be a linear functional. Let I_r denote the set of all elements $x \in R$ such that $f(xa) = 0$ for all $a \in R$. Obviously, I_r is a right ideal in R. According to [87] (see also [80]), there exists a positive functional $f_1(x)$ such that

$$f_1(e) = 1, \qquad f_1(xx^*) = 0 \quad \text{for} \quad x \in I_r. \tag{2}$$

The set of all functionals satisfying these conditions is a weakly closed bounded convex set in the space conjugate to R. By the theorem of Krein and Milman [85], this convex set contains at least one extreme point f_0; f_0 is an indecomposable positive functional satisfying the conditions (2). Hence it follows that $f_0(x) = 0$ for all $x \in I_r$, i.e., $f_0(x)$ is a weak limit point of the functionals $f_0(xa)$.

8. It is not difficult to show that if Beurling's Theorem in the formulation [84] holds in a group, then the Generalized Beurling Theorem holds in the group ring. It would be interesting to find out whether the converse is true.

From what has been shown above, it follows that *in the unimodular group of the second order, Beurling's Theorem does not hold.* Arguments similar to those of this section show that *Beurling's Theorem also fails to hold in every complex semi-simple Lie group.* This is connected with the existence of the so-called supplementary series of representations of these groups. (See the remark in § 52.5.)

Remark: Some of the results of this paper were obtained independently by I. E. Segal (see Bull. Amer. Math. Soc., Vol. 53 (1947), pp. 73-88).

CHAPTER IX

THE DECOMPOSITION OF A COMMUTATIVE NORMED RING INTO A DIRECT SUM OF IDEALS[1]

§ 53. Introduction

1. The following facts concerning the decomposition of a commutative normed ring into a direct sum of ideals are known thus far ([89], § 10).

Let R be a commutative normed ring and $\mathfrak{M} = \mathfrak{M}(R)$ the space of its maximal ideals. We assume that R is decomposed into the direct sum of two non-trivial ideals I_1 and I_2. In particular, the unit element e of R is represented as a sum $e_1 + e_2$, where $e_1 \in I_1$, $e_2 \in I_2$. It is easy to verify that the elements e_1 and e_2 play the role of unit elements in their ideals, which thus are also normed rings with unit element. Since $e_1{}^2 = e_1$, $e_2{}^2 = e_2$, the functions $e_1(M)$ and $e_2(M)$ assume only the values 0 and 1 on the maximal ideals of R. If $F_1 = \{e_1(M) = 1\}$, $F_2 = \{e_2(M) = 1\}$, then F_1 and F_2 are disjoint closed sets whose union is the whole space $\mathfrak{M}(R)$. Thus, the space $\mathfrak{M}(R)$ is *disconnected.*

Conversely, suppose that $\mathfrak{M}(R)$ is disconnected and splits into the sum of two disjoint closed sets F_1 and F_2; we shall also assume that there exists an element $e_1 \in R$ for which the function $e_1(M)$ is equal to 1 on F_1 and to 0 on F_2. Without loss of generality, we may assume that $e_1{}^2 = e_1$. Then it turns out that R splits into the direct sum of ideals I_1 and I_2; the ideal I_1 is generated by e_1, and I_2 by the element $e_2 = e - e_1$. Each of these ideals forms a normed ring, with F_1 and F_2, respectively, as its set of maximal ideals.

2. The results quoted, as we can see, are not yet definitive. The fundamental question—*can a ring R be split into the direct sum of two ideals if the set of maximal ideals of R is disconnected?*—remains open.

---

[1] The present chapter is a translation of a paper by G. E. Shilov [64]. It is added in translation and does not form part of the original Russian text of *Commutative Normed Rings.*

In what follows, a complete—and affirmative—answer is given to this question.

3. In § 54, we give a characterization of the space of maximal ideals of R as a subset of the topological product of discs $|\lambda_\alpha| \leqq \| z_\alpha \|$, where $\{z_\alpha\}$ is a set of generators of R.

In this connection, we mention that a similar characterization has already been given for a ring with a single generator [90], namely: The set of maximal ideals of a ring with one generator is a closed bounded set in the complex plane, with a connected complement. But even for a ring with two generators there cannot be a purely topological characterization of the set of maximal ideals as a subset of the complex two-dimensional space $\{z_1, z_2\}$; for example, the circle $|z_1| = 1$, $z_2 = 0$ is not the set of maximal ideals of any normed ring with generators z_1, z_2, whereas the circle $z_1 = \exp(i\varphi)$, $z_2 = \exp(-i\varphi)$ $(0 \leqq \varphi \leqq 2\pi)$ coincides with the set of maximal ideals of the ring of all continuous functions on this circle whose generators are z_1 and z_2. Thus, the required characterization in this case cannot be stated merely in topological terms, but must take the algebraic nature of the complex space essentially into account.

In § 55, by applying this characterization to the case of a ring with a finite number of generators, we show that every analytic function $f(\lambda_1, \lambda_2, \ldots, \lambda_n)$ that is defined in a domain $G \supset \mathfrak{M}(R)$ coincides on $\mathfrak{M}(R)$ with a function $x(M)$, where $x \in R$. This proposition generalizes a well-known theorem of Gelfand ([89], § 9) to the case of a ring with a finite number of generators. Instead of the Cauchy integral used by Gelfand, we apply the integral representation of Weil [88] (see also § 13.7) to the function $f(\lambda_1, \lambda_2, \ldots, \lambda_n)$.

In § 56, the fundamental problem on the decomposition of R into a direct sum of ideals under the assumption that the set $\mathfrak{M}(R)$ is disconnected will be reduced to the case where R is finitely generated; we shall prove that if $\mathfrak{M}(R)$ is disconnected, then we can find a finitely generated subring in R in which the set of maximal ideals is also disconnected. This enables us later (§ 57), by using the theorem on analytic functions of § 55, to construct an idempotent element $e_1 \in R$, different from the unit element. The problem is thus reduced to the special case in which it has already been solved [89]; this then also gives the general solution.

In § 58, we give some corollaries of our theorem. Apart from corollaries within the theory of rings, we mention a corollary (§ 58.2) concerning the character of convergence of polynomials of several complex variables, which contains a result that appears to be new.

§ 54. Characterization of the Space of Maximal Ideals of a Commutative Normed Ring

1. Let R be a commutative normed ring and $\mathfrak{M} = \{M\}$ the set of all its maximal ideals. If we know a system $\{z_\alpha\}$ of generators of R, then, as is well known ([89], § 5), the space $\mathfrak{M}$ can be realized in the form of a closed subset of the topological product T, with Tichonov topology, of the complex discs $|\lambda_\alpha| \leqq \| z_\alpha \|$. The maximal ideal M is mapped into the point having the number $z_\alpha(M)$ as its α-coordinate. Since the mapping of M so obtained is one-to-one and bicontinuous, we shall identify the image of $\mathfrak{M}$ in the topological product T with the space $\mathfrak{M}$ itself.

Suppose that a point $\{\mu_\alpha\} \in T$ exists that does not belong to $\mathfrak{M}$. This means that there is no homomorphism of R into the field of complex numbers that carries each generator z_α into μ_α or, what is the same, that there is no maximal ideal in R containing all the differences $z_\alpha - \mu_\alpha e$, where e is the unit element of the ring. But the set of all finite sums of the form

$$\sum_{j=1}^{m} (z_{\alpha_j} - \mu_{\alpha_j} e) q_{\alpha_j}, \tag{1}$$

where q_{α_j} are arbitrary elements of R, obviously forms an ideal in R. Since this ideal is not contained in any maximal ideal, it must coincide with the entire ring R; in particular, for some choice of the number m, of the subscripts α_j, and of the elements q_{α_j}, the sum (1) yields the unit element of the ring:

$$e = \sum_{j=1}^{m} (z_{\alpha_j} - \mu_{\alpha_j} e) q_{\alpha_j}. \tag{2}$$

The elements q_{α_j} that occur in (2) can be approximated in norm with arbitrary accuracy by polynomials P_j in the generators z_α. In particular, these polynomials can be chosen so that the norm of the difference

$$e - \sum_{j=1}^{m} (z_{\alpha_j} - \mu_{\alpha_j} e) P_j \tag{3}$$

is less than 1.

The expression (3) is, of course, a polynomial in a finite number of generators from the set $\{z_\alpha\}$. By construction, this polynomial has the following properties:

a) It assumes the value 1 at the point $\{\mu_\alpha\}$;

b) It is less than 1 in absolute value at every point $M \in \mathfrak{M}$.

We have thus obtained the following proposition.

THEOREM 1: *For every point* $\{\mu_\alpha\} \in T$ *that does not belong to* $\mathfrak{M}$ *we can find a polynomial in the generators of* R *that assume the value* 1 *at* $\{\mu_\alpha\}$ *and are less than* 1 *in absolute value at every point* $M \in \mathfrak{M}$.

This theorem will enable us presently to give a characterization of the space $\mathfrak{M}$ as a subset of R.

2. In what follows, a *W-set* shall mean a closed subset of T that is the intersection of open sets of the form

$$\{|P(\lambda_{\alpha_1}, \lambda_{\alpha_2}, \ldots, \lambda_{\alpha_k})| < 1\},$$

where $P(\lambda_{\alpha_1}, \lambda_{\alpha_2}, \ldots, \lambda_{\alpha_k})$ is a polynomial in the given arguments.

Theorem 1 shows that, in particular, the set $\mathfrak{M} \subset T$ is a W-set. We shall now show that every W-set is the space of maximal ideals of a suitable ring R with generators $\{z_\alpha\}$.

Let a W-set $F \subset T$ be given. In the ring of polynomials $P(z_{\alpha_1}, z_{\alpha_2}, \ldots, z_{\alpha_k})$ we introduce a norm by the formula

$$\| P(z_{\alpha_1}, z_{\alpha_2}, \ldots, z_{\alpha_k}) \| = \max_F |P(\lambda_{\alpha_1}, \lambda_{\alpha_2}, \ldots, \lambda_{\alpha_k})| \tag{4}$$

and construct the ring R by completion in this norm.

R can be represented as a ring of continuous functions given on F. Obviously, every point $\{\lambda_\alpha\} \in F$ determines a maximal ideal of R corresponding to the homomorphism $z_\alpha \to \lambda_\alpha$. Since R is generated by the $\{z_\alpha\}$, its set of maximal ideals is a subset of T. Suppose that a point $\{\mu_\alpha\} \in T$ does not belong to F. Since F is a W-set, there exists a polynomial

$$P(\lambda_{\alpha_1}, \lambda_{\alpha_2}, \ldots, \lambda_{\alpha_k})$$

that is less than 1 in absolute value on F and not less than 1 in absolute value at $\{\mu_\alpha\}$. By the definition (4), the norm of this polynomial is less than 1 (F, as a closed subset of the compactum P, is a compactum; and every function continuous on F assumes the upper bound of its values); therefore the polynomial $P(z_{\alpha_1}, z_{\alpha_2}, \ldots, z_{\alpha_k})$ is less than 1 in absolute value on every maximal ideal. Hence the point $\{\mu_\alpha\}$ cannot be in $\mathfrak{M}(R)$.

Thus, the W-set F coincides with the set $\mathfrak{M}(R)$, as stated.

The result obtained can be restated in the form of the following proposition.

THEOREM 2: *A set* $F \subset T$ *is the set of maximal ideals of a normed ring with generators* $\{z_\alpha\}$ *if and only if* F *is a* W-*set.*

§ 55. A Problem on Analytic Functions in a Finitely Generated Ring

1. Let $\mathfrak{M} = \mathfrak{M}(R)$ be the set of maximal ideals of a ring with a finite number of generators $z_1, z_2, \ldots, z_n$. We may identify $\mathfrak{M}$ with a closed subset of a complex n-dimensional space $K_n = \{\lambda_1, \lambda_2, \ldots \lambda_n\}$, by assigning to every maximal ideal $M \in \mathfrak{M}$ the point $\{z_1(M), z_2(M), \ldots, z_n(M)\}$, in a way similar to that of the general case in § 54.

We shall use the term *Weil domain* for every domain $U \subset K_n$ that is the intersection of a finite number of domains of the form

$$\{|P(\lambda_1, \lambda_2, \ldots, \lambda_n)| < 1\},$$

where $P(\lambda_1, \lambda_2, \ldots, \lambda_n)$ is a polynomial in $\lambda_1, \lambda_2, \ldots, \lambda_n$.

Every closed bounded set $F \subset K_n$ that is the intersection of a set of Weil domains will be called a *Weil set*. As was shown in § 54, the set $\mathfrak{M}(R)$ is a Weil set.

THEOREM 3: *For every function $f(\lambda_1, \lambda_2, \ldots, \lambda_n)$ that is analytic in a domain $G \supset \mathfrak{M}$ we can find an element ω in R for which*

$$\omega(M) = f(z_1(M), z_2(M), \ldots, z_n(M))$$

for every $M \in \mathfrak{M}$.

This theorem is already known for a ring with a single generator (see, for example, [90]). In this special case, it is proved by an application of the classical Cauchy formula. In the general case under discussion, we shall use Weil's integral in place of Cauchy's formula.

Proof: We begin by constructing a Weil domain contained in the interior of G and containing $\mathfrak{M}$ in its interior.

Since $\mathfrak{M}$ is a Weil set, it is the intersection of a set of domains of the form

$$G_\nu = \{|P_\nu(\lambda_1, \lambda_2, \ldots, \lambda_n)| < 1\}.$$

Since $\mathfrak{M}$ is bounded, we have to add to these domains another n domains of the form

$$Q_j = \{|\lambda_j/c_j| < 1\} \qquad (j = 1, 2, \ldots, n),$$

where the c_j are positive constants. Let m_ν denote the maximum absolute value of the polynomial $P_\nu(\lambda_1, \lambda_2, \ldots, \lambda_n)$ on $\mathfrak{M}$. The number m_ν is less than 1; we choose a number θ_ν between m_ν and 1 and examine the closed sets

$$F_\nu = \{|P_\nu(\lambda_1, \lambda_2, \ldots, \lambda_n)| \leqq \theta_\nu\}$$

and also the closed sets

$$S_j = \{|\lambda_j| \leqq \alpha_j\},$$

where

$$\max_{\mathfrak{M}} |z_j(M)| < \alpha_j < c_j \qquad (j = 1, 2, \ldots, n).$$

Since $\mathfrak{M}\subset F_\nu\subset G_\nu$, $\mathfrak{M}\subset S_j\subset Q_j$, the intersection of all the sets F_ν and S_j obviously also coincides with $\mathfrak{M}$. The intersection of all the sets F_ν, S_j, and the closed set that is the complement of G is empty; therefore we can choose a finite number of subscripts $\nu_1, \nu_2, \ldots, \nu_m$ such that the intersection of the sets $F_{\nu_1}, F_{\nu_2}, \ldots, F_{\nu_m}$, all the sets S_j, and $K_n - G$ is empty; but this means that *the intersection of the sets* $F_{\nu_1}, F_{\nu_2}, \ldots, F_{\nu_m}, S_1, S_2, \ldots, S_n$ *is contained in* G. A fortiori, the domain G contains the intersection of all the domains

$$\{|P_{\nu_i}(\lambda_1, \lambda_2, \ldots, \lambda_n)| < \theta_{\nu_i}\} \qquad (i=1, 2, \ldots, m)$$

and

$$\{|\lambda_j| < \alpha_j\} \qquad (j=1, 2, \ldots, n).$$

This intersection is obviously a Weil domain contained, by what has already been proved, in G and containing $\mathfrak{M}$.

We now proceed to the conclusion of the proof of Theorem 3. For a ring with uniform convergence, the theorem follows directly from a theorem of Weil that states that every analytic function in a Weil domain G can be represented as a series of polynomials uniformly convergent in the interior of G (i.e., on every closed subset of G) ([88], p. 329; see also [92], p. 300).

In the general case, we consider the integral representation of Weil (which Weil constructed for the sake of the proof of the very theorem quoted above). It is obtained in the following way ([88], p. 312; see also [92], p. 292).

The Weil domain constructed above is defined by a finite number of inequalities of the form

$$|P_\mu(\lambda_1, \lambda_2, \ldots, \lambda_n)| < 1 \qquad (\mu=1, 2, \ldots, N),$$

where the P_μ are polynomials in $\lambda_1, \lambda_2, \ldots, \lambda_n$. Let s_μ be the set of points at which (for a fixed μ) the condition

$$|P_\mu(\lambda_1, \lambda_2, \ldots, \lambda_n)| = 1$$

is satisfied. Since the P_μ are polynomials, we may assume that no system of the n manifolds $s_{\mu_1}, s_{\mu_2}, \ldots, s_{\mu_n}$ intersects in a surface of dimension greater than n.

We denote by $\sigma_{\mu_1\mu_2\ldots\mu_n}$ the intersection of the manifolds $s_{\mu_1}, s_{\mu_2}, \ldots, s_{\mu_n}$, orientated in a definite manner.

For every point $\{\tau_1, \tau_2, \ldots, \tau_n\}\in\mathfrak{M}$ we can write down the expansion

$$P_\mu(\lambda_1, \lambda_2, \ldots, \lambda_n) - P_\mu(\tau_1, \tau_2, \ldots, \tau_n) = \sum_\nu (\lambda_\nu - \tau_\nu)\, Q_{\mu\nu}(\lambda_1, \lambda_2, \ldots, \lambda_n),$$

where the $Q_{\mu\nu}$ are polynomials. We put

$$D_{\mu_1\mu_2\ldots\mu_n} = \det\|Q_{\mu_\alpha j}\| \qquad (j,\ \alpha = 1, \ldots, n).$$

Then the function $f(\lambda_1, \lambda_2, \ldots, \lambda_n)$, analytic in G, can be represented on $\mathfrak{M}$ in the form of the integral

$$f(\tau_1, \ldots, \tau_n) =$$

$$= \frac{1}{(2\pi i)^n} \sum_{(i_1, \ldots, i_n)} \int_{\sigma_{i_1 \ldots i_n}} \frac{D_{i_1 \ldots i_n} f(\lambda_1, \ldots, \lambda_n)\, d\lambda_1 \ldots d\lambda_n}{\prod_{\nu=1}^{n} [P_{i_\nu}(\lambda_1, \ldots, \lambda_n) - P_{i_\nu}(\tau_1, \ldots, \tau_n)]}, \tag{5}$$

The summation is taken over all the combination of subscripts $i_1, i_2, \ldots, i_n$.

Since the polynomial $P_{i_\nu}(\lambda_1, \lambda_2, \ldots \lambda_n) - P_{i_\nu}(\tau_1, \tau_2, \ldots, \tau_n)$ does not vanish when $(\lambda_1, \lambda_2, \ldots, \lambda_n)$ ranges over the set $\sigma_{i_1 i_2 \ldots i_n}$ and $(\mu_1, \mu_2, \ldots, \mu_n)$ over $\mathfrak{M}$, the integral (5) can be interpreted in the ring. In fact, since $\mathfrak{M}$ is the set of maximal ideals of R and the polynomial

$$P_{i_\nu}(\lambda_1, \lambda_2, \ldots, \lambda_n) - P_{i_\nu}(\tau_1, \tau_2, \ldots, \tau_n)$$

as a function of $M = (\tau_1, \tau_2, \ldots, \tau_n)$ does not vanish, there is an inverse element in R

$$[P_{i_\nu}(\lambda_1, \lambda_2, \ldots, \lambda_n) - P_{i_\nu}(\tau_1, \tau_2, \ldots, \tau_n)]^{-1},$$

which obviously depends continuously on the parameters $\lambda_1, \lambda_2, \ldots, \lambda_n$. Since $D_{i_1 i_2 \ldots i_n}$, as a polynomial, also depends continuously on the parameters $\lambda_1, \lambda_2, \ldots, \lambda_n$, we may integrate the ring elements

$$D_{i_1 i_2 \ldots i_n} f(\lambda_1, \lambda_2, \ldots, \lambda_n) \prod_{\nu=1}^{n} [P_{i_\nu}(\lambda_1, \lambda_2, \ldots, \lambda_n) - P_{i_\nu}(\tau_1, \tau_2, \ldots, \tau_n)]^{-1}$$

with respect to the parameters $\lambda_1, \lambda_2, \ldots, \lambda_n$ ranging over $\sigma_{i_1 i_2 \ldots i_n}$. The results obtained can be summed over the $i_1, i_2, \ldots, i_n$ and multiplied by $1/(2\pi i)^n$. We again obtain an element of R as a result of these operations.

Since the homomorphism $R \to R/M$ is continuous, the value of this element on the maximal ideals of R must coincide with the values of the function $f(\tau_1, \tau_2, \ldots, \tau_n)$ at the corresponding points of $\mathfrak{M}$. This completes the proof of Theorem 3. //

2. In connection with the proof of Theorem 3, we raise the following question.

We shall use the term *combined spectrum* of the elements $x_1, x_2, \ldots, x_n$ of an arbitrary normed ring R for the set of all points of the complex n-dimensional space K_n of the form

$$\{x_1(M), x_2(M), \ldots, x_n(M)\}$$

for all $M \in \mathfrak{M}(R)$.

It is easy to verify that the combined spectrum S of a given system of n elements of a ring R is a bounded and closed subset of K_n.

Let an analytic function $f(\lambda_1, \lambda_2, \ldots, \lambda_n)$ be given in a domain $G \supset S$. Is there an element x in R such that

$$x(M) = f[x_1(M), x_2(M), \ldots, x_n(M)]$$

for every $M \in \mathfrak{M}(R)$?

Theorem 3 gives an affirmative answer to this question in the case where the elements $x_1, x_2, \ldots, x_n$ are generators of R; the assumption that this is so enables us to establish that $S = \mathfrak{M}(R)$ is a Weil set and to use the corresponding integral representation. But in the general case, the combined spectrum S is an arbitrary closed bounded set in K_n and, in particular, need not be a Weil set; therefore the solution of the question stated will apparently require new methods. However, for $n=1$, the —affirmative—solution only requires the Cauchy integral ([89], § 9).

§ 56. Construction of a Special Finitely Generated Subring

1. In this section we shall prove the following proposition:

THEOREM 4: *If the set $\mathfrak{M}$ of maximal ideals of R is disconnected, then we can find a subring $\tilde{R} \subset R$ that also has a disconnected set of maximal ideals.*

Proof: As in § 54, we shall assume that $\mathfrak{M}$ is a subset of T. Let $\mathfrak{M} \in \Phi_1 + \Phi_2$ be a decomposition of $\mathfrak{M}$ into the union of two closed disjoint sets. We construct two domains $G_1 \supset \Phi_1$ and $G_2 \supset \Phi_2$ whose closures $\overline{G}_1$ and $\overline{G}_2$ are also disjoint. Every point M_0 of Φ_1 can be separated from $\overline{G}_2$ by a neighborhood of the form

$$U(z_1, z_2, \ldots, z_k, \varepsilon, M_0) = \{| z_i(M) - z_i(M_0) < \varepsilon\} \qquad (i = 1, 2, \ldots, k); \quad (6)$$

moreover, since the neighborhoods of the form (6) form a defining system, we may assume that each of them is contained in G_1. The set of all neighborhoods so obtained covers the entire set Φ_1; and since Φ_1 is closed, we can choose from this a finite covering

$$U_j = U(z_1^{(j)}, z_2^{(j)}, \ldots, z_{k_j}^{(j)}, \varepsilon_j, M_j) \quad (j = 1, 2, \ldots, l).$$

Let U denote the set of points of G_1 that are covered by the neighborhoods U_j.

A similar system of neighborhoods covering Φ_2 and not intersecting $\overline{G_1}$ can be constructed in G_2. Let these be the neighborhoods

$$V_j = V(z_1^{(j)}, z_2^{(j)}, \ldots, z_{k_j}^{(j)}, \varepsilon_j, M_j) \quad (j = l+1, \ldots, m).$$

We denote the union of the neighborhoods V_j by V.

Since $\mathfrak{M}$ is a W-set, by what we have shown it is the intersection of a number of domains of the form

$$G_\nu = \{|P_\nu(z_{\alpha_1}^{(\nu)}, z_{\alpha_2}^{(\nu)}, \ldots, z_{\alpha_\nu}^{(\nu)})| < 1\}.$$

Let m_ν denote the maximum modulus of the polynomial P_ν on $\mathfrak{M}$. The number m_ν is less than 1; we choose a number θ_ν between m_ν and 1 and examine the closed sets

$$F_\nu = \{|P_\nu(z_{\alpha_1}^{(\nu)}, z_{\alpha_2}^{(\nu)}, \ldots, z_{\alpha_\nu}^{(\nu)})| \leqq \theta_\nu\}.$$

Since $\mathfrak{M} \subset F_\nu \subset G_\nu$, it is obvious that the intersection of the sets F_ν also coincides with $\mathfrak{M}$. The intersection of the sets F_ν and of the closed set complementary to $U+V$ is empty; since the space T is compact, we can therefore choose a finite number of subscripts $\nu_1, \nu_2, \ldots, \nu_n$ such that the intersection of the finite number of sets $F_{\nu_1}, F_{\nu_2}, \ldots, F_{\nu_n}$ and $T-(U+V)$ is empty. This means that the intersection of the sets $F_{\nu_1}, F_{\nu_2}, \ldots, F_{\nu_n}$ is contained in $U+V$.

Let us examine all the elements from the set of generators $\{z_\alpha\}$ that occur in the definitions of the neighborhoods U_j, V_j and in the construction of the polynomials P_{ν_k} $(j = 1, 2, \ldots, m; k = 1, 2, \ldots, n)$. The number of such elements is finite; we denote the set of these elements by S.

Let us consider the subring $\tilde{R}$ generated by S in R. We claim that this subring satisfies the condition of our theorem, i.e., that the set of maximal ideals $\tilde{\mathfrak{M}}$ of $\tilde{R}$ is disconnected. In order to prove this, we have to construct the set $\tilde{\mathfrak{M}}$. Every maximal ideal $\tilde{M}$ of $\tilde{R}$ is determined by the values $z_\alpha(\tilde{M})$ of the generators of this ring. Observe that the maximum modulus of the values of an arbitrary function $x(\tilde{M})$ $(x \in \tilde{R})$ is the same as the maximum modulus of the values of the function $x(M)$ on $\mathfrak{M}$, because each of them can be expressed by a well-known formula in terms of the norms of the powers of x ([89], p. 64). Hence for every $\tilde{M} \subset \tilde{R}$ we can find a point $\{\lambda_\alpha\}$ in T at which the coordinates corresponding to the generators $z_\alpha \in S$ are equal to the numbers $z_\alpha(\tilde{M})$. Let us show that every such point falls into the domain $U+V$.

Suppose that a point $\{\lambda_\alpha\}$ is not contained in $U+V$. This means that $\{\lambda_\alpha\}$ does not belong to one of the sets F_{ν_k}—for the sake of definiteness, say to F_{ν_1}—so that at this point we have the inequality

$$|P_{\nu_1}(\lambda_{\alpha_1}^{(\nu_1)}, \lambda_{\alpha_2}^{(\nu_1)}, \ldots, \lambda_{\alpha_{\nu_1}}^{(\nu_1)})| > \theta_{\nu_1}.$$

But then the maximum modulus of the polynomial P_{ν_1} on the maximal ideals of $\tilde{R}$ turns out to be greater than m_ν—the maximum modulus of this polynomial on the maximal ideals of R; but as we have seen, this is impossible.

Thus, every point $\{\lambda_\alpha\}$ at which the coordinates corresponding to the generator $z_\alpha \in S$ are equal to the numbers $z_\alpha(\tilde{M})$ falls into the domain $U+V$. We denote the set of all such points by Z; Z is a closed set contained in $U+V$ and is therefore representable in the form of a union Z_U+Z_V, where $Z_U \subset U$, $Z_V \subset V$ and Z_U and Z_V are closed. Two distinct points of Z may correspond to the same maximal ideal of $\tilde{R}$. But we shall show that two such points $M' \in Z$ and $M'' \in Z$ of which the first belongs to U and the second to V necessarily correspond to *distinct* maximal ideals of $\tilde{R}$.

Indeed, the point M' occurs in one of the neighborhoods whose union, by construction, forms the domain U; hence for the point M' and a certain j the inequality

$$|z_i^{(j)}(M') - z_i^{(j)}(M_j)| < \varepsilon_j \qquad (i = 1, 2, \ldots, k_j)$$

is satisfied. Since the point M'' lies in V, it cannot belong to this neighborhood; consequently, for at least one i and a suitably chosen j we have the inequality

$$|z_i^{(j)}(M'') - z_i^{(j)}(M_j)| \geqq \varepsilon_j.$$

But in that case $z_i^{(j)}(M') \neq z_i^{(j)}(M'')$, i.e., the points M' and M'' correspond to distinct maximal ideals of $\tilde{R}$, the distance between which is therefore not less than $\varepsilon = \min \varepsilon_j$ $(j = 1, 2, \ldots, m)$.

The set of all maximal ideals of $\tilde{R}$ is obtained as the continuous image of Z formed by identifying all the points $\{\lambda_\alpha\} \in Z$ on which the generators $z_\alpha \in S$ assume equal values. The set Z has representatives both in U and in V, because every point $M \in \mathfrak{M}$ determines a maximal ideal of $\tilde{R}$. Since the points of the domains U and V are not identified with each other, by what we have proved, but must necessarily go over into points whose distance from each other is not less than ε, the required image $\tilde{Z}$ of Z is composed of two non-trivial closed disjoint parts—the continuous images of the sets Z_U and Z_V. Therefore, the set $\tilde{Z} = \tilde{\mathfrak{M}}$ is disconnected. //

§ 57. Proof of the Theorem on the Decomposition of a Ring into a Direct Sum of Ideals

1. Let $\mathfrak{M}$ be the set of maximal ideals of a ring R (with an arbitrary number of generators). We shall assume that $\mathfrak{M}$ is disconnected: $\mathfrak{M} = F_1 + F_2$, where F_1 and F_2 are closed. By what has been shown in § 56, we can then find a finitely generated subring $\tilde{R}$ in R, also with a disconnected set of maximal ideals $\tilde{\mathfrak{M}}$.

But $\tilde{\mathfrak{M}}$ lies in the complex finite-dimensional space K_n and the theorem on analytic functions proved in § 55 can be applied to it. Since $\tilde{\mathfrak{M}}$ is disconnected, it splits into the union of two disjoint closed sets $\tilde{\Phi}_1$ and $\tilde{\Phi}_2$. The function $f(\lambda_1, \lambda_2, \ldots, \lambda_n)$ that is equal to 1 in the neighborhoods of $\tilde{\Phi}_1$ and to zero in the neighborhoods of Φ_2 is analytic on $\tilde{\mathfrak{M}} = \tilde{\Phi}_1 + \tilde{\Phi}_2$; therefore R contains an element ω such that $\omega(M) = 1$ for $M \in \tilde{\Phi}_1$ and $\omega(M) = 0$ for $M \in \tilde{\Phi}_2$. Therefore, by the theorem quoted in the introduction, the ring $\tilde{R}$ splits into the direct sum of two ideals. In particular, we can find in $\tilde{R}$ an idempotent element e_1, different from the unit element ([89], § 10). Since $\tilde{R}$ is a subring of R, e_1 also belongs to R. But then R *also splits into the direct sum of non-trivial ideals.* Thus, our theorem is proved.

2. It is useful to verify that our construction leads to a decomposition of R into the direct sum of ideals I_1 and I_2 that, when regarded as normed rings, have the sets F_1 and F_2 as their spaces of maximal ideals. We have embedded the set $\mathfrak{M}(R) = F_1 + F_2$ in disjoint domains $U \supset F_1$ and $V \supset F_2$. The set of maximal ideals of $\tilde{R}$ is the image of a closed set $Z \supset \mathfrak{M}(R)$ contained in the union $U + V$ and is therefore representable in the form $Z_U + Z_V$, where $Z_U \subset U$, $Z_V \subset V$, and Z_U and Z_V are closed.

Under a continuous mapping into the n-dimensional space the closed sets Z_U and Z_V go over into the closed sets $\tilde{Z}_U$ and $\tilde{Z}_V$, which, as we have already proved, are disjoint. The function $\omega(\tilde{M})$ constructed above is equal to 1 on $\tilde{Z}_U$ and to 0 on $\tilde{Z}_V$. Since, in the mapping $Z \to \tilde{Z} = \tilde{\mathfrak{M}}$, only an identification of certain points of Z took place, namely those for which the functions of the subring $\tilde{R}$ assume equal values, the function $\omega(M)$ assumes the value 1 *on the entire inverse image* of $\tilde{Z}_U$, and, in particular, on F_1, and the value 0 *on the entire inverse image* of the set $\tilde{Z}_V$, and, in particular, on F_2. But then it follows immediately, as was shown in § 10 of [89], that the set of maximal ideals of I_1 coincides with F_1, and the set of maximal ideals of I_2, with F_2.

§ 58. Some Corollaries

We shall now give several corollaries of the general theorem proved above, which, in our view, are of independent interest.

1. Apart from the fundamental method of topologizing the set of maximal ideals that was used in our construction there exists another method according to which a maximal ideal M_0 is a point of adherence of the subset $\mathfrak{A} \subset \mathfrak{M}$ if it contains the intersection of all the maximal ideals that occur in $\mathfrak{A}$ ([89], § 12).

The closure $\tilde{\mathfrak{A}}$ of $\mathfrak{A}$ in this topology is, in general, wider than the closure $\underset{\sim}{\mathfrak{A}}$ of $\mathfrak{A}$ in the fundamental topology. Let us show that this closure at any rate *does not contain new connected components*; in other words, *every connected component of* $\tilde{\mathfrak{A}}$ *contains points of* $\mathfrak{A}$.

Proof: We examine the intersection I of all maximal ideals that occur in $\mathfrak{A}$. The residue-class ring R/I has $\tilde{\mathfrak{A}}$ as its set of maximal ideals. If $\tilde{\mathfrak{A}}$ had a connected component $\tilde{F}$ not containing points of $\mathfrak{A}$, then by the theorem proved above, we could construct a function $\tilde{x}(M) \in R/I$ that is equal to 1 on $\tilde{F}$ and to 0 on $\tilde{\mathfrak{A}} - \tilde{F}$. The corresponding function $x(M) \in R$ would be equal to 1 on F (the inverse image of $\tilde{F}$) and to 0 on $\mathfrak{A}$. But then no point of F could occur in $\tilde{\mathfrak{A}}$. This contradiction proves our statement. //

2. Let a sequence of polynomials in n complex variables $P_m(\lambda_1, \lambda_2, \ldots, \lambda_n)$ $(m = 1, 2, \ldots)$ converge uniformly to a function $f(\lambda_1, \lambda_2, \ldots, \lambda_n)$ on a closed bounded set F of the complex n-dimensional space K_n. In general, there exists a closed set $\tilde{F} \supset F$ on which every such sequence continues to be uniformly convergent. (This set does not depend on the special choice of the sequence P_m; for example, if the P_m are polynomials in a single variable λ_1 and F is a circle in the λ_1-plane, then $\tilde{F}$ is the disc bounded by F.) Let us show that, in general, $\tilde{F}$ *does not contain new connected components*; in other words, *every connected component of* $\tilde{F}$ *contains points of* F.

Proof: We examine the ring R defined as the closure of the set of all polynomials $P(\lambda_1, \lambda_2, \ldots, \lambda_n)$ with respect to uniform convergence on F. The set $\tilde{F}$ is obviously the space of maximal ideals of R. If $\tilde{F}$ had a new connected component Q, then by our theorem, R would contain a function that is equal to 1 on Q and to 0 on $\tilde{F} - Q$, and, in particular, equal to 0 on F. The norm of this function in R must obviously be equal to zero; but then it cannot assume the value 1 on any maximal ideal. This contradiction proves our statement. //

3. The same fact can be formulated for a general normed ring: If we denote by $\tilde{F} \subset \mathfrak{M}(R)$ the set on which every sequence of elements $x_n(M)$ of R remains uniformly convergent if it is uniformly convergent on $F \subset \mathfrak{M}(R)$, then $\tilde{F}$ does not contain new connected components. The proof follows the same pattern as under 2.

4. The *boundary* of the space $\mathfrak{M}(R)$ is, of course, defined as the minimal set of points $M \in \mathfrak{M}$ on which every function $x(M)$ assumes its maximum modulus ([89], § 24). Let us show that *every connected component of* $\mathfrak{M}(R)$ *has boundary points.*

Proof: This easily follows from the fundamental theorem of § 57. Indeed, for every connected component $Q \subset M$ we can find a function $e(M)$ that is equal to 1 on Q and to 0 on $\mathfrak{M} - Q$; its maximum is assumed at points of Q, and hence the boundary of $\mathfrak{M}(R)$ contains points of Q. //

5. Theorem (Generalization of a Theorem of Stone for Zero-Dimensional Compacta): *Let the space of maximal ideals of a ring* R *be a zero-dimensional compactum* (in other words, for any two points $M' \in \mathfrak{M}(R)$ and $M'' \in \mathfrak{M}(R)$ there exists a decomposition of $\mathfrak{M}(R)$ into two disjoint closed sets of which the first contains M' and the second, M''). *Then every continuous function* $f(M)$ *is the limit of a uniformly convergent sequence of functions* $x_n(M)$, $x_n \in R$. (Note that in the statement of this theorem the symmetry of R is not required.)

Proof: The proof follows from the fact that R contains an ample set of *real* functions $x(M)$; indeed, in virtue of the assumption that $\mathfrak{M}(R)$ is a zero-dimensional compactum and of the fundamental theorem of § 56, for any two points M' and M'' there exists a function $\omega(M)$ $(\omega \in R)$ (more accurately, assuming only the values 0 and 1 on all of $\mathfrak{M}$) that is equal to 0 at M' and to 1 at M''. Hence the theorem is obtained as an application of Stone's Theorem [89]. //

HISTORICO-BIBLIOGRAPHICAL NOTES

HISTORICO-BIBLIOGRAPHICAL NOTES

CHAPTER I

The foundation of the general theory of commutative normed rings as expounded in Chapter I was laid by Gelfand [Gel'fand] [11, 14]. Theorem 2 of § 4 was first published (without proof) by Mazur [38]. Theorem 1′ of § 7 is due to Stone ([70], pp. 466 ff.). The results of § 8 go back essentially to Raikov [Raĭkov] [46, 50, 19]; normed rings with an involution, both commutative and non-commutative, were first discussed, under certain special assumptions, by Gelfand and Naimark [Naĭmark, Neumark] [17]; in particular, they obtained Theorem 1; the proof of the theorem given here follows Arens [1].

CHAPTER II

The results of § 9 are due to Gelfand [11, 14] and the application of these results to rings of infinitely differentiable functions, to functions, to Shilov [19, 61]. The results of §§ 10-12 are due to Shilov [Šilov] [57, 19]. Case A) of Theorem 1 of § 13 is due to Shilov [64], B) to Arens and Calderon [3], and the general statement of the theorem to Shilov [65]; the construction of an operational calculus in topological linear rings was first given by Waelbroeck [72]. The results of § 14, in the case of a symmetric ring, were established by Gelfand [14], and they were extended to the general case by Shilov [64]. The abstract form of Beurling's Theorem [5, 6], as given in § 15, is due to Gelfand.

CHAPTER III

The results of Chapter III essentially go back to Gelfand [12, 15]. The propositions formulated as Corollaries 1 and 2 to Theorem 1 of § 17 were first obtained by Wiener and Pitt [78].

CHAPTER IV

Theorem 2 of § 20 is due to Gelfand and Raĭkov [18]; the properties of an involution on locally compact groups were first discussed by A. Weil [73]. The results of § 21 are due to Gelfand and Raĭkov [18]. Theorems 3 and 4 of § 22 were also established in the same note [18]; the proof of Theorem 3 given in § 22 and the derivation of topological properties of the group of characters in § 23 go back to Raikov [46, 47, 49]. The construction of an invariant integral on the group of characters explained in § 24 is based on the same idea as the proof of the 'Abstract Plancherel Theorem' in the book by Loomis [35] but, in contrast to the latter, is not based on the representation theorem for positive-definite functions. The results of § 25 are due to A. Weil [73] and M. Krein [Kreĭn] [31] and the proof, to Raikov. The proof of the Pontrjagin

Duality Law in locally compact groups given in § 26 follows Raikov [47]. The results of § 27 are due to A. Weil [73] and Raĭkov [46].

CHAPTER V

Theorem 1 of § 29 is contained as a special case in the results of § 21. Theorem 2 of § 29 is due to Gelfand [13]; the proof of this theorem given in § 29.7 is due to Raĭkov [45]. The propositions formulated as Corollaries 1 and 2 to Theorem 2 were first obtained by Cameron [8] and Pitt [44]. Theorem 3 of § 29 is due to Gelfand [13]. The construction of singular maximal ideals of the ring $V^{(b)}$ given in § 31 follows Raĭkov [19]. Theorems 1 and 1′ of § 32 are due to Raĭkov; somewhat earlier, W. Rudin [53] proved the existence of perfect sets with independent points on every 'I-group' (I-groups are locally compact commutative groups in which every neighborhood of 0 contains an element of infinite order); the proof of Theorems 1 and 1′ is based on the same idea, but was obtained independently. Theorems 2 and 3 of § 32 are due to Shreider [Šreĭder] [67]; recently, Hewitt [28] has extended them to arbitrary I-groups. Theorem 4 of § 32 is due to Wiener and Pitt [78]. § 33 is a summary of the paper [67] by Shreider.

CHAPTER VI

The results of Chapter VI, except where stated in the text, are due to Shilov [56, 59]. Gelfand [16] had previously proposed another approach to the investigation of primary ideals based on a study of the behavior of the resolvent $(x-\lambda e)^{-1}$ in the neighborhood of a spectral value $\lambda = \lambda_0$.

CHAPTER VII

The results of § 43 are due to Gelfand and Shilov [20]. The results of §§ 44 and 45, except where stated in the text, are due to Shilov [62, 63].

BIBLIOGRAPHY

BIBLIOGRAPHY

Items marked with an asterisk * are in Russian

[1] ARENS, R., "On a theorem of Gelfand and Neumark," Proc. Nat. Acad. Sci. USA, vol. 32 (1946), 237-239.

[2] ARENS, R., "Inverse-producing extensions of normed algebras," Trans. Amer. Math. Soc., vol. 88 (1958), 536-548.

[3] ARENS, R. and CALDERON, A. P., "Analytic functions of several Banach algebra elements," Ann. Math. (2), vol. 62 (1955), 204-216.

[4] BANACH, S., "Sur la mesure de Haar." [Appendix to the book: S. SAKS, *Théorie de l'integrale* (Monografje Matematyczne, vol. II), Warsaw, 1933; also, appendix to the book: S. BANACH, *Théorie des Operations Linéaires,* 2nd ed., New York, 1964.

[5] BEURLING, A., "Sur les intégrales de Fourier absolument convergentes et leur application à une transformation fonctionelle," Congrès de Mathématiques à Helsingfors, 1938.

[6] BEURLING, A., "Un théorème sur les fonctions bornées et uniformement continues sur l'axe réel," Acta Math., vol. 77 (1945), 127-136.

[7] BOCHNER, S., *Vorlesungen über Fourier'sche Integrale.* Leipzig, 1932. *Repr.,* New York, 1948.

[8] CAMERON, R. H., "Analytic functions of absolutely convergent generalized trigonometric sums," Duke Math. J., vol. 3 (1937), 682-688.

[9] CARLEMAN, T., *Les fonctions quasi-analytiques.* Paris, 1926.

[10] ČECH, E., "On bicompact spaces," Ann. Math., vol. 38 (1937), 823-844.

*[11] GELFAND, I. M., "On normed rings." Dokl. Akad. Nauk, vol. 23 (1939), 430-432.

*[12] GELFAND, I. M., "On absolutely convergent trigonometrical series and integrals," Dokl. Akad. Nauk, vol. 25 (1939), 571-574.

*[13] GELFAND, I. M., "On the ring of almost periodic functions," Dokl. Akad. Nauk, vol. 25 (1939), 575-577.

[14] GELFAND, I., "Normierte Ringe," Mat. Sb., vol. 9 (51) (1941), 3-23.

[15] GELFAND, I., "Über absolut convergente trigonometrische Reihen und Integrale," Mat. Sb., vol. 9 (51) (1941), 51-65.

[16] GELFAND, I., "Ideale und primäre Ideale in normierten Ringen," Mat. Sb., vol. 9 (51) (1941), 41-47.

[17] GELFAND, I. and NEUMARK, M., "On the imbedding of normed rings into the ring of operators in Hilbert space," Mat. Sb., vol. 12 (54) (1943), 197-213.

*[18] GELFAND, I. M. and RAĬKOV, D. A., "On the theory of characters of commutative topological groups," Dokl. Akad. Nauk, vol. 28, (1940), 195-198.

*[19] GELFAND, I. M., RAĬKOV, D. A., and ŠILOV, G. E., "Commutative normed rings," Uspehi Mat. Nauk 1, No. 2 (12) (1946), 48-146.

[20] GELFAND, I. and ŠILOV, G., "Über verschiedene Methoden der Einführung der Topologie in die Menge der maximalen Ideale eines normierten Ringes," Mat. Sb., vol. 9 (51) (1941), 25-38.

[21] GOFFMAN, C. and SINGER, I. M., "On a problem of Gelfand," Uspehi Mat. Nauk, vol. 14, no. 3 (87) (1959), 99-114.

*[22] GRUŠIN, V. V., "On the structure of closed ideals in the ring of doubly periodic vectorially smooth functions," Vestnik MGU, Ser. Mat., Mech., Astr., no. 3 (1960).

[23] HAAR, A., "Der Massbegriff in der Theorie der kontinuierlichen Gruppen," Ann. Math., vol. 34 (1933), 147-169.

[24] HAMEL, G., "Eine Basis aller Zahlen und die unstetigen Lösungen der Funktionalgleichung: $f(x+y) = f(x) + f(y)$," Math. Ann., vol. 60 (1905), 459-462.

[25] HARTOGS, F. and ROSENTHAL, A., "Über Folgen analytischer Funktionen," Math. Ann., vol. 104 (1931), 606-610.

[26] HELSON, H. and QUIGLEY, F., Existence of maximal ideals in algebras of continuous functions," Proc. Amer. Math. Soc., vol. 8 (1957), 115-119.

[27] HERGLOTZ, G., "Über Potenzreihen mit positivem, reelem Teil in Einheitskreis," Leipzig. Berichte, vol. 63 (1911), 501-511.

[28] HEWITT, E., "The asymmetry of certain algebras of Fourier-Stieltjes transforms," Michigan Math. J., vol. 5 (1958), 149-158.

[29] KATZNELSON, W., "Sur les fonctions opérant sur l'algèbre des séries de Fourier absolument convergentes," C. R. Acad. Sci. Paris, vol. 247 (1958), 404-406.

*[30] KORENBLUM, B. I., "Generalization of Wiener's Tauberian Theorem and harmonic analysis of rapidly growing functions," Trudy Moskov. Mat. Obšč., vol. 7 (1958), 121-148.

*[31] KREĬN, M. G., On a generalization of Plancherel's Theorem to the case of Fourier integrals on a commutative topological group," Dokl. Akad. Nauk, vol. 30 (1941), 482-486.

*[32] KREĬN, M. G., "Integral equations on a half-line with a kernel depending on the difference of the arguments," Uspehi Mat. Nauk, vol. 13, no. 5 (83) (1958), 3-120.

*[33] LEĬBENZON, Z. L., "On the ring of functions with an absolutely convergent Fourier series," Uspehi Mat. Nauk, vol. 9, no. 3 (61) (1954), 157-162.

[34] LÉVY, P., "Sur la convergence absolue des séries de Fourier," Compositio Math., vol. 1 (1934), 1-14.

[35] LOOMIS, L. H., *An Introduction to Abstract Harmonic Analysis.* New York, 1953.

[36] MALLIAVIN, P., "Sur l'impossibilité de la synthèse spectrale sur la droit," C. R. Acad. Sci. Paris, vol. 248 (1959), 2155-2157.

[37] MARKOFF, A., "On mean values and exterior densities," Mat. Sb., vol. 4 (46) (1938), 165-191.

[38] MAZUR, S., "Sur les anneaux linéaires," C. R. Acad. Sci. Paris, vol. 207 (1938), 1025

*[39] MERGELJAN, S. N., "Uniform approximations to functions of a complex variable," Uspehi Mat. Nauk, vol. 7, no. 2 (48) (1952), 31-122.

*[40] MITJAGIN, B. S., "Maximal ideals of certain normed rings," Sibirsk. Mat. Ž., vol. 1 (1960).

[41] MOORE, E. H. and SMITH, H. L., "A general theory of limits," Amer. J. Math., vol. 44 (1922), 102-121.

[42] NAĬMARK, M. A., *Normed Rings,* Groningen, 1963. [Trans. of [42′] below.]

*[42′] NAĬMARK, M. A. *Normed rings.* Moscow, 1956.

[43] PALEY, R. and WIENER, N., *Fourier Transforms in the Complex Domain.* New York, 1934.

[44] PITT, H. R., "A theorem on absolutely convergent trigonometrical series," J. Math. Phys., vol. 16 (1938), 191-195.

*[45] RAĬKOV, D. A., "Positive-definite functions on discrete commutative groups," Dokl. Akad. Nauk, vol. 27 (1940), 325-329.

*[46] RAĬKOV, D. A., "Positive-definite functions on commutative groups with an invariant measure," Dokl. Akad. Nauk, vol. 28 (1940), 296-300.

*[47] RAĬKOV, D. A., "A generalized duality law for commutative groups with an invariant measure," Dokl. Akad. Nauk, vol. 30 (1941), 583-585.

*[48] RAĬKOV, D. A., "A new proof of the uniqueness of the Haar measure," Dokl. Akad. Nauk, vol. 34 (1942), 231-233.

*[49] RAĬKOV, D. A., *Harmonic analysis on commutative groups with a Haar measure and theory of characters.* Trudy MIAN, vol. 14 (1945).

*[50] RAĬKOV, D. A., "On the theory of normed rings with an involution," Dokl. Akad. Nauk, vol. 54 (1946), 391-394.

[51] RIESZ, F., "Über Sätze von Stone und Bochner," Acta Univ. Szeged, vol. 6 (1933), 184-198.

[52] RUDIN, W., "Les idéaux fermés dans un anneau de fonctions analytiques," C. R. Acad. Sci. Paris, vol. 244 (1957), 997-998.

[53] RUDIN, W., "Independent perfect sets in groups," Michigan Math. J., vol. 5 (1958), 159-161.

*[54] ŠATUNOVSKIĬ, S. O., *Introduction to analysis.* Odessa, 1923.

*[55] ŠILOV, G. E., "On certain normed rings," Sb. naučn. stud. rabot, MGU, vol. 18 (1940), 5-25.

*[56] ŠILOV, G. E., "On the theory of ideals in normed rings of functions," Dokl. Akad. Nauk, vol. 27 (1940), 900-903.

*[57] ŠILOV, G. E., "On extensions of maximal ideals," Dokl. Akad. Nauk, vol. 29 (1940), 83-85.

*[58] ŠILOV, G. E., "On the Fourier coefficients of a certain class of continuous functions," Dokl. Akad. Nauk, vol. 35 (1942), 3-8.

*[59] ŠILOV, G. E., *On regular normed rings.* Trudy MIAN, vol. 21 (1947).

*[60] ŠILOV, G. E., "On normed rings with one generator," Mat. Sb., vol. 21 (63) (1947), 25-37.

*[61] ŠILOV, G. E., "On a property of rings of functions," Dokl. Akad. Nauk, vol. 58 (1947), 985-988.

*[62] ŠILOV, G. E., "On a boundary property of analytic functions," Uč. Zap. MGU, ser. mat., vol. 146, no. 3 (1949), 126-128.

*[63] ŠILOV, G. E., "On rings of functions with uniform convergence," Ukrain. Mat. Ž., vol. 4 (1951), 404-411.

*[64] ŠILOV, G. E., "On the decomposition of a commutative normed ring into a direct sum of ideals," Mat. Sb., vol. 32 (1953), 353-364. [A translation of this paper constitutes Chapter IX of the present book.]

*[65] ŠILOV, G. E., "On analytic functions in a normed ring," Uspehi Mat. Nauk, vol. 15, no. 3 (93) (1960).

*[66] ŠNOL', E. E., "The structure of ideals in the rings R_a," Mat. Sb. vol. 27, (69) (1950), 143-146.

*[67] ŠREĬDER, JU. A., "The structure of maximal ideals in rings of measures with an involution," Mat. Sb., vol. 27 (69) (1950), 297-318.

[68] SCHWARTZ, L., "Sur une propriété de synthèse spectrale dans les groupes non compacts," C. R. Acad. Sci. Paris, vol. 227 (1948), 424-426.

[69] SIERPIŃSKI, W., "Sur la question de la mesurabilité de la base de M. Hamel," Fund. Math., vol. 1 (1920), 105-111.

[70] STONE, M., "Applications of the theory of Boolean Rings to General Topology," Trans. Amer. Math. Soc., vol. 41 (1937), 375-481.

[71] TYCHONOFF, A., "Über die topologische Erweiterungen von Räumen," Math. Ann., vol. 102 (1929), 544-561.

[72] WAELBROECK, L., "Le calcul symbolique dans les algèbres commutatives," J. Math. Pures Appl., vol. 33 (1954), 147-186.

[73] WEIL, A., *L'intégration dans les groupes topologiques et ses applications.* Paris, 1940.

[74] WERMER, J., "Function rings and Riemann surfaces," Ann. Math. (2) vol. 67 (1958), 45-71.

[75] WHITNEY, H., "On ideals of differential functions," Amer. J. Math., vol. 70 (1948), 635-658.

[76] WIENER, N., "Tauberian theorems," Ann. Math., vol. 33 (1932), 1-100.

[77] WIENER, N., *The Fourier Integral and Certain of its Applications.* Cambridge [Mass.], 1933.

[78] WIENER, N. and PITT, H. R., "On absolutely convergent Fourier-Stieltjes transforms," Duke Math. J., vol. 4 (1938), 420-436.

BIBLIOGRAPHY FOR APPENDIX

[79] GELFAND, I., "Normierte Ringe," Mat. Sb., vol. 9 (51) :1 (1941), 3-66.

[80] GELFAND, I. and NEUMARK, M., "On the imbedding of normed rings into the ring of operators in Hilbert space," Mat. Sb., vol. 12 (54) :2 (1943), 197-217.

*[81] GELFAND, I. M. and NAĬMARK, M. A., "Unitary representations of the Lorentz group," Izv. Akad. Nauk SSSR., ser. mat., vol. 11 (1947), 411-504.

*[82] GELFAND, I. M. and RAĬKOV, D. A., "Continuous unitary representations of locally compact groups," Mat. Sb., vol. 13 (55) :2-3 (1943), 301-316.

*[83] GELFAND, I. M., RAĬKOV, D. A., and ŠILOV, G. E., "Commutative normed rings," Uspehi Mat. Nauk, vol. 1, part 2 (12) (1946), 48-146.

[84] GODEMENT, R., "Extension à un groupe abélien de quelconque des théorèmes taubériens de N. Wiener et d'un théorème de A. Beurling," C. R. Acad. Sci. Paris (1946), 16-18.

[85] KREĬN, M. and MILMAN, D., "On extreme points of regular convex sets," Studia Math., vol. IX (I) (1940), 133-138.

[86] NEUMANN, J. VON, "Über adjungierte Funktionaloperatoren," Ann. Math., vol. 33 (1932), 294-310.

*[87] RAĬKOV, D. A., "On the theory of normed rings with an involution," Dokl. Akad. Nauk, vol. 54, 5 (1946), 391-394.

BIBLIOGRAPHY TO SECOND APPENDIX

*[88] FUKS, D. A., *Theory of analytic functions of several complex variables.* Moscow-Leningrad, 1948. [See also [91] and [92].]

*[89] GELFAND, I. M., RAĬKOV, D. A. and ŠILOV, G. E., "Commutative normed rings," Uspehi Mat. Nauk, vol. 1, part 2 (12) (1946), 48-146.

*[90] ŠILOV, G. E., "On normed rings with one generator," Mat. Sb., 21 (63) (1947), 25-46.

*[91] FUKS, D. A., *Introduction to the Theory of Analytic Functions of Several Complex Variables,* Moscow, 1962.

[92] FUKS, D. A., *Introduction to the Theory of Analytic Functions of Several Complex Variables,* Providence, 1963. [Trans. of [91] above.]

INDEX

INDEX

THEORY OF FUNCTIONS

By C. CARATHÉODORY

Translated by F. STEINHARDT. The recent, and already famous textbook, *Funktionentheorie.*

Partial Contents: **Part One.** Chap. I. Algebra of Complex Numbers. II. Geometry of Complex Numbers. III. Euclidean, Spherical, and Non-Euclidean Geometry. **Part Two.** Theorems from Point Set Theory and Topology. Chap. I. Sequences and Continuous Complex Functions. II. Curves and Regions. III. Line Integrals. **Part Three.** Analytic Functions. Chap. I. Foundations. II. The Maximum-modulus principle. III. Poisson Integral and Harmonic Functions. IV. Meromorphic Functions. **Part Four.** Generation of Analytic Functions by Limiting Processes. Chap. I. Uniform Convergence. II. Normal Families of Meromorphic Functions. III. Power Series. IV. Partial Fraction Decomposition and the Calculus of Residues. **Part Five.** Special Functions. Chap. I. The Exponential Function and the Trigonometric Functions. II. Logarithmic Function. III. Bernoulli Numbers and the Gamma Function.

Vol. II.: **Part Six.** Foundations of Geometric Function Theory. Chap. I. Bounded Functions. II. Conformal Mapping. III. The Mapping of the Boundary. **Part Seven.** The Triangle Function and Picard's Theorem. Chap. I. Functions of Several Complex Variables. II. Conformal Mapping of Circular-Arc Triangles. III. The Schwarz Triangle Functions and the Modular Function. IV. Essential Singularities and Picard's Theorems.

"A book by a master . . . Carathéodory himself regarded [it] as his finest achievement . . . written from a catholic point of view."—*Bulletin of A.M.S.*

—Vol. I. Second edition. 1958. 310 pp. 6x9. [97] **$4.95**
—Vol. II. Second edition. 1960. 220 pp. 6x9. [106] **$4.95**

ALGEBRAIC THEORY OF MEASURE AND INTEGRATION

By C. CARATHÉODORY

Translated from the German by FRED E. J. LINTON. By generalizing the concept of point function to that of a function over a Boolean ring ("soma" function), Prof. Carathéodory gives an algebraic treatment of measure and integration.

CONTENTS: CHAP. I. Somas (Axiomatic method, somas as elements of a Boolean ring, . . .). II. Set of Somas. III. Place Functions (Functionoids). IV. Computation with Place Functions. V. Measure Functions. VI. The Integral. VII. Application of Integration to Limit Processes. VIII. Computation of Measure Functions. IX. Regular Measure Functions. X. Isotypic Regular Measure Functions. XI. Content Functions. APPENDIX: Somas as Elements of Partially Ordered Sets.

—1963. 378 pp. 6x9. [161] **$7.50**

ALMOST PERIODIC FUNCTIONS

By H. BOHR

Translated by H. COHN. From the famous series *Ergebnisse der Mathematik und ihrer Grenzgebiete*, a beautiful exposition of the theory of Almost Periodic Functions written by the creator of that theory.

—1951. 120 pp. Lithotyped. [27] **$2.75**

LECTURES ON THE CALCULUS OF VARIATIONS

By O. BOLZA

A standard text by a major contributor to the theory. Suitable for individual study by anyone with a good background in the Calculus and the elements of Real Variables. The present, second edition differs from the first primarily by the inclusion within the text itself of various addenda to the first edition, as well as some notational improvements.

—2nd (corr.) ed. 1961. 280 pp. 5⅜x8. [145] Cloth **$3.25**
[152] Paper **$1.19**

VORLESUNGEN UEBER VARIATIONSRECHNUNG

By O. BOLZA

A standard text and reference work, by one of the major contributors to the theory.

—1963. Corr. repr. of 1st ed. ix+715 pp. 5⅜x8. [160] **$8.00**

THEORIE DER KONVEXEN KÖRPER

By T. BONNESEN and W. FENCHEL

"Remarkable monograph."

—*J. D. Tamarkin, Bulletin of the A. M. S.*

—1934. 171 pp. 5½x8½. Orig. publ. at $7.50 [54] **$3.95**

THE CALCULUS OF FINITE DIFFERENCES

By G. BOOLE

A standard work on the subject of finite differences and difference equations by one of the seminal minds in the field of finite mathematics.

Numerous exercises with answers.

—Fourth edition. 1958. xii + 336 pp. 5x8. [121] Cloth **$3.95**
[148] Paper **$1.39**

A TREATISE ON DIFFERENTIAL EQUATIONS

By G. BOOLE

Including the Supplementary Volume.

—Fifth edition. 1959. xxiv + 735 pp. 5¼x8. [128] **$6.00**

CAJORI, "History of Slide Rule," see Ball

HISTORY OF THE THEORY OF NUMBERS

By L. E. DICKSON

"A monumental work . . . Dickson always has in mind the needs of the investigator . . . The author has [often] expressed in a nut-shell the main results of a long and involved paper *in a much clearer way than the writer of the article did himself.* The ability to reduce complicated mathematical arguments to simple and elementary terms is highly developed in Dickson."—*Bulletin of A. M. S.*

—Vol. I (Divisibility and Primality) xii + 486 pp. Vol. II (Diophantine Analysis) xxv + 803 pp. Vol. III (Quadratic and Higher Forms) v + 313 pp. [86] Three vol. set **$19.95**

THE INTEGRAL CALCULUS

By J. W. EDWARDS

A leisurely, immensely detailed, textbook of over 1,900 pages, rich in illustrative examples and manipulative techniques and containing much interesting material that must of necessity be omitted from less comprehensive works.

There are forty large chapters in all. The earlier cover a leisurely and a more-than-usually-detailed treatment of all the elementary standard topics. Later chapters include: Jacobian Elliptic Functions, Weierstrassian Elliptic Functions, Evaluation of Definite Integrals, Harmonic Analysis, Calculus of Variations, etc. Every chapter contains many exercises (with solutions).

—2 vols. 1,922 pp. 5x8. Originally published at $31.50 the set. [102], [105] Each volume **$7.50**

TRANSFORMATIONS OF SURFACES

By L. P. EISENHART

Many of the advances in the differential geometry of surfaces in the present century have had to do with transformations of surfaces of a given type into surfaces of the same type. The present book studies two types of transformation to which many, if not all, such transformations can be reduced.

—2nd (Corr.) ed. ix+379 pp. 5⅜x8. [167] **$4.95**

INTRODUCTION TO THE ALGEBRA OF QUANTICS

By E. B. ELLIOTT

—2nd ed. 1913. xvi+416 pp. 5⅜x8. [184] Prob. **$6.00**

AUTOMORPHIC FUNCTIONS

By L. R. FORD

"Comprehensive . . . remarkably clear and explicit."—*Bulletin of the A. M. S.*

—2nd ed. (Cor. repr.) x + 333 pp. 5⅜x8. [85] **$6.00**

ASYMPTOTIC SERIES

By W. B. FORD

TWO VOLUMES IN ONE: *Studies on Divergent Series and Summability* and *The Asymptotic Developments of Functions Defined by MacLaurin Series.*

PARTIAL CONTENTS: I. MacLaurin Sum-Formula; Introduction to Study of Asymptotic Series. II. Determination of Asymptotic Development of a Given Function. III. Asymptotic Solutions of Linear Differential Equations. . . . V. Summability, etc. *I.* First General Theorem. . . . *III.* MacLaurin Series whose General Coefficient is Algebraic. . . . *VII.* Functions of Bessel Type. *VIII.* Asymptotic Behavior of Solution of Differential Equations of Fuchsian Type. Bibliography.

—1916; 1936-60. x + 341 pp. 6x9. [143] Two vols. in one. **$6.00**

THE CALCULUS OF EXTENSION

By H. G. FORDER

—1941-60. xvi + 490 pp. 5⅜x8. [135] **$4.95**

RUSSIAN MATHEMATICAL BIBLIOGRAPHY

By G. E. FORSYTHE

A bibliography of Russian Mathematics Books for the quarter century 1920-55. Added subject index.

—1956. 106 pp. 5x8. [111] **$3.95**

CURVE TRACING

By P. FROST

This much-quoted and charming treatise gives a very readable treatment of a topic that can only be touched upon briefly in courses on Analytic Geometry. Teachers will find it invaluable as supplementary reading for their more interested students and for reference. The Calculus is not used.

Seventeen plates, containing over 200 figures, illustrate the discussion in the text.

—5th (unaltered) ed. 1960. 210 pp. + 17 fold-out plates. 5⅜x8. [140] **$3.95**

THE THEORY OF MATRICES

By F. R. GANTMACHER

This treatise, by one of Russia's leading mathematicians gives, in easily accessible form, a coherent account of matrix theory with a view to applications in mathematics, theoretical physics, statistics, electrical engineering, etc. The individual chapters have been kept as far as possible independent of each other, so that the reader acquainted with the contents of Chapter I can proceed immediately to the chapters that especially interest him. Much of the material has been available until now only in the periodical literature.

Partial Contents. VOL. ONE. I. Matrices and Matrix Operations. II. The Algorithm of Gauss and Applications. III. Linear Operators in an n-Dimensional Vector Space. IV. Characteristic Polynomial and Minimal Polynomial of a Matrix (Generalized Bézout Theorem, Method of Faddeev for Simultaneous Computation of Coefficients of Characteristic Polynomial and Adjoint Matrix, . . .). V. Functions of Matrices (Various Forms of the Definition, Components, Application to Integration of System of Linear Differential Eqns, Stability of Motion, . . .). VI. Equivalent Transformations of Polynomial Matrices; Analytic Theory of Elementary Divisors. VII. The Structure of a Linear Operator in an n-Dimensional Space (Minimal Polynomial, Congruence, Factor Space, Jordan Form, Krylov's Method of Transforming Secular Eqn, . . .). VIII. Matrix Equations (Matrix Polynomial Eqns, Roots and Logarithm of Matrices, . . .). IX. Linear Operators in a Unitary Space. X. Quadratic and Hermitian Forms.

VOL. TWO. XI. Complex Symmetric, Skew-symmetric, and Orthogonal Matrices. XII. Singular Pencils of Matrices. XIII. Matrices with Non-Negative Elements (Gen'l and Spectral Properties, Reducible M's, Primitive and Imprimitive M's, Stochastic M's, Totally Non-Negative M's, . . .). XIV. Applications of the Theory of Matrices to the Investigation of Systems of Linear Differential Equations. XV. The Problem of Routh-Hurwitz and Related Questions (Routh's Algorithm, Lyapunov's Theorem, Infinite Hankel M's, Supplements to Routh-Hurwitz Theorem, Stability Criterion of Liénard and Chipart, Hurwitz Polynomials, Stieltjes' Theorem, Domain of Stability, Markov Parameters, Problem of Moments, Markov and Chebyshev Theorems, Generalized Routh-Hurwitz Problem, . . .). BIBLIOGRAPHY.

—Vol. I. 1960. x + 374 pp. 6x9. [131] **$6.00**
—Vol. II. 1960. x + 277 pp. 6x9. [133] **$6.00**

LECTURES ON ANALYTICAL MECHANICS

By F. R. GANTMACHER

Translated from the Russian by PROF. B. D. SECKLER, with additions and revisions by Prof. Gantmacher.

Partial Contents: CHAP. I. Differential Equations of Motion of a System of Particles. II. Equations of Motion in a Potential Field. III. Variational Principles and Integral-Invariants. IV. Canonical Transformations and the Hamilton-Jacobi Equation. V. Stable Equilibrium and Stability of Motion of a System (Lagrange's Theorem on stable equilibrium, Tests for unstable E., Theorems of Lyapunov and Chetayev, Asymptotically stable E., Stability of linear systems, Stability on basis of linear approximation, . . .). VI. Small Oscillations. VII. Systems with Cyclic Coordinates. BIBLIOGRAPHY.

—Approx. 300 pp. 6x9. [175] **In prep.**

ARITHMETISCHE UNTERSUCHUNGEN

By C. F. GAUSS

The German translation of his *Disquisitiones Arithmeticae.*

—Repr. of 1st German ed. 860 pp. 5⅜x8. [190] **In prep.**

COMMUTATIVE NORMED RINGS

By I. M. GELFAND, D. A. RAĬKOV, and G. E. SHILOV

Translated from the Russian.

Partial Contents: CHAPS. I AND II. General Theory of Commutative Normed Rings. III. Ring of Absolutely Integrable Functions and their Discrete Analogues. IV. Harmonic Analysis on Commutative Locally Compact Groups. V. Ring of Functions of Bounded Variation on a Line. VI. Regular Rings. VII. Rings with Uniform Convergence. VIII. Normed Rings with an Involution and their Representations. IX. Decomposition of Normed Ring into Direct Sum of Ideals. HISTORICAL-BIBLIOGRAPHICAL NOTES. BIBLIOGRAPHY.

—Approx. 360 pp. 5⅜x8. [170] **In prep.**

THEORY OF PROBABILITY

By B. V. GNEDENKO

This textbook, by Russia's leading probabilist, is suitable for senior undergraduate and first-year graduate courses. It covers, in highly readable form, a wide range of topics and, by carefully selected exercises and examples, keeps the reader throughout in close touch with problems in science and engineering.

The translation has been made by PROF. B. D. SECKLER from a revised and amplified version of the second Russian edition.

Within a few months after publication it has been adopted as a text at a number of leading universities and colleges and seems destined to be a leading textbook at American and British universities.

"extremely well written . . . suitable for individual study . . . Gnedenko's book is a milestone in the writing on probability theory."—*Science.*

Partial Contents: I. The Concept of Probability (Different approaches to the definition. Field of events. Geometrical Probability. Statistical definition. Axiomatic construction . . .). II. Sequences of Independent Trials. III. Markov Chains. IV. Random Variables and Distribution Functions (Continuous and discrete distributions. Multidimensional d. functions. Functions of random variables. Stieltjes integral). V. Numerical Characteristics of Random Variables (Mathematical expectation. Variance . . . Moments). VI. Law of Large Numbers (Mass phenomena. Tchebychev's form of law. Strong law of large numbers . . .). VII. Characteristic Functions (Properties. Inversion formula and uniqueness theorem. Helly's theorems. Limit theorems. Char. functs. for multi-dimensional random variables . . .). VIII. Classical Limit Theorem (Liapunov's theorem. Local limit theorem). IX. Theory of Infinitely Divisible Distribution Laws. X. Theory of Stochastic Processes (Generalized Markov equation. Continuous S. processes. Purely discontinuous S. processes. Kolmogorov-Feller equations. Homogeneous S. processes with independent increments. Stationary S. process. Stochastic integral. Spectral theorem of S. processes. Birkhoff-Khinchine ergodic theorem). XI. Elements of Statistics (Some problems. Variational series and empirical distribution functions. Glivenko's theorem and Kolmogorov's compatibility criterion. Two-sample problem. Critical region. Comparison of two statistical hypotheses . . . Confidence limits). **TABLES. BIBLIOGRAPHY. ANSWERS TO THE EXERCISES.**

—2nd ed. 1963. iv+471 pp. 6x9. [132] **$8.75**

LES INTEGRALES DE STIELTJES et leurs Applications aux Problèmes de la Physique Mathématique

By N. GUNTHER

The present work is a reprint of Vol. I of the publications of the V. A. Steklov Institute of Mathematics, in Moscow. The text is in French.

— 932-49. 498 pp. 5⅜x8. [63] **$6.50**

LECONS SUR LA PROPAGATION DES ONDES ET LES ÉQUATIONS DE L'HYDRODYNAMIQUE

By J. HADAMARD

"[Hadamard's] unusual analytic proficiency enables him to connect in a wonderful manner the physical problem of propagation of waves and the mathematical problem of Cauchy concerning the characteristics of partial differential equations of the second order."—*Bulletin of the A. M. S.*

—viii + 375 pp. 5½x8½. [58] **$4.95**

REELLE FUNKTIONEN. Punktfunktionen

By H. HAHN

—426 pp. 5½x8½. Orig. pub. at $12.80. [52] **$4.95**

INTRODUCTION TO HILBERT SPACE AND THE THEORY OF SPECTRAL MULTIPLICITY

By P. R. HALMOS

Prof. Halmos' recent book gives a clear, readable introductory treatment of Hilbert Space. The multiplicity theory of continuous spectra is treated, for the first time in English, in full generality.

—1957. 2nd ed. (c. repr. of 1st ed.). 120 pp. 6x9. [82] **$3.25**

SET THEORY

By F. HAUSDORFF

Now for the first time available in English, Hausdorff's classic text-book has been an inspiration and a delight to those who have read it in the original German. The translation is from the Third (latest) German edition.

"We wish to state without qualification that this is an indispensable book for all those interested in the theory of sets and the allied branches of real variable theory."—*Bulletin of A. M. S.*

—2nd ed. 1962. 352 pp. 6x9. [119] **$6.50**

VORLESUNGEN ÜBER DIE THEORIE DER ALGEBRAISCHEN ZAHLEN

By E. HECKE

"An elegant and comprehensive account of the modern theory of algebraic numbers."

—*Bulletin of the A. M. S.*

—1923. 264 pp. 5½x8½. [46] **$3.95**

INTEGRALGLEICHUNGEN UND GLEICHUNGEN MIT UNENDLICHVIELEN UNBEKANNTEN

By E. HELLINGER and O. TOEPLITZ

"Indispensable to anybody who desires to penetrate deeply into this subject."—*Bulletin of A.M.S.*

—With a preface by E. Hilb. 1928. 286 pp. 5¼x8. [89] **$4.50**

Grundzüge Einer Allgemeinen Theorie der LINEAREN INTEGRALGLEICHUNGEN

By D. HILBERT

—306 pp. 5½x8¼. [91] **$4.50**

PRINCIPLES OF MATHEMATICAL LOGIC

By D. HILBERT and W. ACKERMANN

The famous *Grundüge der Theoretischen Logik* translated into English, with added notes and revisions by PROF. R. E. LUCE.

"The best textbook in a Western European language for a student wishing a fairly thorough treatment."—*Bulletin of the A. M. S.*

—1950-59. xii + 172 pp. 6x9. [69] **$3.95**

DETERMINANTENTHEORIE EINSCHLIESSLICH DER FREDHOLMSCHEN DETERMINANTEN

By G. KOWALEWSKI

"A classic in its field."—*Bulletin of the A.M.S.*

—Third edition. 1942. 328 pp. 5½x8. [39] **$4.95**

GROUP THEORY

By A. KUROSH

Translated from the second Russian edition and with added notes by PROFESSOR K. A. HIRSCH.

Partial Contents: PART ONE: The Elements of Group Theory. Chap. I. Definition. II. Subgroups (Systems, Cyclic Groups, Ascending Sequences of Groups). III. Normal Subgroups. IV. Endomorphisms and Automorphisms. Groups with Operators. V. Series of Subgroups. Direct Products. Defining Relations, etc. PART TWO: Abelian Groups. VI. Foundations of the Theory of Abelian Groups (Finite Abelian Groups, Rings of Endomorphisms, Abelian Groups with Operators). VII. Primary and Mixed Abelian Groups. VIII. Torsion-Free Abelian Groups. Editor's Notes. Bibliography.

Vol. II. PART THREE: Group-Theoretical Constructions. IX. Free Products and Free Groups (Free Products with Amalgamated Subgroup, Fully Invariant Subgroups). X. Finitely Generated Groups. XI. Direct Products. Lattices (Modular, Complete Modular, etc.). XII. Extensions of Groups (of Abelian Groups, of Non-commutative Groups, Cohomology Groups). PART FOUR: Solvable and Nilpotent Groups. XIII. Finiteness Conditions, Sylow Subgroups, etc. XIV. Solvable Groups (Solvable and Generalized Solvable Groups, Local Theorems). XV. Nilpotent Groups (Generalized, Complete, Locally Nilpotent Torsion-Free, etc.). Editor's Notes. Bibliography.

—Vol. I. 2nd ed. 1959. 271 pp. 6x9. [107] **$5.50**
—Vol. II. 2nd ed. 1960. 308 pp. 6x9. [109] **$5.50**

LECTURES ON GENERAL ALGEBRA

By A. G. KUROSH

Translated from the Russian by PROFESSOR K. A. HIRSCH, with a special preface for this edition by PROFESSOR KUROSH.

Partial Contents: CHAP. I. Relations. II. Groups and Rings (Groupoids, Semigroups, Groups, Rings, Fields, . . . , Gaussian rings, Dedekind rings). III. Universal Algebras. Groups with Multioperators (. . . Free universal algebras, Free products of groups). IV. Lattices (Complete lattices, Modular lattice, Schmidt-Ore Theorem, . . . , Distributive lattices). V. Operator Groups and Rings. Modules. Linear Algebras (. . . Free modules, Vector spaces over fields, Rings of linear transformations, . . . , Derivations, Differential rings). VI. Ordered and Topological Groups and Rings. Rings with a Valuation. BIBLIOGRAPHY.

—1963. 335 pp. [168] **$6.95**

ELEMENTS OF ALGEBRA

By HOWARD LEVI

"This book is addressed to beginning students of mathematics. . . . The level of the book, however, is so unusually high, mathematically as well as pedagogically, that it merits the attention of professional mathematicians (as well as of professional pedagogues) interested in the wider dissemination of their subject among cultured people . . . **a closer approximation to the right way to teach mathematics to beginners than anything else now in existence."**—*Bulletin of the A. M. S.*

—4th ed. 1962. 189 pp. 5⅜x8. [103] **$3.50**

THE THEORY OF MATRICES

By C. C. MacDUFFEE

"No mathematical library can afford to be without this book."—*Bulletin of the A. M. S.*

—(Ergeb. der Math.) 2nd edition. 116 pp. 6x9. Orig. publ. at $5.20. [28] **$2.95**

COMBINATORY ANALYSIS, Vols. I and II

By P. A. MACMAHON

TWO VOLUMES IN ONE.

A broad and extensive treatise on an important branch of mathematics.

—xx + 300 + xx + 340 pp. 5⅜x8. [137] Two vols. in one. **$7.50**

MACMAHON, "Introduction . . . ," see Klein

FORMULAS AND THEOREMS FOR THE FUNCTIONS OF MATHEMATICAL PHYSICS

By W. MAGNUS and F. OBERHETTINGER

Gathered into a compact, handy and well-arranged reference work are thousands of results on the many important functions needed by the physicist, engineer and applied mathematician.

Translated by J. WERMER.

—1954. 182 pp. 6x9. [51] **$3.90**

THEORY OF NUMBERS

By G. B. MATHEWS

CHAPTER HEADINGS: I. Elementary Theory of Congruences. II. Quadratic Congruences. III. Binary Quadratic Forms; Analytical Theory. IV. Binary Quadratic Forms; Geometrical Theory. V. Generic Characters of Binary Quadratics. VI. Composition of Forms. VII. Cyclotomy. VIII. Determination of Number of Improperly Primitive Classes for a Given Determinant. IX. Applications of the Theory of Quadratic Forms. X. The Distribution of Primes.

A reprint of the first edition, with correction of errata and some improvements of notation.

—2nd ed. 1892-1962. xii+323 pp. 5⅜x8. [156] **$3.95**

SET TOPOLOGY

By R. VAIDYANATHASWAMY

In this text on Topology, the first edition of which was published in India, the concept of partial order has been made the unifying theme.

Over 500 exercises for the reader enrich the text.

CHAPTER HEADINGS: I. Algebra of Subsets of a Set. II. Rings and Fields of Sets. III. Algebra of Partial Order. IV. The Closure Function. V. Neighborhood Topology. VI. Open and Closed Sets. VII. Topological Maps. VIII. The Derived Set in T_1 Space. IX. The Topological Product. X. Convergence in Metrical Space. XI. Convergence Topology.

—2nd ed. 1960. vi + 305 pp. 6x9. [139] **$6.00**

LECTURES ON THE GENERAL THEORY OF INTEGRAL FUNCTIONS

By G. VALIRON

—1923. xii + 208 pp. 5¼x8. [56] **$3.50**

GRUPPEN VON LINEAREN TRANSFORMATIONEN

By B. L. VAN DER WAERDEN

—(Ergeb. der Math.) 1935. 94 pp. 5½x8½. [45] **$2.50**

THE LOGIC OF CHANCE

By J. VENN

One of the classics of the theory of probability. Venn's book remains unsurpassed for clarity, readability, and sheer charm of exposition. No mathematics is required.

CONTENTS: PART ONE: Physical Foundations of the Science of Probability. CHAP. I. The Series of Probability. II. Formation of the Series, III. Origin, or Causation, of the Series. IV. How to Discover and Prove the Series. V. The Concept of Randomness. PART TWO: Logical Superstructure on the Above Physical Foundations. VI. Gradations of Belief. VII. The Rules of Inference in Probability. VIII. The Rule of Succession. IX. Induction. X. Causation and Design. XI. Material and Formal Logic . . . XIV. Fallacies. PART THREE: Applications. XV. Insurance and Gambling. XVI. Application to Testimony. XVII. Credibility of Extraordinary Stories. XVIII. The Nature and Use of an Average as a Means of Approximation to the Truth.

—Repr. of 3rd ed xxix+508 pp. 5⅜x8. [173] Cloth **$4.95**
[169] Paper **$2.25**